THE EVOLUTION AND PALAEOBIOLOGY OF LAND PLANTS

THE EVOLUTION AND PALAEOBIOLOGY OF LAND PLANTS

Barry A. Thomas and Robert A. Spicer

CROOM HELM
London & Sydney

DIOSCORIDES PRESS
Ecology, Phytogeography & Physiology Series,
   Volume 2
T.R. DUDLEY, General Editor
Portland, Oregon

© 1987 Barry A. Thomas and Robert A. Spicer
Croom Helm Ltd, Provident House, Burrell Row,
Beckenham, Kent BR3 1AT
Croom Helm Australia, 44–50 Waterloo Road,
North Ryde, 2113, New South Wales

British Library Cataloguing in Publication Data

Thomas, Barry A.
    The evolution and palaeobiology of land
    plants.
    1. Palaeobotany
    I. Title   II. Spicer, Robert A.
    561   QE905
    ISBN 0-7099-2434-8
    ISBN 0-7099-2476-3 Pbk

First published in 1986 in the U.S.A. by
Dioscorides Press, 9999 S.W. Wilshire, Portland,
Oregon 97225, U.S.A.

ISBN 0-931146-06-2 Cloth
      0-931146-07-0 Paper

Filmset by Mayhew Typesetting, Bristol, England
Printed and bound in Great Britain by Mackays of Chatham Ltd, Kent

CONTENTS

INTRODUCTION

A great number of major advances have been made recently in those fields of study encompassing fossil plants. The increasing amount of knowledge, brought about by an expanding research interest, has permitted and indeed stimulated a re-evaluation of many of our former ideas. Palaeobotany is the study of plant fossils, and plant fossils, like living plants, can be studied for many purposes. Morphology, taxonomy, systematics, ecology, plant geography, stratigraphy and evolution are the main interests of palaeobotanists, but the approaches used by the research worker are many and varied.

The unpredictable and chance events involved both in the initial preservation of the plant fragments and in the eventual collecting conspire to thwart our efforts. The evidence is virtually certain to be fragmentary and it is only through the efforts and skills of the investigators that we make progress towards understanding the total biology of the many and varied plants that once lived on the Earth.

Textbooks invariably reflect the interests of the authors, and this can be very apparent in palaeobotany texts. We have attempted to introduce the reader to the evolutionary history of the vascular land plants in the context of plant fossils and what can be deduced of the overall biology of the original living plants. The flowering plants usually receive rather scant coverage in this kind of book, giving the impression that palaeobotany has little to offer towards our understanding of their evolutionary history. This is unacceptable if we consider the present-day importance of the angiosperms in the world's vegetation. To that end we have tried to give a more rounded picture of plant evolution by devoting more space to the angiosperms.

Plant fossils are too often ignored by neobotanists, and palaeobotanists can become too engrossed in their own extinct plants to see them in the wider context of an evolving ecosystem. We have attempted to get away from the strait-jacket of writing a text entirely on plant fossils, for they must be seen in the context of living plants. For this purpose we have included some of the theoretical ideas of evolution together with the associated evidence obtained from studying living plants. Plant fossils are the remains of once living plants, and it is their total biology that we are aiming to bring to the reader rather than a catalogue of the evidence.

A GEOLOGICAL TIME SCALE

ERA	SYSTEM/ PERIOD			SERIES / EPOCH		STAGE / AGE	Ma B.P.
P A L A E O Z O I C				PERMIAN			248
						– – – –	–286
	C A R B O N I F E R O U S	PENNSYLVANIAN	C₂	Gzelian	Stephanian		
				Kasimov –ian			
				Moscov –ian	Coal Measures	Cantabrian	– – – – –296
						Westphalian	
				Bashkir –ian			– – – – –315
					Namurian — (Millstone Grit)		
		MISSISSIPPIAN		Serpukhov –ian			
			C₁				– – – – –333
				Visean	Dinantian — — (Carboniferous Limestone)		
							– – – – – –352
				Tournaisian			
	DEVONIAN		D₃	Upper	Famennian		360
					Frasnian		367
							374
			D₂	Middle	Givetian		380
					Eifelian		387
			D₁	Lower	Emsian		394
					Siegenian		401
		D			Gedinnian		408
	SILURIAN			Pridoli			
							– – – – – –414
				Ludlow			– – – – – –421
				Wenlock			– – – – – –428
				Llandovery			
		S					– – – – – –438
	ORDOVICIAN			O			– – – – – –505
	CAMBRIAN			Є			590
	PRECAMBRIAN						

viii

ERA	SYSTEM/ PERIOD	SERIES/ EPOCH	STAGE / AGE	Ma B.P.
CENOZOIC Cz	Quaternary TT	Holo.		0.01
		Pleisto.		2
	TERTIARY	Pliocene		5.1
	NEOGENE Ng	Miocene		
				24.6
		Oligocene		32.8
	PALAEOGENE Pg	Eocene		
				54.9
		Palaeocene		65
MESOZOIC Mz	CRETACEOUS K	K₂ Seno-nian	Maastrichtian	73
			Campanian	83
			Santonian	87.5
			Coniacian	88.5
			Turonian	91
			Cenomanian	97.5
		K₁	Albian	113
			Aptian	119
			Barremian	125
		Neoc-omian	Hauterivian	131
			Valanginian	138
			Berriasian	144
	JURASSIC J	Upper (Malm) J₃		163
		Middle (Dogger) J₂		188
		Lower (Lias) J₁		213
	TRIASSIC Tr	Tr₃		231
		Tr₂		243
		Scythian Tr₁		248

Source: Modified from W.B. Harland et al. (1982) A geological time scale. Cambridge Univ. Press 131 pp.

PART I: THE FIRST LAND PLANTS

1 PRE-LAND PLANTS

Land has been covered with large multicellular plants for approximately only the last 400 million years. As the Earth is generally taken to be about 4700 million years old, it clearly took plants an immense period of time to evolve to the extent that some were able to leave the oceans and abandon an aquatic existence.

Historically the first 4100 million year period was thought to be devoid of life. It was called the Precambrian era on the basis that the first evidence of life had been recorded in the Cambrian of Wales. The Precambrian does, however, represent nearly 90 per cent of the Earth's history, and it is during this time that the Earth's mantle solidified, the oceans were formed and life first began.

It was only in 1954 that indisputable evidence was published of Precambrian life (Tylor and Barghoorn, 1954). Since then it has become firmly established that microbial fossils range back to over 3000 million years ago in a wide variety of Precambrian rocks. Much more attention is now being paid to the Precambrian, and our ideas of its fossil record are becoming much clearer. However, although an ever increasing number of forms are being described, it is often very difficult to compare them with living forms and especially to relate them to extant taxa. Comparisons are almost always restricted to morphology, and the simple structures found do not have many diagnostic characters.

For the last 2500 million years of the Precambrian the world was dominated by simple non-nucleated cells, the prokaryotes. They existed for the most part in the sea, but also colonised the shallow, relatively flat areas of the intertidal coastal regions. Although the evidence is not yet conclusive, it is inferred that the most primitive forms were spheroidal and perhaps comparable to the extant clostridial forms of bacteria. They probably derived their energy from fermenting organic materials that were being synthesised in the anoxic early atmosphere. Later, aerobic prokaryotes developed the ability to photosynthesise and most likely gave rise to the modern photosynthetic bacteria (Schopf, 1978). By the late Precambrian, prokaryotes had evolved that were strikingly similar in size and organisation to extant blue-green algae, even to the extent of having enclosing sheaths.

Although such prokaryotes are still extremely successful in terms of numbers and diversity and in the range of ecological niches that they now occupy, their pre-eminence was clearly in the Precambrian. Some of them evolved the ability to create the large numbers of stromatolites that have been found in so many Precambrian rocks. These are not strictly fossils but are organo-sedimentary structures that seldom preserve the organisms that were responsible for their formation. Their interpretation therefore rests on studies of living stromatolite surfaces where blue-green algae are known to build these structures through a combination of carbonate precipitation and sediment trapping. Stromatolites

3

form today in a few arid coastal plains such as along the shores of Western Australia and the Persian Gulf. Here the growing mats of microbes dominated by cyanobacteria become covered with sediment. Then, by abandoning their buried sheaths, the filamentary forms migrate upwards through the sediment to form a new surface mat. Repetition of this gives a characteristically laminated stromatolite. Sectioning of fossil stromatolites often reveals structural features that allow an identification of the original microbial communities.

The oldest stromatolites are from rocks dated as 3500 million years old in Western Australia. They rapidly became abundant about 2000 million years ago and it is possible that such stromatolites dominated most of the available shoreline during the late Precambrian and the succeeding early Cambrian. But it is also during this period that two vitally important evolutionary changes were occurring that were to shape plant evolution and to result in the land vegetation as we know it today.

Nucleated cells appeared and commenced the evolutionary line leading to the thalloid algae and the land plants, and large numbers of herbivorous animals adapted to feeding on the stromatolites. The effects of this grazing may not at first sight seem all that important, but the eventual effect was to be devastating for the algae. Early Cambrian croppers appear to have maintained a natural balance with the stromatolites (Edhorn, 1977). A series of variations in shape and size of the stromatolites resulted in forms that might not have been suitable as food for the early croppers. Then other varieties of animals appeared, which made use of the new algal forms. This self-regulating ecosystem was not disrupted until the late Cambrian when molluscs evolved that overcropped the algae. From then on the stromatolites became more and more reduced in numbers and restricted in distribution. This must have cleared widespread areas of shallow coastal waters that were ideal sites for colonisation by other groups of evolving algae.

Schopf (1978) has suggested that the first widespread appearance of stromatolites, about 2300 million years ago, possibly marked the earliest diversification of oxygen-producing cyanobacteria. For several hundred million years the iron dissolved in the oceans reacted with the generated oxygen and precipitated as ferric oxides. Only when all the unoxidised iron and similar materials had been swept from the oceans would the concentration of oxygen in the atmosphere have begun to rise towards modern levels. Then, about 1400 million years ago, the first of the nucleated cells, the eukaryotes, made their appearance (Schopf and Oehler, 1976). They were probably fully aerobic from the start, and the known diversity of cell types by 1000 million years ago suggests that some kind of sexual reproduction had evolved by then. This allowed the eukaryotes to accelerate the rate of evolutionary change so that within the next 400 million years multicellular forms of life existed that were recognisable ancestors of modern plants and animals.

One important and intriguing feature of eukaryotic cells is that they contain membrane-bound organelles such as mitochondria and chloroplasts. It has been suggested that these organelles evolved from free-living prokaryotes that invaded and became successful symbionts with ancestral eukaryotes. If this is true, the symbiosis was an extremely effective one, for mitochondria are present in all eukaryotes that carry out aerobic respiration, and chloroplasts allow the organisms that possess them to photosynthesise.

With the ability to photosynthesise and carry out aerobic respiration there was the potential for large multicellular plants to evolve, and it was during the latter part of the Precambrian that thalloid algae made their first appearance. Some of them have been referred to the red algae — the Rhodophyta. The problem here is that current classification principles of extant algae are based on the colour of the pigments, the chemical nature of their reserves, the type of reproductive cells and to a lesser extent their morphologies. Fossil algae instead are classified mainly on morphological characters and to a lesser extent on reproductive organs.

The evolutionary history of the green algae — the Chlorophyta — is of course much more important as it is generally accepted that the land plants were derived from them. It is possible that the first Chlorophyta existed in the Precambrian. They were definitely present in the Cambrian, where six genera have been shown to be comparable in morphology to the extant Dasycladaceae. These are radially symmetrical algae with a non-septate central axis and whorls of lateral, sometimes branched, appendages. Their gametes were produced in operculate cysts. Preservation of these algae has been enhanced by their ability to secrete lime around their thallus, thereby facilitating their incorporation into reefs of algal limestone. The Dasycladaceae have been thought to be linked in some way to the evolution of the land plants. Unfortunately these early ancestral forms all lived in marine conditions, and we have no record as yet of any pre-Devonian freshwater forms.

The other major group of green algae that have been thought to be connected in some way to the evolution of the earliest land plants is the Charophyta. The living forms are distinguished from other green algae by having multicellular axes divided into nodes and long internodes with whorls of branches arising from the nodes. Sexual reproduction occurs readily with the female oogonium consisting of an egg surrounded by several coiled tubes. Such oogonia are commonly found calcified in the fossil record and are known as gyrogonites. Such gyrogonites can be found in rocks of Upper Silurian (Downtonian) age onwards. As we will see later, this suggests that the charophytes and the earliest land plants evolved at roughly the same time, which would seem to preclude the charophytes from being the ancestral group.

2 THE FIRST LAND PLANTS

The environment on dry land to which previously aquatic plants had to become adapted differed considerably from almost all modern terrestrial situations. Our reconstructions of this alien environment are based on diverse sources of evidence, including sedimentological and mineralogical studies and the observed effects of modern deforestation.

Try to imagine all the land surfaces of the Earth totally devoid of any vegetation. Furthermore, because plant life is responsible for their formation, there were no soils. The only modern analogues might be deserts, but even these would appear benign compared with the terrestrial environments of 500 million years ago.

Recent deforestation studies show that plant growth affects the amount of water vapour circulated through the atmosphere. Where vegetation is removed, not only does the surface environment become more arid but cloud cover is reduced which in turn increases insolation and daytime ground-level temperatures. Temperatures during the night, however, are lowered because clear skies allow the radiation back into space of heat accumulated during daytime.

Before the presence of vegetative cover the land surfaces represented an extremely hostile environment. Precipitation must have occurred because water vapour would have entered the atmosphere by evaporation from the oceans. Either the rainwater would quickly have soaked away from the surface through unconsolidated sands or there would have been rapid runoff from more impervious surfaces. Fine-grained material such as clays would have formed hard crusts when dry, and would have been relatively impervious to water when wet. Rapid runoff would have resulted in highly fluctuating 'flashy' stream discharge.

Rock weathering due to diurnal temperature changes and wind and water erosion was undoubtedly rapid, and without a mantle of vegetation and the binding action of roots, the products of weathering — rock fragments varying in size from boulders to the finest clay particles — would form unstable, constantly shifting surfaces. There would have been frequent dust storms and large dune fields would have existed.

Where permanent pools of water became established, the high rates of weathering and general aridity may well have led to varying degrees of salinity. The level of salinity would have fluctuated depending on precipitation and evaporation, and this would have imposed a considerable selective pressure on any organism living in such waters or attempting to colonise them.

Rivers would have had high suspended, bed and solute loads and their courses would have changed constantly as they cut down and eroded unconsolidated products of weathering. Braided rivers would have been the norm. For a more detailed account of the pre-vegetation terrestrial environment see Beerbower (1985). This then was probably the harsh, unstable terrestrial environment to which aquatic plants had to become adapted if colonisation of the land was to take place. The problems confronting early land plants were primarily those associated with desiccation.

Before examining the evidence for the evolution of early terrestrial plant life it is necessary to clarify what we mean by the term 'land plants'. Until recently the term was considered by some to be synonymous with vascular plants, but this is incorrect because many plants have a terrestrial existence but do not possess a vascular conducting system. Perhaps the most appropriate concept of land plants is that which concerns those photosynthetic organisms that customarily live on land and whose relations are primarily to other plants living on land (Gray and Boucot, 1977). Therefore not only vascular plants but bryophytes and bryophyte-like plants are included. Excluded, however, are terrestrial algae and fungi (Gray and Boucot, 1977), but this idea may be unnecessarily restrictive in view of the possible early succession of plant forms that may initially have colonised land surfaces.

The fossil record has been criticised for failing to provide evidence covering the evolutionary 'gap' between the presumed green algal ancestors of land plants and land plants themselves, and in particular those with a vascular system. It is becoming increasingly evident that the gap exists not so much in the record itself but in our examination and interpretation of it. It is true that there is a dearth of megafossil evidence for critical phases of land colonisation: fossils of anything like whole plants are exceedingly rare. But this is not surprising in view of the likely fragile nature of the plants themselves and the kinds of environments they were likely to have inhabited. A more profitable avenue of research has been the examination of dispersed fragmentary remains and microfossils, and it now appears that there is evidence for the existence of land plants somewhat earlier than the Upper Silurian, the previously accepted date for the first colonisation of land plants based on the occurrence of undisputed vascular remains.

Algal Ancestors

It has long been accepted that the progenitors of land plants belonged to the green algae or Chlorophyta. Their shared similarities with modern land plants include the type of stored food, cell wall chemistry, pigments, and method of reproduction (Stewart and Mattox, 1975, 1978; Picket-Heaps, 1976). More specifically, the Charophyta are regarded as being the most advanced of the algae and possibly related, but not necessarily ancestral, to land plants.

Largely due to their capacity to precipitate calcium carbonate, the charophytes have a good fossil record, currently beginning in the Pridolian (Upper Silurian). They typically grow in shallow brackish water, but some

Figure 2.1: Diagram of Simplified Plant Life Cycle Showing Alternation of Phases. The sporophyte phase is diploid ($2n$) and the gametophyte phase is haploid (n). The relative dominance of the sporophyte and gametophyte phases differs not only among algal species but also among the various groups of homosporous land plants. For a more complex explanation see text

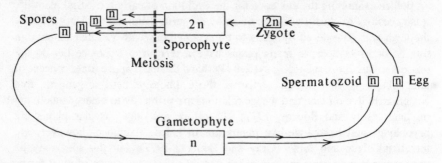

species can tolerate a degree of drying. They have a complex plant body with an axial growth habit, nodes and internodes, corticated axes, parenchymatous organisation and specialised complex reproductive organs. In fact they are so different from the other green algae that they are sometimes treated as a separate class.

The morphological resemblance between the axial growth habit of the charophytes and that of higher land plants is striking, but there is no reason to believe that early terrestrial plants had erect branching axes. On the contrary, they were more likely to have had a ground-hugging thallus, which, because it was prostrate, was not exposed to the full effect of drying winds.

The Sexual Life Cycle

The algae display a wide variety of reproductive methods. Vegetative reproduction and asexual reproduction are common, and to date sexual reproduction has been observed in members of all algal groups except the Euglenophyta.

Sexual reproduction involves the combination of nuclear material, and sometimes cytoplasm, from two organisms usually of the same species. Its profound evolutionary importance is that it provides the opportunity for the exchange and formation of new combinations of genes (for the significance of this see Chapter 5). Although some algal sexual cycles are extremely complex, most may be reduced to the fundamental form shown in Figure 2.1. The reproductive cells, the gametes, are haploid; that is to say they contain a single set of n chromosomes. When these haploid gametes fuse, they form a zygote with a double set of chromosomes ($2n$), one set from each 'parent' gamete. This diploid zygote germinates, without undergoing dormancy, to produce a diploid plant. Before sexual fusion of nuclear material can occur again, the chromosome number must be halved. In the type of reproductive cycle we are considering here, specialised cells of the diploid plant undergo meiotic divisions and give rise to haploid spores. These spores are dispersed from the diploid sporophyte (spore-producing plant) and, perhaps after being carried some distance from the sporophyte, eventually germinate to give rise to a haploid gametophyte plant which produces the gametes, so completing the sexual cycle.

In different species of algae different parts of this basic cycle are emphasised while other parts are reduced. In *Ulva* both the gametophytes and sporophytes are multicellular plants that look outwardly the same. *Ulva* is therefore said to have an isomorphic sexual cycle. Other algae display a heteromorphic life cycle in which the sporophytes and gametophytes are morphologically distinguishable. In the case of the brown alga *Laminaria*, the sporophyte is a large strap-like plant, whereas the gametophyte consists of a small branched thread of cells. In another brown alga, *Fucus*, the sporophyte is the obvious plant, and the gametophyte phase has been reduced entirely to the gametes themselves. Halving the genetic complement by meiosis occurs when the gametes are produced within specialised organs known as gametangia.

In *Chara* the vegetative plant is haploid, and in this case the gametes are formed not by meiosis but by mitotic divisions within the gametangia. Meiosis occurs immediately after fertilisation during germination of the zygote. Therefore in this plant the diploid sporophyte has been practically eliminated.

All sexual life cycles where a haploid phase is followed by a diploid phase again are said to display alternation of phases — a phenomenon which is also a feature of green land plants.

The free-living plant body of most bryophytes is the haploid gametophyte; the diploid sporophyte (the seta and capsule) is dependent to a large extent on the gametophyte for its nourishment.

In vascular plants, however, the sporophyte forms the more conspicuous phase of the life cycle, and, as we shall see in subsequent chapters, improved adaptation to a terrestrial existence is paralleled by a reduction in the gametophyte phase.

Within the algae the gametes also display a degree of differentiation into two distinct forms. In some unicellular algae such as *Chlamydomonas* the organisms themselves may function as gametes and may be morphologically identical, or isogamous. In some cases one gamete of a uniting pair may be morphologically slightly different from the other, and the pair are described as being anisogamous.

The most extreme case of dimorphy, and one which is considered to be the most advanced, is when one gamete (the egg) is considerably larger than the other and non-motile, and the smaller one (the sperm) is free swimming. This condition is known as oogamy and is a feature of, among others, the charophytes, coleochaetes, bryophytes and vascular plants.

Adaptations to Life on Land

Spores

In many respects the potentially most vulnerable time of a plant's existence is during the dispersal phase of its life cycle, and there are distinct selective advantages in favour of those plants which have the capacity to protect the disseminule from adverse environmental conditions.

Spores with durable protective walls are known to be produced by a number of different marine algae, notably those belonging to the Rhodophyta (red algae) and Phaeophyta (brown algae), but none is reported to be resistant to desiccation or the acid treatment (acetolysis) necessary to extract them from rock and prepare them for microscopy. Among modern freshwater and terrestrial green algae, resistant spores are formed, but none is known to possess a persistent triradiate (trilete) mark (Figure 2.2B) indicative of their having been formed in a tetrahedral tetrad (see discussion in Gray and Boucot, 1977).

Although acid-resistant trilete spores are not found in all bryophytes or even ferns, they do seem to be produced only by green land plants, and until it can be demonstrated otherwise the occurrence of trilete spores in the fossil record is cited as evidence for some degree of adaptation to a terrestrial existence.

It has been suggested (Heslop-Harrison, 1971; Gray and Boucot, 1977) that the evolution of the desiccation-resistant spore was a fundamental event which allowed the spread of plant life over land surfaces. Such spores may initially have been produced by aquatic or semi-aquatic green algae in response to periodic desiccation. However, the capacity to tolerate drying enabled the spores to be dispersed by wind and therefore allowed colonisation of isolated or ephemeral water bodies to take place. The establishment of isolated communities in this way under extreme selection pressures from desiccation probably gave direction and impetus to the evolution of a fully terrestrial existence.

Support for the idea that resistant trilete spores evolved prior to other specialisations for life on land is found in the fossil record. At the time of writing the oldest fossil remains that can be used as evidence for some kind of terrestrial adaptation are trilete spores from Upper Ordovician (Ashgillian) and early Silurian (Llandoverian) near-shore shallow marine sediments (Corna, 1970).

The plants that produced these spores are unknown and it would be premature to postulate that they are derived from a flora fully adapted to a terrestrial environment. However, from early through to late Silurian times, when the first undisputed vascular plant remains appear, an increasing number and type of trilete spores occur in the fossil record from both Wales and Libya (Richardson and Lister, 1969; Richardson and Ionnides, 1973).

Cuticles and Stomata

A surface cuticle is a feature of the exposed parts of most vascular plants. The composition and thickness of the cuticle is highly variable, but its function is primarily to help prevent desiccation. In addition to limiting water loss the

Figure 2.2: A, *Spongiophyton nanum*; Portion of Thallus. B, A Spore with a Trilete Mark. C, Portion of the Thallus of *Protosalvinia*. D, *Sporogonites exuberans*. E, Reconstruction of *Parka decipiens*

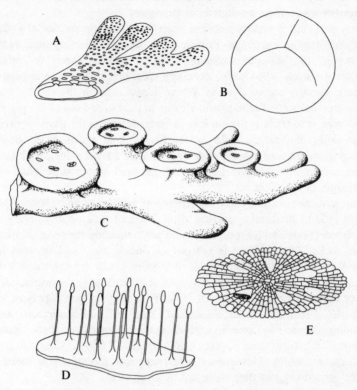

cuticle is also resistant to gas exchange, chemical substances, microbial attack, abrasion and mechanical injury. In an aquatic environment a plant is bathed in a nutrient solution and any barrier to the free exchange of water, solutes and the waste products of metabolic processes across the surface of the plant would be detrimental to its survival. It is no surprise then that aquatic algae do not have a cuticularised epidermis (Church, 1968). In sub-aerial environments, however, the opposite is true and an impervious coat becomes an advantage. This is borne out by the fact that a cuticle is a common feature not only of vascular plants but also of those bryophytes which absorb water primarily by rhizoids rather than over their entire surfaces.

The evolution of a cuticle in marine algae is unlikely not just because of the deleterious effect it would have on the free passage of substances across the plant surface. According to Lee and Priestly (1924) the fatty acids of which cutin is composed would readily saponify in seawater. It appears then that a cuticle probably evolved after the initial land colonisation had taken place or on the emergent distal branches of non-marine aquatic plants. Once the capability of synthesising a cuticle had arisen, a direct positive relationship between the extent of cutinisation and the extent and frequency of exposure to air would have been a relatively simple evolutionary development, but one that allowed plants to grow further away from an aquatic environment.

Although a cuticle greatly restricts water loss from the surface of a plant, it also inhibits gaseous exchange. Provided that the cuticle is very thin, as it is in some bryophytes, gaseous exchange vital to photosynthesis is apparently unimpeded (Watson, 1964), but, generally speaking, cuticles are impermeable to gases, especially carbon dioxide. Where the cuticle is in any way substantial, some method has to be found to allow gases to pass in and out of the plant. The simplest way is to trade off some loss in water retention for gaseous exchange by perforating the cuticle with holes. Living land plants have, however, improved upon this and have evolved the variable aperture device we know as a stoma which controls the passage of gas and water depending on the requirements of the plant.

Fossil evidence for the early occurrence of cuticles is ambiguous. Durable sheets of cells or material appearing to be sheets of cells occur as early as the Precambrian (Muir and Sutton, 1970) and are frequently found in Silurian and younger rocks. The problem is not one of finding the fossil material but of relating it to land plants. The cuticles of modern plants are extremely variable in chemistry, thickness and durability, so it is difficult to generalise about early forms or identify them with any particular type of plant (see papers in Martin and Juniper, 1970, and Cutler, Alvin and Price, 1982). Furthermore, animals can produce cuticle-like coverings which in the dispersed fossil state might be indistinguishable from those of plant origin (Banks, 1975b, c; Gray and Boucot, 1977; Taylor, 1982b). However, the resistant nature of many of these latter 'cuticles' would suggest they were not of algal origin.

The cuticles that occur in pre-Devonian rocks are devoid of stomata, but that does not mean to say that these cuticles do not represent land, or perhaps even, vascular plants. Late Silurian plants proven to have vascular tissue apparently lacked stomata, and it is not until the early Devonian (Gedinnian-Siegenian) that stomata appear in the fossil record. The earliest stomata are reported to occur in *Zosterophyllum myretonianum* (Lele and Walton, 1962) and have been likened not to those typical of modern vascular plants but to those found in moss sporophytes. The stomatal type more usually found in vascular plants apparently did not occur until another 4 to 5 million years later. As Gray and Boucot (1977) point out, stomata are not a prerequisite for a terrestrial existence because functional stomata, even today, are lacking in bryophyte gametophytes.

With the limited competition likely to have been experienced by the earliest land plants, relatively inefficient photosynthesis caused by restricted gas diffusion rates through a thin imperforate cuticle could no doubt be tolerated. As crowding of the land surfaces, especially those near water, increased, so would the premium on more efficient photosynthetic productivity, and the co-evolution of such features as a cuticular coat and stomata became advantageous.

Tubes and Tracheids

While plants remained in an aquatic environment there was little requirement for a translocation system within the plant body to distribute water, solutes and the products of photosynthesis. Passive diffusion through the tissues was sufficient for all metabolic processes. All cells were within a short distance of water, and without the need for structural competence most cells had the capacity to photosynthesise. In a situation where most or all of the plants exist for long periods of time out of the water, the loss of water through evaporation, even with a cuticle, and increasing specialisation of cells away from photosynthesis towards structural and reproductive roles resulted in the need for more fluids, solutes and metabolites throughout the plant. The nutrient medium which previously had bathed the entire plant had now to be contained within the plant body.

To some people the colonisation of the land is synonymous with the development of a vascular system. This clearly need not be the case, however, as many plants, for example mosses and liverworts, successfully grow on land and yet lack true vascularisation. Provided a plant grows on a relatively moist substrate and grows close to the ground, a vascular system is not required. A conducting system only becomes essential when the plant adopts an upright habit, grows away from the humid environment close to the ground surface and exposes itself to the desiccating effect of the wind. It then also requires the development of extensive structural support.

There were probably many selective pressures that led to the evolution of an erect habit. Perhaps the strongest involved the advantages gained in releasing spores into the higher wind energies encountered at greater distances from the substrate surface. This increased the likelihood of longer distance dispersal and therefore the chance of colonising new areas. The initial evolutionary trends in this direction may have begun with the distal positioning of sporangia in emergent aquatic plants and were carried over as a terrestrial adaptation. Another stimulus following the initial colonisation of land may have been the necessity to avoid being grown over by a neighbouring thallus. Associated with an erect growth habit is the possession of a vascular system of some kind, and the first undisputed vascular plant, *Cooksonia*, was apparently small but erect.

Specialised cells forming the function of tracheids (i.e. conduction and mechanical support) are lacking in the algae. Mosses, however, do possess fluid-conducting tubes, called hydroids, that share a number of characters with tracheids of true vascular plants (Hebant, 1977). It would seem likely that early land plants similarly produced systems which, while not necessarily being morphologically similar to tracheids, or even evolutionarily related to them, nevertheless performed the same function. Equally, it would be surprising if the tracheid arose *de novo* without passing through some kind of evolutionary grade. All this leads us to expect to find a variety of tubular structures in early land plants that may, to some degree or another, have fulfilled the role of tracheids.

A characteristic of tracheids which predisposes them to preservation is the lignification of their walls. Lignin is a major constituent of many vascular plant cell walls but is apparently absent in all aquatic plants (Gray and Boucot, 1977) with the possible exception of the green alga *Staurastrum* (Gunnison and Alexander, 1975). It is thought that lignin is a metabolic by-product of photosynthesis and is the result of a detoxification process (Reznik, 1960; Stewart, 1960, 1966; Neish, 1968; Freudenbert, 1968). Land plants are unable to excrete waste products, and typically store them, chemically transformed, in their own tissues. Lignin stored in this way undoubtedly proved to be advantageous to the more erect early land plants because it bonds cells into a composite structure that has improved mechanical integrity, it is resistant to attack by micro-organisms, and it increases the efficiency of fluid conduction in tracheids by decreasing the permeability of their walls (Sarkanen and Ludwig, 1971). Aquatic plants have little need to be mechanically self-supporting because an erect habit can be achieved through buoyancy. On land, however, increasing height requires considerable structural strength of the axis. The constructive use of lignin appears to have been the key to the continued evolution of the upright habit.

A variety of tubes and tracheid-like structures, some of which may be of plant origin, have been recovered from Silurian and Devonian rocks. In 1978 Pratt, Phillips and Dennison described an assemblage of well-preserved dispersed spores, cuticles and tubes, together with some macrofossil remains from siltstone lenses of early Silurian age (Llandoverian). The tubes occur as two forms. The most common type is simple, smooth, imperforate and aseptate, up to 45 μm in length and 20 μm wide. The other form is up to 200 μm long, with internal ring-like thickenings, and often occurs aligned in mats. The ends of the tubes appear rounded or papillate, and in no cases have the tubes been observed arranged end to end.

Niklas and Pratt (1980) claim they have detected lignin or lignin-like degradation products in these fossils and envisage the tubes as being more or less functionally equivalent to a tracheid or a hydroid. The problem here is that the same lignin products can also be obtained from living bryophytes so this biochemical analysis is of no diagnostic value for vascular plants (Logan and Thomas, 1985). Mierzejewski (1982) has pointed out that tubes morphologically like these in Silurian sediments can be produced by graptolites. Therefore, with or without the presence of lignin derivatives, the likely affinities and function of these tubes must remain highly speculative. Until they can be demonstrated to form an integral part of plant bodies their nature must continue to be questioned. It is to macrofossil remains that we must turn for reliable evidence of early land-plant evolution.

Macroremains

As well as dispersed spores, cuticles and tubes, Silurian sediments have yielded a variety of larger pieces of plants. Some are readily identifiable as belonging to bryophytes or even vascular plants, but others remain problematical and are collectively referred to as nematophytes.

Thalloid Land Plants

This group of plants, of unknown affinities, have little in common except that they typically have a robust cuticle and a filamentous pseudoparenchymatous organisation. Some taxa are known to have produced spores. It is not known whether these plants represent specialised near-shore algae or early 'attempts' at terrestrial colonisation.

A thalloid Middle Devonian plant with some interesting features is *Spongiophyton* (Figure 2.2A). Normally preserved only as a thick cuticular envelope, the tubular dichotomously branched thallus may reach up to 2–5 cm long and 2–5 mm wide. The cuticle, on one side, is perforated by irregularly distributed subcircular pores 200–300 μm in diameter. The cuticle of the other surface of the thallus is quite smooth and about one-third as thick as the porate side. Chaloner, Mensah and Crane (1974) consider *Spongiophyton* to be a terrestrial plant. There is currently no evidence to suggest that the plant was vascular, and Taylor (1982b) regards it as being at a bryophyte level of organisation. He suggests the pores may have provided continuity between the air and intercellular spaces and 'may represent a structure only slightly removed from a regulating transport system involving stomata'.

Protosalvinia (Figure 2.2C) is a thalloid late Devonian plant about which there has been some controversy in recent years. It has been variously considered as having affinities with the brown algae or with the bryophytes, but at the present time it would seem appropriate to regard it as an extinct member of a group that has no extant relatives but which displays a number of features found in land plants (Taylor, 1982b). Some people claim to have isolated organic molecules, possibly of cuticular origin (Niklas and Phillips, 1976) from *Protosalvinia* remains, which would support the idea that a waxy covering was present (White and Stadnichenko, 1923). Schopf (1978) on the other hand found no evidence of a cuticle and was of the opinion that the plant was a fully marine organism (Schopf and Schwietering, 1970). Gray and Boucot (1977) envisage the plant as being an emergent aquatic which had only the tips of its thallus cutinised.

Cuticle or no, *Protosalvinia* bore sporocarps with hypodermal conceptacles which contained thick-walled trilete spores in tetrads. The spores measure approximately 200 μm in diameter and are ornamented by minute pits. The large size of these spores had led to the suggestion that they may have functioned as ova (Schopf and Schwietering, 1970), but their thick desiccation-resistant wall is inconsistent with this interpretation. Niklas and Phillips (1976) interpret the tetrahedral tetrads as resulting from meiosis, thereby implying that the thallus represents a diploid sporophyte. There is, however, no real justification for this viewpoint, and it remains unknown whether the spores were meiotic or mitotic derivatives (Taylor, 1982a).

Parka (Figure 2.2E) was an Upper Silurian/Lower Devonian thalloid plant that appears to have lived in shallow marine environments but which was subject to periodic drying (Niklas, 1976b). The thallus was oval or lobed, 5–7 cm in diameter and constructed of radiating files of cells extending from a structure that might have represented a holdfast. The spores were about 42 μm in diameter and were borne in sporangia which were uniformly distributed over the surface of the thallus. They apparently did not occur in tetrads nor did they have a trilete mark. There is a striking similarity between these fossil remains and the modern green alga *Coleochaete* (Coleochaetales).

Another enigmatic Lower Devonian plant is *Orestovia*. Compressed remains of *Orestovia* form coal beds which extend over 100 km along the Baryas River in Western Siberia, indicating not only that the plant was non-marine but also that it grew in enormous quantities over large areas of land. Krassilov (1981) regards *Orestovia* as a leafless vascular plant rather than an alga, but one that was related in some way to *Protosalvinia*.

The stem of *Orestovia* was apparently erect and radially symmetrical, and contained a thin vascular cylinder composed of long tracheid-like tubes with usually spiral, but sometimes annular or reticulate, wall thickenings. It is reported to have possessed three cuticles. The external and middle cuticles were relatively thick, but the inner cuticle was thin. The outer cuticle bore simple stomata. Abundant spores, 150–190 μm in diameter, have been found between the cortex cuticles, presumably indicating some kind of sporogenous tissue rather than a well-defined sporangium. Two other types of spore, one 75 μm in diameter and one 150 μm in diameter, have also been found in association with *Orestovia*, but these would appear to represent contaminants. Many questions remain concerning the precise morphology of this plant and the way in which the spores were produced. It does, however, demonstrate that even in early Devonian times land surfaces were being aggressively and successfully colonised by forms that appear in some respects to bridge the gap between the algae and those plants that we regard as being truly vascular.

Bryophytes

The recent discovery of lower Middle Devonian fossil plants from the USSR that can be assigned to the Marchantiales is regarded as evidence that some major bryophyte divisions were well established by the Lower Devonian and that the bryophytes in general must have played a part in the initial colonisation of land (Ishchenko and Shylakov, 1979).

Pallaviciniites devonicus is an Upper Devonian (Frasnian) thalloid form that resembles the living liverwort *Pallavicinia* in the Jungermanniales (Schuster, 1966). The fossils consist of fragments of thallus, but enough is known to indicate that it was dichotomously branched, had minute marginal teeth, and bore rhizoids.

A Lower Devonian (Emsian) plant with an apparently widespread distribution is *Sporogonites exuberans* (Figure 2.2D). This plant consisted of an irregularly shaped thallus from which arose numerous narrow erect stalks, each terminating in an elongate sporangium. The sporangium contained small (20–25 μm) smooth-walled spores arranged around a central sterile columella. This feature, together with the overall aspect of the plant and the apparent lack of vascular tissue, has led to *Sporogonites* being regarded as bryophytic.

As more Silurian and Devonian plant fossil assemblages are described, it is becoming apparent that there was a richness and diversity of form among the early land plants and their progenitors that hitherto has not been appreciated. It is clear that we must retain an open mind when attempting to interpret these early forms, and critical detailed observation is essential. However, we are still a long way from understanding the sequence of evolutionary changes that took place as algae became adapted to a harsh terrestrial environment and gave rise to the erect vascular forms we will be discussing in the next chapter.

PART II: THE INITIAL DIVERSIFICATION

9

3 EARLY EVOLUTIONARY TRENDS

It had taken nearly 3000 million years for plant life to evolve the simplest vascular design as found in *Cooksonia*. Yet within about 30 million more years the plant form had evolved and diversified to give a multitude of different types, displaying a wide range of vegetative and reproductive characters. The first vascular plants to invade the land had little competition for space, although, as we have seen in the previous chapter, the ground was barren with virtually no organic content. Once the plants had mastered survival on the land they were free to spread and colonise large tracts of the Earth's surface with almost no constraints other than their need for water and a substrate suitable for their growth form.

Evolution at this level in a virgin landscape could have been simply a succession of adaptive radiations related to colonisation of diverse environments. Evolutionary diversification at the lowest levels of plant taxa (species and subspecies) is thought to be due to changes in the pool of genetic variability. It has been stated (Stebbins, 1971) that in today's advanced and genetically sophisticated flowering plants only a small proportion of the adaptive radiation shifts that occur are associated with an increase in genetic information and only a small percentage of such increases lead to further evolutionary opportunity. In contrast, a population of early land plants, with both the simplest of morphology and presumably correspondingly simple genetic make-up, must have had immeasurably more potential for variation and diversification in both structure and genes. Even relatively minor changes in the simple genotypes would have been likely to have resulted in morphological change. Further, the low level of competition would have 'allowed' most changes to survive and many were unlikely to have been selected against. Similarly, the opportunities for genetic interchange via hybridisation and polyploidy must have been considerably higher in such genetically simple plants.

The early changes in form must therefore be seen in the context of the simplest of all plant interactions, allowing an almost free range of developmental change to evolve. The problem that we have to face here is how to make meaningful interpretations of our limited evidence. Fossil plants must be considered in the context of our knowledge of the time in which they existed, without basing all our interpretations upon preconceived ideas from living plants, or upon wild speculation. It is also vital to keep in mind that fossil plants are fragmentary and that they do not represent the total number of plants that existed in any particular past period of time. Evidence is still accumulating so our ideas will no doubt themselves require subsequent modification.

Acceptance of *Cooksonia* (Figure 3.1) as the first vascular land plant is based on Lang's (1937) studies on the Downton floras of the Welsh Borderland and

21

Figure 3.1: Reconstructions of *Cooksonia caledonica* (A) and *Steganotheca striata* (B). Their respective heights must have been at least 7 cm and 5 cm. (After Edwards, 1970)

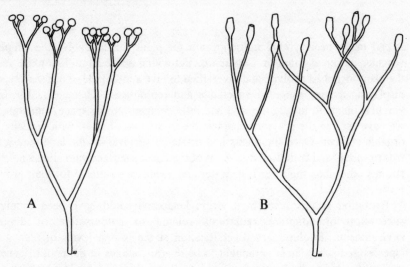

South Wales. He described branching fertile specimens and associated axes with central vascular strands, and concluded that there were two species, *Cooksonia pertonii* and *C. hemisphaerica*; although the latter is now known to be of Ditto- nian (Gedinnian) age. Even though Lang's deductions were based upon this association of characters, his belief that *Cooksonia* was a tracheophyte has found general acceptance.

Several species of *Cooksonia* have subsequently been described from widely separated localities in Ireland, South Wales, Scotland, Germany, Czechoslovakia, Podolia in the USSR, North Africa and North America. They range in age from Wenlock strata (Wenlockian Series, Silurian) in Ireland to Emsian (Lower Devonian). The earliest fertile specimens are very small indeed, clearly demonstrating how susceptible these first land plants must have been to physical changes of the environment. One fertile specimen of *C. pertonii* dichotomised twice to terminate in sporangia, and yet is only 13 mm in overall height with an axis diameter of between 0.3 and 0.4 mm. In contrast, the later *C. hemisphaerica* was possibly a much more conspicuous and probably robust plant having dichotomising axes up to 6.5 cm long and 1.5 mm wide. Apart from size the recognised species also vary in their reproductive organs. The sporangia exhibit various shapes, from rounded in *C. hemisphaerica* to broader and longer in *C. pertonii* and often reniform in *C. caledonica* and *C. crassiparietilis*. Spores have been recovered from only two species. In *C. pertonii* they are 25–35 μm in diameter, trilete and smooth-walled, and in *C. crassiparietilis* they are 50–65 μm in diameter, trilete and covered with spines 1 μm long. Unfortunately there is no evidence available at present of the basal parts of these plants, so we do not know if each axis was an individual plant or part of a creeping mat-like plant. Indeed, we are making an assumption, however justified it may seem, in thinking that they are not branches of a much larger plant.

This simple type of plant with dichotomising vertical axes and terminal sporangia persisted and diversified throughout the late Silurian and early Devonian, although other basic designs in plant form were also becoming recognisable. Our understanding of these primitive plants has developed over the years with discoveries of many diverse forms and of structurally preserved specimens of outstanding interest and botanical value. The accumulated knowledge eventually led Banks in 1968 to propose that the simple Silurian and early Devonian plants could be split into three main forms. These he then used as the basis for establishing the Rhyniophytina, Zosterophyllophytina and Trimerophytina as groups, thereby abandoning the catch-all order Psilophytales which had been used for all simple leafless fossil plants until that time. This threefold grouping is widely accepted today, although there is some disagreement over the level of classification at which the groups should be recognised. Banks originally suggested three subdivisions, but Taylor (1981) has since raised each to the level of divisions.

Another early land plant does however create a very large problem for our understanding of the initial stages of vascular-plant evolution. On the present evidence a flora, with the lycophyte *Baragwanathia*, occurs in Ludlow strata of Victoria, Australia (Garrett, Tims, Rickards, Chambers and Douglas, 1984). The plants from Australia are therefore more complex than are those so far described from the Ludlow of the Northern Hemisphere. It seems that it took up to 40 million years for the Southern Hemisphere plants to migrate northwards to what is now North America. Garrett *et al.* suggest that the Urals geosyncline may have acted as a barrier preventing the dispersal of the floras.

This apparently simultaneous appearance of rhyniophytes and lycophytes in different hemispheres could be taken as evidence for the polyphyletic origin of vascular plants, albeit from within a common pool of archegoniates. However, as suggested in Edwards and Fanning (1985), it may be a little premature to come to any firm conclusions on this. We all wait eagerly for more data on Silurian plants.

The Rhyniophytina, as understood today, is based largely upon the studies of the extraordinary beautifully preserved plants of the Rhynie Chert. This small deposit of volcanic chert found near the village of Rhynie in Scotland was first studied by Robert Kidston and William H. Lang who published their interpretations in a series of papers between 1917 and 1921. Since then there have been several other published works which together have increased our understanding of these remarkable plants.

Kidston and Lang included two species in *Rhynia*, which has since always been thought of as their best-understood genus. However, David Edwards has now conclusively shown them to be very different both morphologically and anatomically. *Rhynia gwynne-vaughanii* (Figure 3.2D) had aerial stems that

Figure 3.2: Devonian Rhyniophytina. A, B: *Aglaophyton major*; A, reconstruction of a plant about 18 cm tall; B, cut-away of a sporangium showing the relative size of the spores. C, D: *Rhynia gwynne-vaughanii*; C, reconstruction of a plant about 20 cm tall; D, cut-away of stem showing its cylindrical protostele of xylem (x) and phloem (p), stomata on its stem surface (s). E, F: *Horneophyton lignieri*; E, reconstruction of a plant about 30 cm tall; F, cut-away of a dichotomising sporangium showing the central columella and the relative size of the spores. (A, after Edwards, 1986; B, D, after Chaloner and Macdonald, 1980; C, after Edwards, 1980; E, after Eggert, 1974)

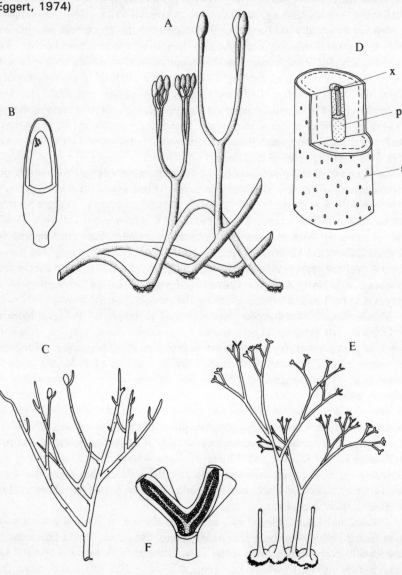

tapered distally to a height of about 20 cm. These arose from prostrate axes that anchored the plants to the substrate by delicate rhizoids, which may also have aided water and nutrient absorption. The vertical axes had both dichotomous and lateral (adventitious) branching, which gave it a monopodial appearance that was accentuated by abscission and overtopping of its terminal fusiform sporangia after they had dehisced and released their spores. Their spores were trilete and all about 40 μm in diameter, showing the plants to be homosporous. The cellular organisation of these plants mirrored the physiological complexity that must have come with increasing size. The entire plant was covered with a thick cuticle to help conserve water. Only the aerial parts had simple stomata in their epidermal layer, suggesting that it was only this region that photosynthesised. The need to transport water and nutrients and the products of photosynthesis presupposes that an efficient conducting system was available. Both rhizome and aerial shoots had a slender central xylem strand of tracheids with broad, annular, rarely spiral thickenings. These were surrounded by a narrow zone that was comparable in position and cellular anatomy to simple phloem, and which presumably functioned in a similar way. The surrounding cortical tissue was two-zoned, the outer zone having more tightly packed cells that extended into small hemispherical projections on the surface. The relative thickness of the cortex and the general paucity of tracheids suggest that it was cortical turgor rather than xylem rigidity that supported the plant. The outer cortex with its tightly packed cells is analogous to a tube having structural strength with minimal use of material. A uniform cortex would have been less efficient; so even here there was specialisation in the economy of tissue.

The other species of *Rhynia* has now been transferred to another genus and called *Aglaophyton major* (Figure 3.2A) on the basis of its having a rather different branching pattern and no central xylem strand (Edwards, 1986). The dichotomously branching naked axes parted at wide angles of over 60° and arose from rather curiously arching rhizoid-bearing axes. Sporangia were terminal and fusiform in shape, and produced spores that were about 65 μm in diameter. The central conducting strand consisted of elongate thin-walled cells surrounding numerous elongate uniformly thickened cells, the central cells of which were narrower and had thinner walls than the peripheral cells. Edwards stressed that the lack of any tracheids in the conducting strand of this plant precluded its assignment to the Tracheophyta and suggested that non-vascular plants with a pteridophytic life cycle existed at the same time as the more highly evolved tracheophytes.

There are other plants that have been considered to be best assigned to the Rhyniophytina even though their affinities are still debatable. Edwards and Edwards (1986) have recently suggested a new broader concept of the Rhyniophytina encompassing a number of levels of organisation that range from isotomously branching forms with simple sporangia to those that have a complexity in both branching pattern and sporangial characteristics. They suggest that this is consistent with the adoption of a variety of strategies in

Figure 3.3: *Renalia hueberi*. A: Reconstruction of a plant at least 20 cm tall. B, C: Cut-aways of dehiscing sporangia (at right angles to one another) showing their resultant two flaps and the relative size of their spores. (A, after Gensel, 1976)

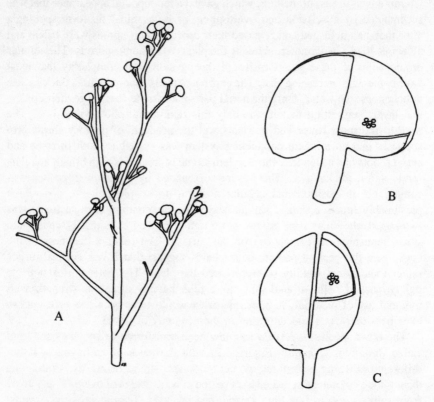

growth and reproductive behaviour by the earliest vascular land plants. In their concept of the Rhyniophytina, plants that did not possess conventional tracheidal thickenings were excluded and those in which evidence was lacking were considered to be 'rhyniophytoid'.

Renalia (Figure 3.3) as described by Gensel (1976) from the Gaspé Sandstone in Quebec can be included in this broad concept of the Rhyniophytina. It had main axes up to 11 cm long and 1.5 mm wide with much smaller lateral branches which dichotomised once or twice before terminating in rounded to reniform sporangia. The sporangia were clearly dehiscent along the distal margin, opening up into two equal halves to shed the several thousand spores that they contained. In contrast, Edwards and Edwards (1986) only include *Cooksonia* and the very similar *Steganotheca* and *Salopella* as 'rhyniophytoid'.

Horneophyton (Figure 3.2E), which also came from the Rhynie Chert, is thought of as a questionable member of the Rhyniophytina. It had dichotomising aerial axes that arose from rather unusual basal parts. Instead of having horizontally growing and dichotomously branching vascular axes, *Horneophyton* possessed a series of corm-like swollen structures that were devoid of vascular tissue. These plants also had unusual sporangia that branched repeatedly with up to three successive dichotomies. These sporangia, like those of *Rhynia*, had stomata; but unlike those in other members of the Rhyniophytina, they had a central column of sterile tissue around which the spores developed. Dehiscence was via a slit at the apex of each sporangial lobe. This condition has been viewed as an intermediate between a typical sporangium with a single spore cavity and the typical synangium consisting of several sporogenous cavities where each is thought to correspond to a sporangium. This would suggest that septation led to synangial evolution (El-Saadawy and Lacey, 1979b).

The development of synangia might, however, have evolved in two different ways. In contrast to *Horneophyton, Nothia* suggests that fusion might have been the cause (El-Saadawy and Lacey, 1979b). *Nothia* sporangia were reniform and dehisced by apical transverse slits. The arrangement of the sporangia on the axes and branches is extremely irregular, being in spirals, pairs, whorls of three, terminal clusters or apparently at random. Occasional fusion seems to have occurred in the terminal clusters where fused sporangia can be seen to have a single spore cavity with two dehiscence lines and fused stalks with two vascular strands. Such an extreme variability in the manner of attachment of *Nothia* sporangia makes comparison possible with many other forms of plants. It also suggests that it was a morphologically unstable condition from which various other arrangements could easily have arisen (El-Saadawy and Lacey, 1979a).

The Zosterophyllophytina is a group of plants that showed increasingly elaborate morphology. They either evolved independently from algal ancestors or developed very rapidly from the earliest members of the Rhyniophytina. Members of the Zosterophyllophytina then gave rise to the lycophytes, a group that has survived to the present day. New plant groups were appearing throughout the early Devonian while their ancestral groups were still surviving. The changes seem to indicate continuous, gradual evolution although none of them was as yet massive. The fossil record clearly shows many plants from both ancestral and descendant groups appearing together in assemblages. Thus plants of the Rhyniophytina grew not only with members of the Zosterophyllophytina and Trimerophytina but also with representatives of the lycophytes which had subsequently evolved. For example, *Trimerophyton* (Trimerophytina) and *Sawdonia* (Zosterophyllophytina) have been found in the same Gaspé flora as *Renalia*. Similarly, the lycophyte *Asteroxylon* is preserved with *Rhynia* in the Rhynie Chert.

The earliest *Cooksonia* species possibly grew in fluviatile areas close to shore or on tidal flats (Edwards, 1979). Later plants clearly extended their range into the hinterlands as they are found in a much wider range of sedimentary rocks. As plants gradually covered the surface of the land, competition was to become a factor of increasing importance and eventually whole plant groups would become extinct. Most of the simpler forms of plants disappeared by the end of

Figure 3.4: Devonian Zosterophyllophytina. A: *Gosslingia breconensis*, reconstruction of a plant about 50 cm tall. B, C: *Zosterophyllum myretonianum*; B, reconstruction of a plant about 15 cm tall; C, portion of fertile spike with sporangia about 2 mm long. D: *Serrulacaulis furcatus*, reconstruction of plants with axes about 60 cm long. E, F: *Sawdonia acanthotheca*; E, reconstruction of a portion of a plant that was up to 1 m tall; F, portion of fertile branch with sporangia about 4 mm in diameter. (A, after Edwards, 1970; B, after Walton, 1964; D, after Hueber and Banks, 1979; E, F, after Gensel, Andrews and Forbes, 1975)

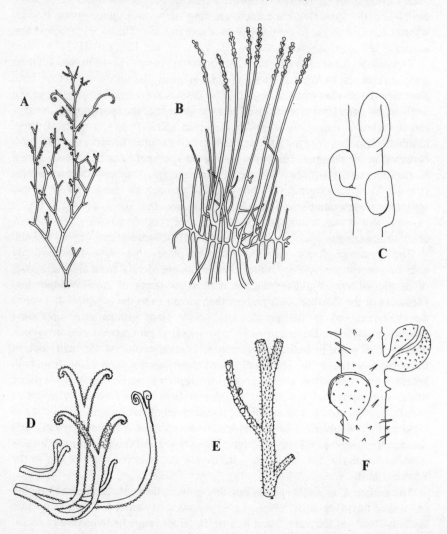

the Lower Devonian. This was the start of a process of vegetational change that would become increasingly apparent in the Upper Devonian and Lower Carboniferous.

Members of the Zosterophyllophytina were relatively small plants which, although branching dichotomously, often exhibited pseudomonopodial patterns (Figure 3.4). *Zosterophyllum* itself had a rather distinctive, so-called H-branching system where a horizontal axis gave off upward- and downward-growing branches, presumably by a series of close 90° dichotomies (Figure 3.4B). This growth pattern led to the suggestion that *Zosterophyllum* was a semi-aquatic plant growing on coastal or estuarine flats. This idea is supported by the discovery of relatively entire plants, suggesting that they grew on mud or sand from which they could be washed out without much damage (Lele and Walton, 1961). Axes of many of the Zosterophyllophytina were often covered with emergences that have been variously described as spines or teeth, and had xylem that was more developed than that of the Rhyniophytina. The central strand was larger, elliptical in cross-section, exarch and consisted mainly of scalariform rather than annular thickened tracheids. However, the main distinguishing feature of these plants was not their stem morphology but their sporangia. They were globose to reniform, dehiscing along their distal edge, but, most important of all, they were borne laterally and not terminally at the stem apices. Various authors have commented on the wisdom of assuming that lateral sporangia evolved from terminally positioned sporangia. Any such change must have occurred very rapidly, for lateral sporangia had already appeared by the Gedinnian — the earliest part of the Devonian. An alternative proposal is that lateral sporangia arose independently in land plants from algal ancestors that had already developed lateral sporangia. The available evidence as yet does not substantiate either hypothesis. The variation in general appearance of the plants assignable to this group has led to their being divided into two families, providing a useful system of grouping similar plants into more manageable units (Banks, 1968). As we will see later, such divisions are largely artificial but become absolutely vital in handling the larger numbers of forms that evolved in the later groups of plants. Plants such as *Sawdonia, Gosslingia, Crenaticaulis* and *Kaulangiophyton* make up the Gosslingiaceae and share the common character of having sporangia scattered along their axes. In contrast, *Zosterophyllum* and *Rebuchia*, with sporangia aggregated into terminal spikes, make up the second family, the Zosterophyllaceae. Although we are not certain how lateral sporangia originated, it does seem highly likely that scattered sporangia would have given rise to others with aggregated spikes of sporangia. But again the fossil record does not give us any suitable confirmatory evidence for such a hypothesis. Indeed *Zosterophyllum*, with its terminal spike, was the first member of the group to appear, being known from the Gedinnian of Europe.

Zdebska (1982) has suggested that her genus *Konioria* from the Emsian of the Western Carpathians of Poland might even merit the establishment of a third family. *Konioria* seems to have been a small plant with dichotomising erect axes that terminated in recurved hook-like points. The axes had up to four longitudinal wings and were covered in spines of variable length. The xylem was exarch with scalariformly thickened tracheids. The sporangia were rounded to reniform in shape and dehisced into two equal valves. But it was their positioning which was so unusual and important. They were borne singly below dichotomies and usually on short stalks, thus being quite unlike the other two families.

Although evidence for the origin of and the evolutionary changes within the Zosterophyllophytina seems to be minimal, evidence for the group giving rise to the lycophytes is virtually incontestable. There are several avenues that we can explore here. Both groups of plants possessed exarch protosteles and reniform sporangia. There is also good evidence to support the notion that lycophyte leaves evolved from the emergences of their ancestors. Presumably such extensions of stem tissue must have given these plants more surface area for gaseous exchange and light-catching ability, coupled with a greater internal volume for photosynthetic activity. These morphological adaptations would presumably have made the plants physiologically more effective and therefore more competitive than their neighbouring rival species.

Asteroxylon, from the Rhynie Chert, is an early lycophyte that can be considered to be primitive in many of its morphological features (Figures 3.5 and 3.6). The plants have leafless horizontally growing rhizomes with upright, monopodially branching leafy shoots that grew up to 50 cm tall. The sporangia were kidney shaped and borne randomly among the leaves on the terminal portions of the upright axes. What was unusual was that the leaf traces extended only to the base of the leaves without entering them, unlike those of all other lycophytes. It is certainly plausible that such morphologies must have existed during the early diversification of lycophytes. *Asteroxylon* seems to represent a stage in the evolution of the group in which vascular tissue had not yet differentiated in the leaves in addition to the cortex of the stem. Any further slight physiological stimulus presumably would have been enough to cause further tissue differentiation leading to vascularisation of the leaves. Such further change must have occurred fairly rapidly, for there are Lower Devonian lycophytes, such as *Drepanophycus* (Figure 3.6A), which had well-developed leaves with vascular strands extending along their entire lengths. These leaves, which may have evolved via a modification of unvascularised emergences, are often called microphylls.

Zdebska (1982), however, has put forward a hypothesis which suggests a rather different origin for sporophylls than from emergences. She has proposed changes from the random lateral arrangement of sporangia in plants like *Gosslingia* via that in *Konioria* to the branching sporophylls of such lycophytes as *Colpodexylon* and *Leclercqia*, then by reduction to the undivided sporophylls of the type shown by the Devonian lycophyte *Cyclostigma*. If this idea is correct, then it confirms the axial origin of the lycopod sporophyll and the explanation by Zimmerman (1930) of the origin of the lycopod sporophyll from sterile and

Figure 3.5: *Asteroxylon mackiei*. A: Cut-away of stem showing stellate xylem (x), phloem (p), leaf traces (t) terminating at the bases of the leaves (l), stomata on the leaves and stem surfaces (s). B: Shoot apex showing leaves and lateral sporangia. C: Cut-away of sporangium showing the relative size of the spores. (After Chaloner and Macdonald, 1980)

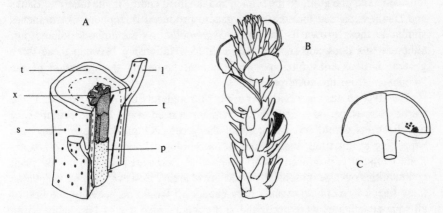

Figure 3.6: Reconstructions of Fertile Devonian Lycophytes. A: *Drepanophycus spinaeformis*. B: *Asteroxylon mackiei*. Both plants stood about 50 cm tall. (A, After Krausel and Weyland, 1935; B, after Kidston and Lang, 1920)

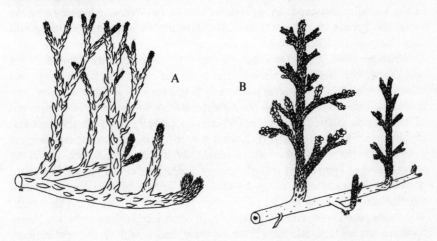

fertile axes by overtopping and reduction. There are clearly contradictory ideas here, so we must wait for further confirmatory evidence before we can be confident that we understand the origin of sporophylls.

It has been suggested that *Crenaticaulis* and *Gosslingia* contribute to the hypothesis that lycophytes descended from the Zosterophyllophytina by virtue of their unusual anatomical features (Banks and Davis, 1969). *Crenaticaulis* had branches attached on one side of the stem. These branches were similar to the main axes except in not possessing their marginal teeth. Scars were sometimes present in their place resembling what are often called axillary tubercles. *Gosslingia* had similar tubercles and a complex associated vascular system. The vascular strand divided twice below a branching point; one strand continuing up the main axis, one going to the branch and the third ending in the tubercle. Banks and Davis suggested that these two genera produced rhizophore-like branches similar to those grown by the living *Selaginella*. As the authors pointed out, although this does not demonstrate a direct relationship between these three genera, it does constitute another argument favouring the evolution of the lycophytes from the zosterophylls.

The lycophytes diversified rapidly throughout the Devonian and the Carboniferous periods, developing in many varied ways. Their evolutionary modifications would seem to parallel the great ecological changes that were happening at this time. Innovations in morphological and reproductive adaptations eventually produced giant plants that dominated vast areas of plant communities over geographically widespread areas. However, before we pursue these lines of lycophyte evolutionary change, it would be better to see first of all what adaptations were occurring in the other group that had probably arisen from the Rhyniophytina, the Trimerophytina.

The Trimerophytina was the group that held the key to successful plant evolution. Although there are few genera included within this group, these plants gave rise to all the major groups of vascular plants that have since dominated most of the world's landscape. From descendants of the Trimerophytina came the ferns, the cycads, the conifers and ultimately the angiosperms; all being still very much in evidence today.

What provided these Devonian plants with the evolutionary potential not possessed by more advanced members of the Rhyniophytina and Zosterophyllophytina? In general it was their morphology and in particular their branching patterns which gave them their capacity for large-scale change and diversification. Their main axes branched monopodially to give lateral axes that rapidly divided many times by dichotomies or trifurcations. Such morphological division into main and lateral axes brought about a fundamental change in the organisation of growth and indeed the very appearance of the plants. Continued growth of the main axes, coupled with limited growth of the laterals, resulted in taller and more bush-like plants than had previously existed in the Devonian. This shape, combined with a relatively large centrarch protostele of scalariform-bordered tracheids, would have given the plants more stability, stimulating them to grow even taller. We know that one species, *Pertica varia*, even reached nearly 3 m in height.

The genus *Psilophyton* was originally used as the basis for the now abandoned group, the Psilophyta. Kidston and Lang, in working on the flora of the Rhynie Chert, recognised an affinity between *Rhynia* and *Psilophyton*, which had been created by Dawson in 1859 for spiny, dichotomously branching vascular plants found in the Devonian of the Gaspé Peninsula in Canada. In

Figure 3.7: Reconstructions of Devonian Species of *Psilophyton*.
A: *P. dapsile* was about 35 cm tall. B: *P. microspinosum* was about 50 cm
tall. C: *P. forbesii* was about 1 m tall. D: *P. princeps* was about 2 m tall.
E: *P. dawsonii* was about 25 cm tall. (A, B, C, D, After Andrews *et al.*,
1977; E, after Banks *et al.*, 1975)

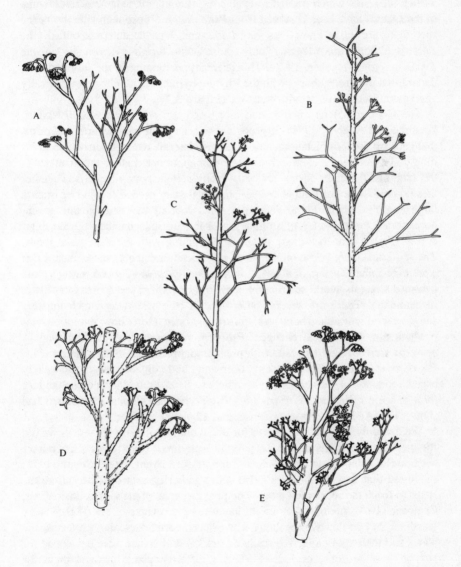

acknowledging the validity of Dawson's work, Kidston and Lang named the group Psilophyta. From that time the group became a repository for all the simplest land plants, concealing within it the affinities that Banks ultimately used as the basis for his tripartite classification. *Psilophyton* itself similarly became a catch-all genus which needed re-evaluating (Banks, 1975a). As a result some of the species have been removed from *Psilophyton*. Those with laterally borne sporangia are now known as *Sawdonia* and have been removed to the Zosterophyllophytina, leaving only those with terminal sporangia in the Trimerophytina (Hueber, 1971). This terminal positioning of the sporangia is a feature that the group shares with the Rhyniophytina, but here they are typically aggregated into fertile lateral branches (Figure 3.7).

The most completely understood member is *Psilophyton dawsonii* (Banks, Leclercq and Hueber, 1975). It grew into a large plant with several main axes that gave off smooth dichotomising vegetative laterals which branched up to six times. The stem had a centrarch protostele of about one-quarter the diameter of the axis, and the outer cortex contained many collenchyma-like cells. Together these would have constituted the supporting tissues needed to hold up such a large complex plant. Photosynthesis, still a role of the stem in this group, necessitated the presence of stomata and sub-stomatal chambers in the outer cortex. Fertile branches arose in zones alternating with the vegetative zones. The sporangia were borne on the repeatedly dichotomising fertile branches that were in terminal clusters of about 32 in number. Each was about 5 mm long and dehisced along its lateral surface to release the smooth-walled spores (40–70 μm in diameter). There are several other species of *Psilophyton* which together show several variable characters illustrating once more how evolution was producing morphological changes. *Psilophyton dawsonii, P. forbesii* and *P. princeps* were at least 1 m tall, with pseudomonopodial main axes. In contrast, *P. dapsile* was under 50 cm tall and had dichotomous to weakly pseudomonopodial branching axes. Whereas *P. dapsile, P. dawsonii* and *P. forbesii* were glabrous, *P. charientos, P. microspinosum* and *P. princeps* had spiny axes (Andrews, Forbes, Gensel and Chaloner, 1977).

We can only guess at the reason for the evolution of these spines, as we did for the larger enations of the Zosterophyllophytina. They might simply have been useful adaptations that helped support scrambling axes. Smart (1971) suggested that they might have aided spore-gathering arthropods to climb the stems to reach the sporangia; thereby helping dispersal of the spores. In contrast, Chaloner (1970) thought they might have had the reverse effect. If they were glandular they could have repelled herbivorous invertebrates, thus protecting the plants and their sporangia. We really do not know if spines were developed for any of these reasons, or indeed whether different plants evolved them for different reasons.

Trimerophyton robustius, from which the group gets its name, was itself originally described as a species of *Psilophyton* by Dawson (1859). Unlike this other genus, *Trimerophyton* has a characteristic trifurcating branching pattern.

Figure 3.8: Reconstructions of *Pertica*. A: *P. varia*; representative portion showing the main axis with sterile and fertile branches. B: *P. varia*; partial restoration of a sterile branch that is essentially pseudomonopodial. C: *P. quadrifaria*; partial restoration of a dichotomous sterile branch. (A, B, After Granoff *et al.*, 1976; C, after Kaspar and Andrews, 1972)

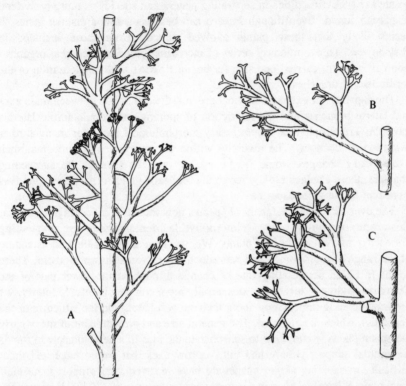

The numerous lateral branches divide trichotomously twice before continuing to branch dichotomously. Sporangia are terminal and are borne erect in groups of three.

Pertica varia (Figure 3.8) was the largest member of the Trimerophytina yet known, standing nearly 3 m tall (Granoff, Gensel and Andrews, 1976). This species, and the smaller *P. quadrifaria* (Kaspar and Andrews, 1972), had distinct main axes with numerous, much branched, laterals. In *P. quadrifaria* the primary branches were in four ranks and alternate, while in *P. varia* they were in subopposite pairs. In both these species lateral branches divided many times: in *P. quadrifaria* by wide-angled dichotomies; in *P. varia* by more or less equal dichotomies to strongly pseudomonopodial divisions. Sporangia, which dehisced longitudinally, were in terminal clusters on fertile branches. In *P. varia* these fertile branches were much smaller than the sterile ones. Most unfortunately we know nothing about the internal anatomy of these plants. This is a great pity because their large dimensions suggest that they might have started to develop a structural feature that we have not yet encountered, namely secondary thickening.

Pertica presents a taxonomic problem since it is so similar to the more advanced species of *Psilophyton*. Granoff *et al.* (1976) and Gensel (1979) have pointed out the difficulties in separating genera and species as more early land plants are found. Even though *Pertica* can be accepted as a distinct genus, it seems likely that these plants evolved directly from those included in *Psilophyton*. An evolutionary series of increasing pseudomonopodial organisation can be seen, carried one step further in *P. varia* with condensation of the fertile lateral branches.

The capacity for extended vertical growth, the distinction between main axes and lateral branches and the restriction of sporangia to certain fertile laterals gave the Trimerophytina the necessary morphological bases for many further evolutionary changes. The diversity within the group also ensured that such evolutionary changes would be varied. As will be seen in the succeeding chapters, these changes rapidly led to the ferns and other forms that themselves gave rise to the gymnosperms.

The ever increasing numbers of species now included in the Rhyniophytina, Zosterophyllophytina and Trimerophytina demonstrate our expanding knowledge of these early land plants. We now have a reasonable understanding of the range of morphology and reproductive structures shown by them. There seems to be an accelerating rate of change throughout the lower part of the Devonian leading to larger and structurally more complex plants. Similarly, we are accumulating information about their growth habits and the structure of the vegetation which they formed. The general size and growth habit of many early Devonian plants would seem to substantiate the idea of a need to grow in stands for mutual support. Many had tall, narrow axes that would have had great difficulty in standing alone, but would have received vital support from their neighbours if they had been in dense stands. Andrews *et al.* (1977), in discussing the ecological setting of the later early Devonian landscape in northern Maine in the USA, came to a similar conclusion. Many slabs of rock are covered with large numbers of parallel aligned axes of single species. This suggests that the plants were growing in dense pure stands that were periodically flattened by the sediments that now enclose them. They likened the growth habit of their plants to stands of modern grasses and sedges, but perhaps a better model might be the branching species of the living *Equisetum*: for example, *E. giganteum* often attains heights of nearly 10 m when growing in dense patches. If the Devonian plants did grow in stands like our living models, it seems logical to suppose that they may have similarly spread by vegetative horizontal axes.

There is, however, one large gap in our knowledge of the early Devonian plants. We know so very little about their life histories. All the plants so far described have been sporophytes. They represent those parts of the plants' life patterns that form sporangia and spores for reproduction. Such *in situ* spores have been recovered from about twenty species of these plants and all show the trilete mark characteristic of meiotically produced spores (Gensel, 1980).

The apparent lack of any gametophyte plants has prompted several people to look for evidence in the best-preserved Devonian material, the Rhynie Chert. Some germinating spores have been described by Lyon (1957), but unfortunately they were isolated so we have no idea of the parent plant. Controversy has, however, settled upon those axes originally described by Kidston and Lang as *Rhynia gwynne-vaughanii*. Merker (1958, 1959) thought some of its axes were gametophytic; Pant (1962) suggested that it was entirely gametophytic with archegonia and young sporophyte plants; and Lemoigne (1971 and earlier papers) believed it to be the gametophyte plant of *R. major*. He, like Pant, described the axes as having archegonia and also thought there to be antheridia. D.S. Edwards (1980) disagrees with these interpretations of *R. gwynne-vaughanii* as entirely gametophytic, convincingly demonstrating that some axes were sporophytic as described earlier in this chapter. The problem really centres upon the interpretation of certain poorly preserved structures either described as sex organs or dismissed as lesions caused by fungal attack or arthropod damage.

Fungi are certainly known from the Rhynie Chert, for hyphae can be found in many of the axes. Kevan, Chaloner and Saville (1975), however, believe that most fungal activity followed rather than caused the lesions. They suggest that the sap-sucking arthropods or other metazoans attacked the Rhynie plants for food, thereby initiating traumatic response to their physical injuries. Some wounds even extended to the phloem-like region of the stele, having been subsequently filled with a black plugging material that presumably exuded from the plant tissue. Once again, it seems highly likely that these were caused by sap-feeding arthropods.

Remy and Remy (1980b,c) have described two other Lower Devonian plants that they believe to be gametophytes (Figure 3.9). *Lyonophyton rhyniensis*, from the Rhynie Chert, is thought to have a rhizome with upright axes, each terminating in a lobed gametangiophore cup. The anatomy of both kinds of axis is virtually identical to that of *Rhynia* and *Horneophyton*. Stalked antheridia were individually attached to the surface of the gametangiophore lobes, and the archegonia are thought to have been centrally attached in groups on a common stalk. Remy and Remy tentatively suggested on the basis of the plant's lobed morphology and on evidence of association that *Horneophyton lignieri* might have been the corresponding sporophyte plant. However, they also make the suggestion that it seems to be the gametophyte of a plant representing an evolutionary stage prior to the separation of land plants into bryophytes and higher land plants (Remy and Remy, 1980a,b). This latter suggestion is completely contrary to the idea that bryophytes and vascular land plants arose independently from the green algae before invading the land.

Sciadophyton is also thought to be a gametophyte (Remy and Remy, 1980c; Schweitzer, 1981). Examples from the Siegenian of West Germany have a variable number of vertical axes which occasionally dichotomise or give off

Figure 3.9: Putative Devonian Gametophytes. A: *Lyonophyton rhyniensis*, about 3 cm tall. B: *Sciadophyton steinmannii*, about 7 cm tall. (A, After Remy and Remy, 1980a, b; B, after Remy and Remy, 1980c)

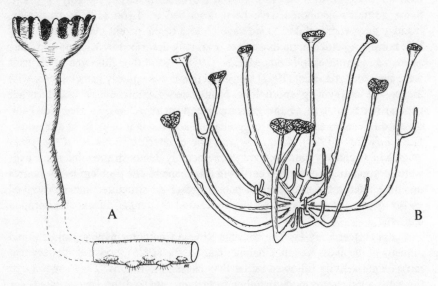

laterals at right angles. These axes, which are described as being stelar, terminate in cup-shaped 'gametangiophores' containing what have been called gametangia.

There is clearly much disagreement over these rather debatable Devonian gametophytes. All our information is from compression fossils and we have no knowledge of their internal anatomy. However, it seems highly probable that the gametophytes of early land plants were long-lived and robust. If this were true, they might easily have had vascular conducting tissue. Part of the supporting argument for this contention is the structure of the gametophyte of the living pteridophyte *Psilotum*. It is very much like the rhizome of the sporophyte plant; both are subterranean and branched. But what is more important is that the larger individuals of the diploid race of gametophytes have central steles.

The present evidence for Devonian gametophytes is still very limited and seemingly rather debatable. Even accepting that *Lyonophyton* and *Sciadophyton* are gametophyte plants, we still have precious little idea about the total biology of the plants or their relationships to other Devonian sporophyte plants. Much fascinating information about the life histories of early land plants is yet to be discovered.

4 GEOLOGICAL TIME SCALE, FOSSILISATION PROCESSES AND EVOLUTION

A lack of appreciation of the enormous lengths of time represented by the layers of the Earth's rock was one of the problems encountered by the biologists of the last century as they tried to get to grips with the revolutionary ideas of Charles Darwin. Changes in geological thinking during the late eighteenth and early nineteenth centuries set the stage for the appearance of a credible theory of evolution.

It was James Hutton who, in his Theory of the Earth communicated to the Royal Society of Edinburgh in 1785, first presented an irrefutable body of evidence showing that the mountains and uplands of the present day, far from being everlasting, were continually being resculptured by erosion. The resulting alluvial sediment was redeposited elsewhere, usually in lakes or the sea, to form the basis for new rocks. He realised, as no one had done before, that the vast thickness of these older sedimentary rocks implied the operation of erosion and deposition throughout a period of time that then could only be described as inconceivably long. Hutton was also the first to realise that the doctrine known as catastrophism, which held that the Earth was subject to violent and super-natural cataclysms (e.g. the Flood of Noah), denied the basic orderliness of nature, a principle upon which science up to that time had been based. Hutton stated that 'the past history of a globe must be explained by what can be seen to be happening now'. This principle was later formalised into the 'Principle of Uniformitarianism' by Sir Charles Lyell. This principle neither implies a uniform rate of deposition nor excludes naturally occurring local catastrophic events (such as volcanic eruptions or flood events) which unfortunately some people thought it did. In short it stated that 'no powers are to be exployed [invoked] that are not natural to the globe, no action to be admitted except to those of which we know the principle'.

In most cases, sedimentary rocks are those that are made from the erosional products of other pre-existing or contemporary rocks. From the time when the first solid crust formed on the Earth under a primitive atmosphere, rock fragments became broken off from their parent rock by physical and chemical processes, were transported by gravitational force and/or fluid flow and began to accumulate elsewhere, thus forming the first sedimentary deposits. A sedimentary deposit may be defined as a body of solid materials accumulated at or near the surface of the Earth under the temperatures and pressures that characterise this environment. Sediment is generally, but not always, deposited from a fluid in which it was contained either in a state of suspension or in solution. This includes, as well as sands, clays and chemical precipitates, the fragmental materials ejected from volcanoes and deposits on the deep-sea floor.

Most sedimentary rocks are stratified. Layer upon layer of sediments are built up gradually and the strata deposited near the top of a section or sedimentary pile are clearly laid down after those below them. Occasionally rocks are bent or fractured and a given rock exposure may exhibit a succession that has been completely overturned (when this happens it can usually be recognised), but most sedimentary rock units can safely be assumed to be younger than those they overlie.

Towards the end of the eighteenth century the Industrial Revolution prompted a large number of civil engineering projects to be undertaken, such as building canals, collieries and, later, railways. William Smith was originally a land surveyor but as a hobby he had collected fossils, then considered only curiosities. In the course of his work he made separate collections from the various sedimentary rock units to which his work led him. He found that each rock unit had a suite of fossils largely unique to itself, and that using fossils alone he could tell the rocks apart. Moreover the fossil types were similar over large geographic areas within the same rock unit. Therefore using fossils Smith could correlate widely separated rock units. The relative ages or positions in the sedimentary sequence could be determined, and by 1808, following similar discoveries by Cuvier and Brongniart in France, it was possible to correlate formations on both sides of the Channel. Biostratigraphy, as it was later known, developed rapidly.

It was noted that, generally speaking, older rocks contained organisms more different from those alive today than did the younger rocks. Furthermore, over and over again, different groups of fossils exhibited increasing complexity of form as they came from younger and younger sediments. Everywhere the sequence of fossils revealed a gradual development of different forms of life and it became possible to divide the whole of the fossiliferous stratified rocks into appropriate divisions or units, each with its distinctive fossils. The oldest fossiliferous rocks known at that time comprised a major division known as the Palaeozoic or 'ancient life'. This was followed by rocks laid down during the era of 'middle life' or Mesozoic, and this was followed by the Cenozoic or 'recent life'. What was missing was an ability to divide the rocks into units of real time.

If the thickness of sediment can be measured and the rate of deposition is known, then the total time interval between the bottom and the top of a rock unit can be determined. Unfortunately the rate of deposition is rarely known, and such computations are impossible for all but a restricted set of glacial lake sediments.

Early attempts to put absolute ages on the relative chronology based on fossil evidence centred around assumptions as to the time taken for the Earth to cool from a molten state to its present temperature. In 1862 the physicist Lord Kelvin arrived at an age for the Earth of between 20 and 400 million years and by 1897 had narrowed the limits to 20 and 40 million years. Unfortunately one error in his calculation (but which he admitted was a possibility) was that he disregarded the fact that the Earth continued to generate heat once the solid crust had formed. It was of course not appreciated when Kelvin computed his figures that radio-activity was responsible for this heat. It was radioactivity, however, that proved the key to a chronology that could be related to a known passage of time.

All radioactive elements decay over time into other elements. Some elements do this extremely rapidly and others extremely slowly. The rate at which this decay occurs for any given element is, however, constant irrespective of the physical and chemical conditions to which it is exposed. Thus if the ratio of the first element to that of its products is measured and the rate at which that change occurs is known, then, provided there are no losses or gains of either the parent or product (daughter) elements, the time over which the decay has been taking place can be measured.

Fortunately there are a number of these radioactive clocks. Uranium 238 decays to lead 206 with a half-life of 4498 my. Other examples are uranium 235, thorium 232, rubidium 87 and potassium 40:

$U_{235} \rightarrow Pb_{207}$ (half-life 713 my)
$Th_{232} \rightarrow Pb_{208}$ (half-life 13900 my)
$Rb_{87} \rightarrow Sr_{87}$ (half-life 50000 my)
$K_{40} \rightarrow A_{40}$ (half-life 11850 my)
$K_{40} \rightarrow Ca_{40}$ (half-life 1470 my)

Elements like rubidium and potassium are found in a number of common minerals with properties suitable for dating. Some minerals are formed in the oceans and some occur in volcanic ashes, so radioactively datable horizons frequently occur interspersed in other fossiliferous sediments providing fixed points in the chronology.

Fossils

A fossil is: 'any specimen that demonstrates physical evidence of occurrence of ancient life (i.e. Holocene or older)' (Schopf, 1975), and is formed when a living organism dies and becomes buried by sediment. Usually only the hard parts of an organism are preserved as these are most resistant to decay. Under exceptional circumstances, however (usually very rapid burial in fine-grained sediment, immersion in mineral-rich water or freezing), soft parts may be preserved. Burial and preservation usually take place in an aqueous environment.

The fossils we find and study represent a minute proportion of organisms that were alive in the past. To illustrate this let us consider an object such as a leaf and the selection processes it is exposed to before becoming a specimen under study in the laboratory (Figure 4.1).

Assuming the leaf completes its useful life on the parent tree and is not eaten

Figure 4.1: Diagram Showing the Possible Ways in which a Potential Leaf Fossil Might be Destroyed before Being Studied by an Evolutionary Plant Biologist. See text for details

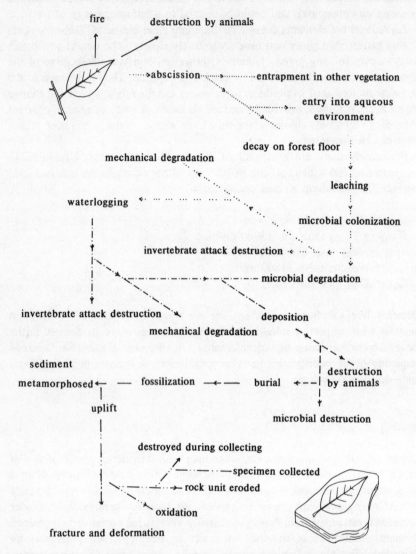

by insects or disintegrates in place (as leaves of herbaceous plants do), eventually it will be abscised. Unless the tree is overhanging water the leaf must be blown to a site of deposition. The leaves at the top of a tree are exposed to higher wind energies than those below (particularly in closed canopy forests) and therefore may be dispersed over a wider area and have a greater chance of entering a depositional environment. The distance which any given leaf will travel under a given wind regime will depend on the inherent aerodynamic properties of the leaf. Its shape, size, weight per unit area (including moisture content), presence/absence of a petiole and whether it is curled or flat all affect the 'flight' properties (Spicer, 1981; Ferguson, 1985). If a leaf does not enter directly into water, it may become entangled in the vegetation and disintegrate over time, or it may fall on the forest floor and rot or be eaten.

Let us assume our leaf does enter a river or lake. It will immediately begin to lose soluble compounds such as sugars and soluble tannins and it will also begin to absorb water. But before it sinks it will float for a while, the actual floating time being dependent upon water chemistry, temperature, agitation, leaf chemistry, anatomy, surface characteristics and degree of damage. While floating the leaf may be transported by water flow or wind or both. Micro-organisms will invade the leaf at a rate that is in part governed by water temperature, chemistry, leaf anatomy and leaf chemistry. The invasion of micro-organisms will then affect the mechanical strength, water absorption properties, and density of the leaf, and predisposes the leaf to attack by both large and small particle invertebrate feeders. Eventually when the leaf sinks it may be transported for a while in suspension by water flow or by rolling (salta-tion) along the river- or stream-bed. As the effect of microbial breakdown weakens the leaf, it may become mechanically damaged by flexing in the fluid flow or being knocked against obstacles. The activity of invertebrate feeders may also destroy it.

If it arrives at a depositional site, the water energy will fall and the leaf will be deposited on a lake- or stream-bed. Here it will be covered by inorganic sedi-ment, if it does not decay and is not eaten. Sometimes sediment build-up on the leaf surface can occur during the transport phase, thereby slightly protecting the leaf but increasing its density. Leaves that are buried rapidly will stand a better chance of being preserved because they are protected from being eaten. As sedi-ment build-up continues, the environment of the leaf will be gradually depleted of oxygen by the activity of micro-organisms, and eventually the microbial activity will be severely slowed by the anaerobic conditions.

The next stages of fossilisation are not well understood and the method by which the leaf may be preserved will depend on the composition of the organic material and its entombing matrix.

Widening our consideration now from leaves to other organic remains such as wood, cones, fruits, bones and shells, preservation may take place by cellular permineralisation, coalified compression/impression, and authigenic preserva-tion. These processes are illustrated in Figure 4.2. Sometimes the original material comprising the organisms is so durable that it survives more or less unchanged. Examples of this 'duripartic preservation' as it is sometimes called are some shells and bones, plant cuticles, spore and pollen walls, and fossil charcoal.

These preservation categories are not mutually exclusive and there is considerable gradation between one form and another. For example, plant

Figure 4.2: Fossilisation Processes. The diagram illustrates the changes that take place during fossilisation of a hollow stem. The left-hand side of each illustration depicts the stem in cross-section in relation to the surrounding sediment, and the right-hand side shows the changes that take place at the cellular level. Illustrations a, f and k show the stem at the time of deposition on a sub-aqueous sediment surface. In b, g and i, the stem has become buried, and in i, intracellular mineral precipitation has begun. In c and d, the plant tissues have rotted away leaving a cavity infilled with sediment. This yields a cast, e, of the external morphology of the stem, the surrounding sediment forming the mould. No internal (cellular) information is preserved. In h and i, compression has begun distorting both cellular and morphological information in the vertical plane. Coalification takes place with the loss of the more volatile organic components. Splitting the rock along the bedding plane (j) reveals a compression and an impression of the stem. In m, mineral precipitation within the cells has given rise to a three-dimensionally preserved permineralisation. If mineralisation continues and the organic cell walls are also replaced, a petrifaction is formed (n) which on recovery (o) reveals a three-dimensional specimen with cellular detail still intact

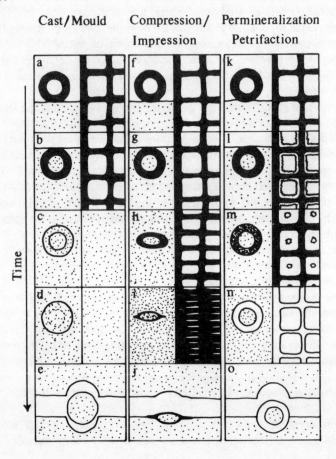

remains may become partly compressed before becoming permineralised. Similarly a spore-bearing cone may become compressed but still yield the relatively unaltered coats of the spores it contains.

Compression fossils can be studied by direct observation with both light and scanning electron microscopes. Oxidative acid treatments are also usually employed to remove the coaly substances and to liberate the relatively unaltered cuticle or spore coats contained within the specimen.

Impression fossils often yield considerable information about the surfaces of plants even though no organic material may be present. Details of the size, shape and arrangement of epidermal cells, stomata, and trichomes may be present. The high quality of preservation exhibited by some impressions may be due to the fact that the impression can begin to form even before the plant organ has been buried (Spicer, 1977). Plant surfaces can become covered with precipitates within days of entering an aqueous environment, possibly even while the plant parts are still being transported. These coatings not only serve as the basis for the impression but they also protect the plant surface from abrasion and biological attack.

Often organic matter remains within the crevices of the impression obscuring the finer details, but this can be oxidised away by burning (Chaloner and Collinson, 1975) and a faithful copy of the original plant surface may be obtained by making a mould of the impression in silicone rubber (Watson and Alvin, 1976).

Similar surface detail can sometimes be preserved in natural casts and moulds. No internal organic structure is present but sometimes sectioning of such specimens reveals sedimentological and/or mineralogical information which may be useful in elucidating the circumstances of burial and the diagenetic history of the specimen.

Traditionally palaeobotanists have concerned themselves with examining the individual specimen and describing its anatomy and morphology. However, such individual specimens usually represent only parts of a whole organism. Furthermore, what we understand as a living plant consists not just of an individual organism but has a complete life cycle. If we just restrict the concept to an organism such as a tree, it will inevitably be represented in the fossil record as a series of detached organs. The flowers, fruits, seeds, leaves, trunk, branches, roots and pollen will all have different 'preservation potentials'. Flowers are ephemeral, delicate organs that usually develop into fruits. They are therefore rather rare in the fossil record. Leaves are produced in large numbers and are relatively robust, and many are shed at the end of their useful life; therefore they are common in the fossil record. The most common fossils of terrestrial plants are, however, pollen and spores. Not only are they produced in vast numbers but their coats are very resistant to degradation. These different organs will inevitably be transported various distances away from the growing plant before being deposited in sediment. An assemblage of plant debris composed of remains that have undergone some transport is termed allochthonous and consists of a jumble of organic debris derived from many different plants; whereas an assemblage deposited essentially *in situ* (e.g. peat

Figure 4.3: Whole Plants are Rarely Fossilised Intact so Reconstructions are Based on Evidence from Detached Plant Parts Each of Which Usually has its Own Name. This reconstructed Carboniferous arborescent lycophyte illustrates the variety of names associated with it. *Knorria* is the name given to impressions of old wrinkled bark, and *Lepidophloios* one name given to impression/compression fossils of leaf-cushion-covered stems. The leaves themselves are referred to as *Lepidophylloides*. Rhizophore or 'root' systems are called *Stigmaria*. The microsporangiate (male) cone is known as *Lepidostrobus* and yields *Lycospora* spores from sporangia borne on *Lepidostrobophyllum* cone scales. The female megasporangiate cone also has *Lepidostrobophyllum* cone scales but when permineralised scales are found with megasporangia containing only one functional megaspore, the name *Lepidocarpon* is applied. The dispersed megaspore is known as *Cystosporites*. Although both male and female cones are shown on the same tree, they may in reality have been borne on separate individuals. For clarity some detail has been omitted from the reconstruction

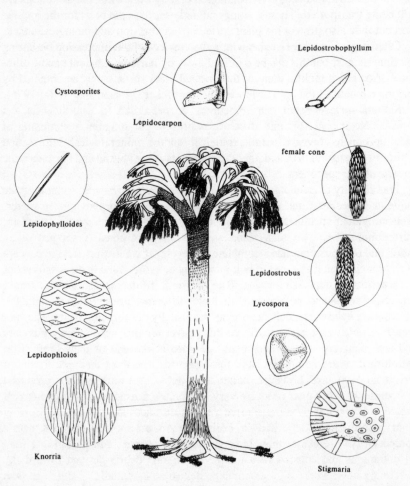

Cystosporites

Lepidostrobophyllum

Lepidocarpon

female cone

Lepidophylloides

Lepidostrobus

Lycospora

Lepidophloios

Knorria

Stigmaria

bog, swamp) is termed autochthonous.

Detached organs or plant parts are given names in the form of binomials in the same way as whole plants, because initially little is known of their relationship to other organs or parts. Therefore when the fossil parts are reconstructed into an ancient organism it is not unusual for it to have a large number of different names (Figure 4.3). Often, however, the reconstructed form commonly adopts the name of its component part that was first described.

If a selection of plant or animal parts are consistently found together, these may be grounds for suspecting that they were once parts of the same type of organism (or community). However, consistent association is the weakest of foundations upon which to reconstruct ancient life. The association may be due, for example, to the original fragments having similar hydrodynamic properties or similar resistance to breakdown. There may have been other fragments deposited that subsequently decayed. Similarity of structure and/or surface characteristics may further suggest an original in-life association, but proof is only obtained where two or more parts are found in organic attachment. If on one occasion organs A and B are found attached, then later B and C are also found connected, A and C can safely be considered to represent different parts of the same plant species. In this way a reliable picture of the original organism may be built up.

Taphonomy is the study of the processes of fossil assemblage formation. It is probably true to say that traditionally there has been comparatively little scientific effort expended to interpret how plant assemblages form and what the relationships are between the patterns of deposited remains and the patterns of organisation within the living plant source communities. Palaeobotanists originally concentrated on understanding the nature of extinct plant forms. Studies of palaeoecology were limited and some reconstructions of past vegetation were often fanciful affairs with more imagination than scientific content. It is not unusual to find all the plants known from a given era pictured as growing in the same community, that is as 'botanic gardens'.

Taphonomic work is now being undertaken and as well as being observational involves significant experimental elements (for example see Spicer, 1981; Scott and Collinson, 1983; Schiehing and Pfefferkorn, 1984; Ferguson, 1985). There have already been some attempts to document patterns of deposition within fossil assemblages by quantitative sampling (e.g. Spicer and Hill, 1979) but ultimately, in spite of sophisticated computer analysis of the resulting data, reliable interpretation must depend upon a detailed knowledge of depositional processes. The rewards for this type of study are enormous, for if it is possible to reconstruct past plant communities reliably, and in some detail, then we can deduce far more about interactive biology of extinct plant forms.

5

TIME, SPACE AND SPECIES

When we consider the palaeobiology of extinct plants, we have to base our interpretations on aspects of morphology, and possibly some environmental data
from the physical and chemical nature of contemporaneous rocks. To fill out this
often sparse information we are obliged to make reasonable assumptions based
on our knowledge of living plants.

There is a danger in this, however, for if we accept that evolutionary change
has taken place, we cannot apply freely the principle of uniformitarianism to
ancient organisms. Nevertheless, we have no reason to believe that the process
of genetic change, diversification, and speciation we observe in modern plants
is any different from that which operated in the past and gave rise to the extant
forms. The fact that, as far as we know, the processes that control the sorting
and mixing of genes during sexual reproduction are similar for all eukaryotic
organisms attests to its antiquity. In this chapter we will be considering the
nature of species and the way in which plant species may arise, based on our
knowledge of living organisms.

Species Concepts

The concept that organisms may be grouped into more or less distinct units
called species is fundamental to the whole of biological science. To Linnaeus,
species were the natural groups of organisms as they had been created. In his
classification Linnaeus was merely seeking out the pattern of design of Creation,
and species were immutable. In fact later in his life Linnaeus himself departed
from a rigid adherence to the immutability of species because of the problem
of hybrids. To Darwin, however, species changed with time as they became
adapted to changing environments and the less fit individuals were selected out.
From generation to generation there was descent with modification. Although
Darwin was able to observe and document variation within and between species,
he was totally unaware of the mechanism of inheritance or of the source of variation upon which natural selection could operate. We now know that the origin
of variation is in gene mutations and rearrangements of the genetic material
within the organisms. Before examining the genetic nature of evolution we must
consider what we mean by the word 'species'.

To some, following the rediscovery of the Mendelian principles of heredity,
species were totally artificial groupings of organisms. Bessey (1908) was of the
opinion that 'Nature produces individuals, nothing more . . . Species have no
actual existence in nature. They have been invented in order that we may refer
to great numbers of individuals collectively.' To a degree this attitude has some

validity. Sexual reproduction involves the exchange of genetic material between individuals, and it is individuals which comprise a common line of descent. However, individuals can only share their genetic material if they are compatible. The parents have to be similar to a high degree in order to produce viable offspring. Furthermore if, as biologists, we treated each individual as totally unique, we could not replicate any experiments, apply statistical analysis, or extrapolate findings from one individual to another. In order for experiments to have any general applicability, we have to group individuals into units within which, in the majority of cases, the response of one or a few individuals to a particular treatment or situation is considered to be typical of the response of all the others in that unit. There is clearly then a practical need to seek out groups of similar organisms. Living things, however, are not just objects: their similarity has a special basis in that they may exchange between themselves the cause of their similarity: their genetic information.

There is a tendency for one's concept of a species to be coloured by one's interest, and it is not unusual to see references to morphological species, genetic species, biochemical species, and so on. Mayr (1957), however, considered that the underlying nature of species may be reduced to three philosophical concepts which embody all of its more empirical expressions.

The Typological Species

The typological species concept is one in which the word species is interpreted to mean 'kind of'. A group of individuals may be summarised as a collection of typical or essential attributes. Taken together these attributes constitute the mental image that we have when we think of a particular species. The observed variation within a species may be thought of as merely the imperfect manifestation of the basic or fundamental type of that species.

The embodiment of a species in a single, albeit perhaps hypothetical, individual has severe limitations. It is difficult to accommodate intraspecific variation, polymorphism between sexes, or phases in the life cycle (gametophyte and sporophyte), or morphologically indistinguishable yet reproductively isolated populations (cryptic or sibling species). Nevertheless, an extrapolation of the concept has found practical application. In many instances when a taxonomist makes a decision as to whether to include a particular specimen in one species or another he often has very restricted information as to the characteristics of the population from which the specimen was extracted. Crudely speaking the taxonomic assignment is based upon mental images of known species to which the specimen might belong. These mental images are ultimately based on a type specimen.

This type individual, or course, comes from a single population at a fixed point in time. In order to properly represent the species, the type specimen should be as representative of all the individuals in a species as possible. To put it another way it should be situated at the centre of gravity of a cloud of individuals distributed in multidimensional space, the position of each individual being determined by measures of the characters it shares with the type specimen. Even here we face problems. If we accept Darwin's concept of evolution as one of common descent with modification, then species are interconnected and so are the clouds of individuals representing the species. If the clouds are interconnected, then strictly speaking the centre of gravity for each cloud cannot be determined because we cannot define, except arbitrarily, the cloud boundaries.

In the case of traditional taxonomy the type specimen was rarely, if ever, designated after the full morphological variation within a species was known. It is possible that the type specimen was an unrepresentative individual from the population that was first discovered, and this population was not necessarily a representative population within the species.

Species are clearly more than a collection of similar objects, and the traditional, primarily morphological, emphasis of the typological concept cannot encompass all aspects of a species. We know that individuals belonging to the same species may reproduce and interact with each other in a variety of ways to form populations and communities. Furthermore these interactions give rise to a structure within populations that cannot be described in purely morphological terms. A population of organisms interacts with its environment, including other populations, in ways which are not entirely dependent on the morphology of the individuals. For example, population structure may be determined to a large extent by behaviour and the timing of reproductive cycles.

Sexual reproduction results in a complex interchange of genes and a mixture of genotypes. The expression of the various genetic complements, in conjunction with environmental influences, gives rise to the observed morphological diversity within populations. The underlying genetic complexity cannot be adequately described by a single archetypal individual. Morphology is derived from, and secondary to, the genetic composition or genotype of an individual. The fundamental role of the genotype in determining even morphological species limits must therefore be recognised in any species concept. Modern taxonomic studies try to take these factors into account where living organisms are concerned. Unfortunately the palaeobotanist is usually restricted by the amount of information available when describing extinct organisms. The palaeobotanist's species cannot therefore be fully equated with those of living plants.

Non-dimensional Species

A second fundamental species concept is that of a naturalist. Mayr (1957) refers to it as a non-dimensional species concept. It deals with the relationship between two co-existing natural populations at a single locality at a single point in time. We know that when we examine a range of organisms that occur in, for example, a woodland at any point in time, it is possible to assign individuals to discrete morphological units. Jordan (1905) summarised the naturalist's viewpoint by referring to species as units of which the fauna (or flora) of a region is composed, and which are separated from each other by gaps which at a given place are not bridged by intermediates. Our experience tells us that this concept of species is apparently real. But let us examine this a little more closely.

The distinctiveness of one group of organisms from another is only valid for a particular place at a particular time. Furthermore the unity of one group depends on its relationship to the other groups; a given group has no intrinsic features that unite it. Mayr likens the concept to that of a relationship, such as being a brother. The term brother, when applied to one individual only has meaning with respect to a second phenomenon; with respect to another individual. Being a brother has no inherent property such as hardness in a stone. Such a relationship is absolute but it is limited to the context of the two individuals, or groups, it refers to. Being only a relationship the concept lacks dimensions of time and space. In a palaeontological situation we may examine two specimens from the same horizon and determine whether they represent the same type of organism or different organisms, but only with respect to one another, and only for that given stratigraphic horizon.

The gaps between the units may be explained in Darwinian terms by saying that, if only the best adapted organisms are able to survive in a given set of circumstances, then the wide variation inherent in one group grading into another will be selected against. Some part of the range of organisms must be less well adapted than others. This species concept is only valid of course for a given situation at a given time and has no general applicability in an evolutionary context.

Hologamodemes and Biological Species

Mayr's (1957) third species concept is that of the polytypic or multidimensional species. This concept considers species merely as groups of populations which are capable of interbreeding. This affords species continuity in both space and time and is the basis of what is known as the biological species concept. Unfortunately by becoming multidimensional it becomes virtually impossible to define species limits and the concept loses objectivity. Nevertheless it does reflect the behaviour of natural populations and is therefore biologically meaningful.

Mayr (1940, 1963) defines species as 'groups of actually or potentially interbreeding natural populations which are reproductively isolated from other such groups', whereas Dobzhansky (1970) considers that species are 'systems of populations; the gene exchange between these systems is limited or prevented by a reproductive isolating mechanism or perhaps by a combination of several such mechanisms'. Although both authors use fertility barriers to define species limits, they also stress that absolute reproductive isolation cannot be taken as a criterion for distinguishing between species. The production of fertile hybrids can, and does, occur between species that are normally genetically isolated and maintain distinct and different gene pools. A species then is a system of populations that normally do not, rather than cannot, interbreed with others. Strictly speaking these definitions cannot apply to populations which reproduce asexually because they cannot interbreed.

Such definitions imply that we cannot delimit species in morphological terms. Some authors maintain that 'the generalization that organisms exist as distinct species is largely invalid' (Ehrlich and Holm, 1962). This conclusion is based to a large degree on the results of phenetic classification where groups of individuals are sought based on observable similarities in the phenotype, the phenotype being the expression of the genotype conditioned by the environment. There is a justification for this in that it is on the phenotype, not the genotype, that natural selection operates. Yet if evolution is taking place, there will be instances where changes in the genotype are not expressed in the phenotype for a given environment or the variation in the phenotype is on such a small scale that separation into distinct morphological species is not justified. All this results in an intergradation of form that frustrates traditional morphological partitioning.

Problems in distinguishing between species are not always the result of phenetic classification. Stebbins (1963) was of the opinion that 70–80 per cent of species of higher plants 'conform well to the biological species definition' and consequently exhibit morphological discontinuity based on reproductive isolation. Most of the difficulties arise, according to Stebbins, due to a combination of hybridisation and polyploidy. These processes occur far more frequently in plants than in animals and therefore plants have a greater number of controversial species. This problem should not disillusion us, however, because it is to be expected if evolution is taking place. At any point in time there are bound to be some overlaps between species as new forms arise.

Clearly there is a case for distinguishing unambiguously between species as delimited by the morphological expression of the genotype and species defined on the basis of the interchange of genetic material within and between populations or organisms. Although it is not universally accepted, there is a terminology proposed by Gilmour and Gregor in 1939 and later elaborated by Gilmour and Heslop-Harrison (1954) which not only facilitates such a distinction but also allows us to refer to different kinds of populations of organisms depending on how we wish to view them for a particular study.

Because it was in use a long time before the genetic basis of inheritance and change was in any way understood, we may retain the word 'species' for a group of individuals showing the same characters. These characters will usually be morphological but may be extended to include biochemical and cytological characters. Thus the word 'species' is applicable to museum collections, fossil specimens and so on and is primarily a taxonomic term.

If we are interested in populations of organisms, however, it is useful to have a separate terminology and the one we are going to consider is based around a common neutral suffix '-deme'. This suffix simply means 'a group of individuals of a specified taxon'. The terms referring to different types of population are made by adding appropriate prefixes.

If, for example, we wish to refer to a group of individuals of a particular taxon that occur within a specified area, we describe them as belonging to the same topodeme. They comprise a 'local population'. If we need to refer to a group of individuals of a specified taxon that exist in a specified kind of environment, then we say that they comprise an ecodeme. An ecodeme may occupy a single area (in which case it is also a topodeme) or the individuals may be found occupying a series of discontinuous areas. So long as the environment is considered to be the same, then all the individuals belong to the ecodeme. Populations may similarly be defined depending on the aspects under consideration. For example, a cytodeme is a group with a particular chromosomal condition and a genodeme is a group which is known to differ genetically from other groups. Perhaps the most useful concepts in respect to our understanding of biological species are the gamodeme and hologamodeme. A gamodeme is a group of individuals of a specified taxon, which are so situated spatially and temporally that within the limits of the breeding system all can interbreed (Briggs and Walters, 1969). This is a subset of a wider population which comprises a hologamodeme. A hologamodeme is defined as 'a group of individuals of a specified taxon which are believed to interbreed with a high level of freedom under a specified set of conditions', and may be considered to be more or less equivalent to the biological species. Unfortunately the -deme terminology has not been used very widely in spite of its usefulness in clarifying population concepts. In the second edition of their book *Plant Variation and Evolution*, Briggs and Walters (1984) have limited their use of this terminology but rightly lament its demise.

One can immediately see that there are likely to be problems associated with identifying the origins and limits of a hologamodeme under natural conditions because far more has to be known about the organisms than morphology and distribution. White (1978) gives thirteen requirements for the full documentation of a single case of speciation, which include the distributions of the organisms both past and present, vegetative and reproductive morphology, geographical variation, ecological data, biochemical and genetic information, reproductive cycles and hybridisation studies. This information has to be known, not only for the original (parent) hologamodeme, but also the derived (daughter) hologamodemes. Because of this complexity there is probably no single case of 'speciation' for which all this information has been collected!

Genetic Variation within and between Populations of the Same Hologamodeme

Fundamental to Darwin's theory of evolution was the maintenance of a reservoir of variation within a species (hologamodeme) upon which natural selection could act, and yet Darwin was completely unaware of the source of this variation or the mechanism for inheritance. Following the recognition of Mendel's work, the genetic basis for inheritance was established and variation was seen to be allied to differences in the frequency of genes within a population. Furthermore it was soon realised that the maximum rate of evolutionary change is limited by the amount of genetic variation in a population (Fisher, 1930). Before we examine in greater detail the key role of genetic variation in evolutionary theory we shall briefly consider how this variation originates.

The Source of Variation

Genetic variation originates by changes taking place in the hereditary material itself; in other words by mutation. There are three basic types of mutation: one involves a change in as little as a single DNA nucleotide pair and is known as a gene or point mutation; another concerns structural chromosomal changes; and the third is the result of changes in the chromosome number.

Gene Mutation. Gene mutations are the result of changes in DNA or RNA molecules brought about by a variety of agents, including chemicals or ionising radiation. The change may involve the substitution, addition or deletion of one or more nucleotides. The degree to which this will affect the organism will depend upon the biological function of the gene. The rates of these mutations vary from organism to organism and from one gene locus to another. Gene mutations in bacteria and other simple organisms typically occur at a rate of 1 in 1 million per gene per division. Such a mutation rate might seem very low but each organism has many genes and there are usually a large number of organisms in each species. Add to this a significant time dimension and gene mutations become a common source of variation.

Structural Chromosome Changes. Most characteristics of an organism are produced by groups of genes, and any structural changes in the chromosomes that produce new groups of linked genes are likely to have an effect on the phenotype.

There are four types of structural chromosomal mutation which usually occur at meiosis. The first of these involves the loss of a portion of a chromosome and is known as a deletion. The loss may involve as little as one gene but usually more are involved. Any loss of genetic information has potentially serious consequences, and deletions tend to reduce fitness or prove fatal in the homozygous condition.

When a section of a chromosome is present both at its normal location and elsewhere on that chromosome, it is said to be duplicated. This results in an increase in the number of genes on a chromosome which may cause an imbalance in gene pairing at meiosis. This may reduce the viability of an organism but many duplications appear to have been incorporated into normal genetic complements in the course of evolution.

Inversions occur when a block of genes within a chromosome become rotated through 180°. This is not as drastic a change as it appears because normal homologous chromosomes pair gene by gene at meiosis. The chromosome with the inverted segment will tend to pair with its normal counterpart in such a way that the pair are transmitted as a single unit. Inversions are quite widespread in natural populations and have significant evolutionary consequences in that they may keep adaptively advantageous gene complexes together. This may also lead to segregation of genes within a population and hence incipient speciation.

Translocations arise when two non-homologous chromosomes break simultaneously and exchange segments. In a sense this gives rise to two 'new' chromosomes. Like inversions, translocations alter the linkage relationships for genes at recombination and may lead to the inheritance of apparently linked groups of genes. This in turn, as with inversions, may lead to genetic segregation and eventually new species.

Changes in Chromosome Number

Aneuploidy: For every organism which possesses them there is a normal set of chromosomes. In the diploid phase of a plant life cycle this set is made up of a number, n, of homologous pairs of chromosomes such that the total chromosome count is $2n$. Sometimes the number of chromosomes in this diploid set becomes altered by one or more: a condition known as aneuploidy. Aneuploidy includes the loss of one (the monosomic condition) or both (nullosomic) chromosomes from a homologous pair, or the addition of extra chromosomes to a pair (trisomic, tetrasomic).

Polyploidy: In the haploid phase of the life cycle there are n single or unpaired chromosomes. If this complete haploid set is added to the diploid set, the plant (now carrying $3n$ chromosomes) is said to exhibit the condition known as polyploidy or more specifically to be a triploid individual. Polyploidy may involve any multiple of the haploid set.

Aneuploidy and polyploidy result in an increase in the amount of genetic material present in organisms, and as such are different from gene mutations and structural changes in the chromosomes which result only in differences in the nature and arrangement of genes. Polyploidy, in particular, plays a critical role in certain types of speciation which will be considered later.

Measurement of Genetic Variation

Mendel's experiments on pea plants (*Pisum sativum*) were based on simple characters, e.g. wrinkled versus smooth seeds. These character states were later found to be controlled by one of a pair of genes (alleles) that resided at a given position or locus on a chromosome. When Mendel carried out his breeding experiments he fortuitously chose characters that were determined by single or very few genes. The characters of most plants, however, are determined by a number of genes, and in such polygenic systems we cannot assess genetic variation simply by examining the phenotype. Ideally we need a way of looking at the genotypes directly. Fortunately, this can in a sense be done, by a technique known as electrophoresis. Different enzyme forms produced by different alleles at the same locus possess different electrical properties. By allowing these enzymes to migrate in a gel across which is applied an electrical charge, we can separate the different enzymes, or allozymes as they are known. At the molecular level dominance is not usually expressed and the heterozygote condition is revealed by the presence of two distinct allozymes; one derived from the dominant allele and one from the recessive allele.

Variation and Selection

Variation between individuals in a population from generation to generation results from the shuffling of genes during sexual reproduction. One might expect that this mixing of genetic information at each generation would soon give rise to a uniform population and rare traits would be swamped. This however is not the case. In 1908 W. Weinberg and G.M. Hardy independently realised that in large populations where there is random mating and where the proportions of genes in a population are not being affected by mutations or natural selection, the original ratio of dominant to recessive alleles will be retained from generation to generation. This principle, the Hardy-Weinberg law, has been demonstrated in the laboratory, but in nature mutation and natural selection lead to changes in the gene frequencies. The difference between the expected and the observed genotypic frequencies is the result of differences in 'fitness' between genotypes.

Fitness. Fitness may be thought of as the relative advantage of one genotype over another in terms of the number of offspring produced. The genotype with a greater degree of fitness contributes more to subsequent generations than a less fit genotype. It is, however, not just simply a measure of the number of offspring produced (fecundity). A genotype with high fecundity may, for instance, produce a large number of offspring many of which die before they are able to reproduce, while a second genotype may produce fewer offspring but a greater number survive to reproduce. In this case the second genotype has the greater fitness.

It is important to realise, however, that a measure of fitness is only valid for the conditions under which a given set of observations were made. In another environment, with other selective forces, a previously less fit genotype may be more fit because it is better adapted.

Fate of Genes under Selection

A dominant allele is usually expressed. If the expression of the allele is manifest in the phenotype, it may be operated on by natural selection and the frequency of that allele in the population may be affected. In the extreme case of a dominant allele being strongly selected against, it will, of course, be rapidly eliminated from the population. This then results in the recessive allele being 'fixed' in the population because only individuals with the homozygote recessive condition will survive and the population becomes entirely composed of homozygous recessive individuals.

If there is the same intensity of selection against the recessive allele instead

Figure 5.1: The Effects of Directional, Stabilising and Disruptive Selection on Genetic Variation within a Population. The shaded portion of the population is selected against in successive generations

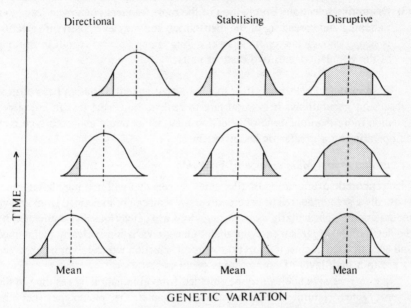

of the dominant, however, the recessive allele will not be eliminated in the same way. A reservoir of recessive alleles will reside in the heterozygote because the dominant allele 'shields' the recessive allele from elimination.

So far we have only been considering alleles at a single gene locus. Obviously, in nature the situation is far more complex, and natural selection operates on systems with more than one gene: so-called polygenic systems. The inheritance of most characters depends on more than one gene and more than two alleles at each locus. Characters may have several forms or states which may not be distinct but merge into one another and so vary continuously. When natural selection operates on these characters several genes will be affected but still the gene frequencies will change just as they do in simple systems. The only real difference is that changes in the phenotype will tend to be more gradual. If we plot the observed genetic variation as expressed in the phenotype against the frequency of individuals with a given trait we get the distribution curve shown in Figure 5.1. Any selection against individuals in part of the curve will, over subsequent generations, result in changes in this distribution.

Three types of selection may be recognised (Figure 5.1).

(1) Directional selection. Selection against one extreme causes the phenotypic frequency to change directionally as generation succeeds generation.
(2) Stabilising selection. Here selection is in favour of the most common

phenotypes. The extremes of the population, as represented by the tails of the distribution, are eliminated. Over successive generations the phenotypes will display less variation; they will become more similar.

(3) Disruptive selection. Elimination of the most frequent phenotype leads to a flattening and spreading of the distribution and may eventually, if selection is strong enough and operates over a long enough period, result in splitting of the population into two distinct units.

These modes of selection affect the amount of genetic variation passed on to subsequent generations. It is important to realise, however, that in any given situation more than one type of selection may be operating and each type may be operating to a greater or lesser extent.

Accumulation and Maintenance of Variation

Electrophoretic examination of allozymes reveals that within a population there is usually a greater degree of genetic diversity than can be explained from known mutation rates, particularly as most mutations are deleterious and eliminated by the death of the organism carrying them. Genetic variation is clearly maintained and accumulated against the forces of natural selection which might be expected to create a high level of homozygous genotypes.

Rare or recessive alleles may be shielded from elimination by residing in the heterozygote. Natural selection operates against the phenotype, not the genotype, and, provided it is not expressed, a recessive allele will not be selected against. In fact in some cases the heterozygote may be fitter than the homozygote dominant. When this occurs it is known as heterozygote advantage, and leads to an increase in the frequency of the recessive allele.

Another way in which recessive alleles may be maintained or increased is by frequency-dependent selection. Here selection varies directly with the frequency of alternative genotypes. When present in low frequencies a genotype has a high fitness and therefore increases as a result of natural selection.

An example of frequency-dependent selection is seen when a population of plants is attacked by a fungal disease. The fungus might be able to attack readily plants with the most frequent genotype but a few individuals might have genes conferring resistance to the fungus. As the plant population is decimated these individuals will reproduce and will eventually form the basis of a new population in which the frequency of the resistant gene is high. However, in the fungal population a mutant strain may arise against which the previously resistant gene is ineffective. The frequency of the mutant fungal gene then increases in the fungal population as the plant population is again decimated until another, previously rare, gene for resistance against the new strain increases in frequency by virtue of its enhanced fitness.

In some ways allied to frequency-dependent selection are the mechanisms employed to avoid crossing with genetically similar organisms. Where there is excessive inbreeding, potentially deleterious alleles are reinforced and progeny tend to have lower survival rates and poorer reproductive success. There are a number of ways of preventing inbreeding depression and increasing the frequency of outcrossing. These include incompatibility mechanisms as well as the consequences of heterospory and the advantages associated with the seed habit. These topics will be developed in later chapters.

Meiotic Drive

Through the processes of segregation and recombination at fertilisation, alleles normally have an equal chance of being passed on to the next generation. Sometimes, however, some alleles are associated with chromosome changes at meiosis which ensure that they have a greater chance of being passed on, whereas other alleles have a reduced chance because they are associated with chromosome abnormalities. Over successive generations this selection of alleles can give rise to changes in gene frequencies within populations and is known as meiotic drive. Under natural situations, however, other forces acting on the genetic composition of populations are so strong that meiotic drive is considered to play a relatively minor role in maintaining genetic variability.

Genetic Drift

Genetic drift is a phenomenon confined to small populations and is defined as an undirected change in gene frequencies found in a population which cannot be ascribed to the action of meiotic drive, natural selection or other processes. The gametes produced by an individual at meiosis are a random assortment of haploid genotypes. If we take a small random sample of these, the sample cannot possibly encompass the full range of gene frequencies present in the population at large, and therefore the sample will be biased. In large populations this bias occurs all the time but produces no net 'error' because the bias in one direction in favour of a given allele in one individual will be cancelled out by bias in the other direction in other progeny. However, in small populations there are few progeny for the sampling errors to cancel each other out. Consequently the gene frequencies drift as the sampling errors build up.

Molecular Drive

As well as by the diverse processes of selection and genetic drift, gene mutations may be fixed and spread throughout a sexually reproducing population by a process recently described as molecular drive. Molecular drive is defined as 'the fixation of variants in a population as a consequence of stochastic and directional processes of family turnover' (Dover, 1982). Families in this context refers to families of genes where multiple copies of genes and gene sequences exist within an individual. There may be many hundreds of copies scattered over different chromosomes but within a species there is a high degree of homogeneity (the copies are very similar to one another) compared with that between species. The degree of homogeneity depends on the relative rates of mutation, selection, and the rate of homogenisation.

Homogenisation is brought about by the frequent molecular exchanges (known as turnover) between the various copies within a family. These exchanges take place by a number of different processes. In gene conversion one copy can alter another so that both copies end up sharing the same DNA sequence. The direction of conversion may be random or directional but either way gene conversion has a tendency to homogenise all copies of a gene family.

The process of transposition, where segments of DNA duplicate themselves and then become inserted into different chromosomes, also has a homogenising effect as do the structural chromosome changes we have already discussed.

Homogenisation of a gene family by these mechanisms takes place only very slowly. The important feature of molecular drive is that in a sexually reproducing population where the rate of homogenisation is much slower than the cycle of reproduction a synchronous genetic change throughout the breeding population may result. While the slow processes of homogenisation are taking place, sexual reproduction at each generation ensures that the chromosomes are being repeatedly shared out among the members of the population with the result that, at any point in time, the composition of a gene family tends to be similar in all individuals.

One of the profound consequences of molecular drive is that genetic variation, differences in fitness, and to some extent phenotypic variation (inasmuch as it is partly dependent on the genotype) will be minimised, but more importantly there would be a simultaneous evolution throughout an entire population. Populations which are isolated from one another would undergo their individual 'in unison' genetic change which would lead to genetic differences between the populations, and to fertility barriers and speciation.

Genetic Variation between Hologamodemes

We have now seen how selection processes, breeding systems, genetic drift and molecular drive alter gene frequencies within a population of a single hologamodeme. With such a range of influences on the genotypes it is easy to appreciate that when two populations are genetically isolated (that is, there is no exchange of genetic information between them), their gene frequencies may diverge over a comparatively small number of generations. The factors that bring about genetic variation within populations should therefore be seen as part of, and continuous with, those that primarily enhance variation between populations.

If we follow the Neo-Darwinist gradualist theory, the variation within a species or hologamodeme is to a large extent continuous with that between species. The accumulation of differences within and between populations belonging to a species eventually gives rise to breeding barriers that delimit species. The key to genetic divergence and therefore speciation is that of the restriction of gene flow between populations. The following is a summary of how this may occur.

combination of factors as might be encountered at different altitudes up a mountainside. The various populations along the cline might, in themselves, have only a restricted range of environmental tolerances. The elimination of populations part way along the cline may fragment the cline into two sets of isolated populations and subsequent genetic divergence could take place.

In any species which is geographically widespread there is a tendency for local populations to undergo genetic differentiation. This is particularly true of plants where the distance travelled between the germination of an individual and the germination of its offspring is limited. Populations with unusual alleles, or frequencies of alleles, may occupy different portions of the total area occupied by the species. When two or more such populations expand their ranges and meet, there may be a sharp change in gene frequencies. The hybrids produced across this boundary could well be less fit than the parents with the better adapted gene complexes. This may lead to further divergence and eventual reproductive isolation. This so-called area-effect speciation does not assume a regular environmental gradient.

A slight variation of the area-effect speciation model is known as stasipatric speciation. This model differs from area-effect speciation in that it involves structural changes (inversions, deletions, translocations, duplications) in the chromosomes rather than gene mutations. A chromosomal rearrangement which reduces fecundity when heterozygous may originate at some point within the area occupied by the ancestral species. Whether by genetic drift, meiotic drive or molecular drive, once such a rearrangement becomes established it may act as an incipient isolating mechanism between the population homozygous for it and the original population.

Like the previous two models, sympatric speciation concerns the formation of new species within the same geographical area. It may involve the separation of a gamodeme by the adaptation of two subpopulations to different environments or by achieving sexual maturity at different times, for example. In plants, sympatric speciation is also likely to occur as a result of polyploidy.

Plants in which the chromosome numbers are some multiple of a basic set are termed polyploids and they arise when division of the chromosomes is not followed by division of the nucleus. The fusion of two gametes each with a haploid chromosome set (A) gives rise to a diploid plant with the genetic complement (AA). This diploid plant may produce tetraploid cells ($AAAA$) if cell division does not occur after duplication of the chromosomes has taken place. Vegetative reproduction involving these cells may result in individual plants which are entirely tetraploid. The tetraploid reproductive structures of these plants may produce gametes which are diploid rather than haploid and fertilisation would produce tetraploid progeny. In reality the situation is not as simple as this, however, because meiosis will be disturbed in tetraploid cells. Instead of pairing, the chromosomes will arrange themselves in fours, and there will be an increased chance of irregular chromosome numbers in the gametes as well as reduced fertility.

The real importance of polyploidy is its role in interspecific hybridisation. Suppose two species, one with a diploid chromosome number of 8 (*A*) and the other of 12 (*B*), produce a hybrid plant: this would have a diploid chromosome set of 10 (*AB*). Four chromosomes would have come from one parent and six from the other. This hybrid would be sterile, however, because the chromosomes cannot pair off at meiosis. A tetraploid reproductive structure produced by doubling of the diploid set (*AABB*) to give 20 chromosomes would be fertile because at meiosis the chromosomes will fall into 10 homologous pairs (see also p. 134).

If self-fertilisation takes place, the progeny will combine the genotypes from the two parent species. Furthermore these plants will be fertile between themselves but reproductively isolated from the normal plants of the parent species because they cannot form fertile hybrids. Whatever criterion we use to define a species, be it reproductive isolation or morphological distinctness, polyploidy can result in 'instantaneous' speciation.

The events just described may sound improbable but they have been observed to occur both naturally and under laboratory conditions, and there can be little doubt that polyploidy has played a major role in plant evolution.

Following genetic divergence between populations, reproductive isolation has to be maintained for the populations to have a valid status as hologamodemes or species.

Maintenance of Hologamodeme Separation

If we accept the biological species concept, it follows that there must be mechanisms which normally prevent free genetic exchanges between species or hologamodemes, when the species coexist in the same area and environment. Such mechanisms may be conveniently separated into two groups: those that prevent or impede hybridisation (pre-zygotic) and those that reduce the viability or fertility of the hybrid (post-zygotic).

Pre-zygotic isolating mechanisms

(1) Ecological or habitat isolation. Here hologamodemes occupy different habitats in the same geographical area. This is perhaps more important in animal populations than in those of plants. Unless highly dependent on a specific animal pollinator with a restricted habitat, plants, through the relatively long-distance dispersal of pollen and spores, can effect genetic exchange between individuals in different local environments.

(2) Seasonal or temporal isolation. Hologamodemes in the same geographical area may become sexually mature at different times. This is an important way in which hologamodeme separation is maintained in plants. Production of the reproductive organs must be more or less synchronous for free exchange of genetic material to occur. However, by desynchronising the production of gametes within an individual the chances of self-fertilisation are reduced, outbreeding is facilitated, and the potential for a greater degree

of gene exchange results. The staggering of gamete production within individuals tends to increase the length of the reproductive period and strict isolation by this means it not as effective as it might otherwise be.

(3) Mechanical isolation. The mechanical structure of sporangia or flowers impedes or prevents cross-fertilisation. This is particularly significant in insect-pollinated plants where flower structure may facilitate pollination by only one specific insect vector. Slight changes in flower form or size may prevent pollination taking place.

(4) Gametic isolation. Male and female gametes of different hologamodemes may fail to attract or unite with each other, or the pollen grains of the hologamodeme may fail to germinate on the styles of flowers of another hologamodeme.

Post-zygotic isolation mechanisms

(1) Hybrid inviability. Zygote development fails to begin or fails to proceed beyond a certain point, or the hybrid dies before reproduction.

(2) Hybrid sterility. The hybrid fails to produce functioning sex cells or gametes.

(3) Hybrid breakdown. The hybrid may be viable but the progenies of hybrids may have reduced viability or fertility.

The outline just given has treated separately the various mechanisms of genetic change, variability, and speciation, and for clarity they have been discussed in isolation from each other. There are many examples where studies have shown a particular mechanism to be operating to a greater degree than the others, and therefore allowed it to be studied in detail. In most natural situations, however, a number of mechanisms will be operating simultaneously, each one affecting the genetic composition of the population to a greater or lesser degree depending on a myriad of variables peculiar to that set of circumstances. The overall effect is a change in the nature of the population, and possibly the origin of new species, that cannot be ascribed to any single process.

Gradualism and Punctualism

Darwin's view of evolution, 'descent with modification', envisaged the gradual change of one species into another. Selection would operate on the small amount of inherent differences between individuals present in a species and in so doing would increase variation. Eventually breeding barriers would become established and therefore new species would arise. Further change would result perhaps in a degree of change worthy of being categorised as a new genus and so on. That is not to say that only one new species could be derived from a single ancestral species, or how would there ever be an increase in diversity and numbers of species? Clearly this gradual evolutionary divergence could give rise to multiple species from a single ancestral population but the process was not

Figure 5.2: Left — the Gradualist View of Evolution. Groups A–F (species or higher taxa) are derived from an ancestral stock by a gradual divergence. The rate of divergence is not uniform but no sudden jumps or saltations are involved. Right — the Punctuational Model Envisages Groups A–F as Arising as a Series of Very Rapid Divergences (Saltations) Followed by Long Periods of Relative Stasis

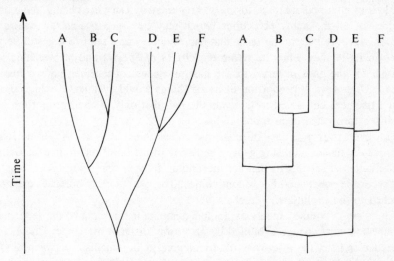

envisaged to be prone to quantum jumps of one form into another (Figure 5.2).

It is easy to see why this was Darwin's preferred model. At the time Darwin made his observation, particularly in the early phase of his career, geology had become revolutionised by the 'Principle of Uniformitarianism'. This was generally taken to mean that gradual processes of erosion and deposition would, if given enough time, wear down mountain ranges and give rise to new rocks. This process could not be observed in its entirety in a man's lifetime because it was a slow process. Similarly speciation events could not be observed directly because a man's lifetime was too short. Variation could be observed but the long-term effects of selection had to be inferred in much the same way as the origin of sedimentary rocks had to be inferred from the movement of sand grains along a stream-bed. If Darwin had proposed a rapid speciation model, his contemporaries might have expected to see direct evidence for it and failing in this would have rejected evolutionary theory altogether. It is interesting to note, however, that in the first edition of the *Origin of Species* Darwin does suggest that rapid speciation might be possible but this speculation was dropped from later editions.

It was far more appropriate to propose that speciation was a slow gradual process. There was, however, an immediate problem. If evolution was a gradual process, one would expect to find that the fossil record would tell a story of gradual change of one form into another. Everywhere there would be intermediate forms connecting one 'species' with another. Unfortunately the fossil record is characterised by an overwhelming absence of such intermediates. 'Missing links' are the rule rather than the exception.

Darwin considered the fossil record as being particularly important and yet it singularly failed to support his gradualist ideas. The gaps between species were glaringly wide and despite continued research they remain so to this day. The lack of intermediate fossil forms was regarded by Darwin as being 'the most obvious of all the many objections which may be urged against my views'. Darwin was fortunate in the sense that palaeontology as a science was still very much in its infancy when he wrote the *Origin of Species* and he was able to account for the gaps in terms of the incompleteness of the then known fossil record. Only a small proportion of living things would ever leave evidence of their existence, but added to this was the fact that only a small proportion of fossilised forms had been found and studied.

However, over a century of investigation has failed in most cases to narrow significantly those tantalising gaps. The nature of the fossil record is essentially one of suites of specimens that are morphologically distinct from one another. The record is punctuated by discontinuities. The evolutionary branches are not attached to the evolutionary tree.

There are of course exceptions. Included among these might be the sequence of morphological change exhibited by *Micraster* shells in the Upper Chalk, or at a coarser level the discovery of *Archaeopteryx* intermediate between dinosaurs and birds. As far as plants are concerned, evidence for gradual morphological change with time within a single sedimentary sequence is exceedingly rare even when the abundance of pollen and spores allows the analysis of larger populations (Stidd, 1981; Müller, 1984). No matter how detailed the work it seems apparent that evolution has progressed by a series of jumps or saltations. The question remains as to how real these jumps are. Are they a product of the evolutionary process or are they the product of the vagaries of fossilisation?

We have seen that the process of speciation often involves genetic change in small populations (strict allopatry with narrow population bottleneck). Furthermore we know that the fossil record only contains specimens representing very few individuals compared with the living source population. If fossilisation is a rare process, then it becomes almost inevitable that the critical phase of speciation will not be preserved. We will therefore never see speciation events preserved in the fossil record except under the rarest of circumstances.

However, this is not the only explanation of the punctuated nature of palaeontological information. Fossils are only formed in sedimentary environments. Suppose the geographic distribution of a species is such that at least some individuals normally live in or near a depositional environment. The species will have a chance of entering the fossil record. Speciation may occur in part of the population away from any depositional site and the only evidence that it has occurred would be when the parent species was replaced by the daughter species.

Similarly if the environment, or competition from other plants, changes, then so too might the normal distribution of a species. If this results in a move away from depositional sites, any record of speciation events will be lost until perhaps a descendant species reoccupies such areas. Here again we would have no record of the process of speciation, not as a result of small population size, but due to changing distribution in relation to preserved sites of deposition. In general, depositional environments most conducive to the preservation of plants (i.e. swamps) tend not to be the environments that exert the strongest selective pressures, and therefore the rate of evolutionary change within them is likely to be low.

There are other possible reasons for a punctuated pattern of evolution and these concern the processes of evolutionary change themselves rather than the properties of selective fossilisation. Evidence for these processes comes from the fossil record itself. It is well known that some ancient fossil forms are virtually indistinguishable from modern living organisms. Examples are numerous, including such plants as *Equisetum, Ginkgo* and cycads. In many cases their fossil record is abundant and continuous, representing little evolution over tens if not hundreds of millions of years. These so-called 'living fossils' offer as much 'proof' as one is ever likely to get that some plant forms exist for very long periods of time. Today such forms can either be widespread (e.g. selaginellas) and therefore exposed to a large number of selective pressures, or may be restricted in distribution (e.g. ginkgo) with perhaps minimal selection but with modern distributions rarely reflecting distributions of the past. The long temporal existence of such forms cannot be directly related to selection alone and argues against the all-embracing gradualistic model of evolutionary change. On the other hand, if evolution proceeds by irregular 'jumps', then the persistence of a form over long periods of time is no surprise. It is to be expected. The punctuational model (Gould and Eldredge, 1977) also says that speciation can be extremely rapid. We have already seen that in certain circumstances plants can naturally develop reproductive isolating mechanisms (and by definition species) almost instantaneously by polyploidy. Polyploidy need not be associated with significant morphological change, but it does serve for the rapid creation of small isolated populations within which genetic divergence, and subsequently morphological change, can occur quickly.

Irrespective of the mode of isolation and initial genetic change, if the alterations in the phenotype wrought by these processes are favourably selected by a changing environment, or confer special competitive advantages, then further speciation will follow. If mutation rates are high, then potentially at least speciation rates can be high, particularly if they are enhanced by suitable selection pressures.

It is not possible to say precisely why some groups of organisms have remained evolutionarily rather static, whereas others have undergone major radiations. It may be due to mutation rates, selection pressures or 'genetic potential'. One thing is certain, however: rates of evolutionary change are not constant. Certain plants in some environments have, at some time, undergone gradual evolution. On the other hand other plants appear to have evolved almost instantaneously against a geological timescale. Because these two modes of evolution intergrade, any distinction between gradual and punctuated evolution is in reality purely artificial.

Mozaic Evolution

So far we have considered selection pressures on populations and individuals. Even though the unit of genetic continuity is the individual, selection also operates at the level of individual organs or even tissues. The fossil record provides overwhelming evidence, presented throughout this book, that different organs evolve at different rates. In angiosperms, for example, leaf morphology and anatomy may remain relatively static whereas floral morphology evolves rapidly in response to selection pressures from the behaviour of and morphological change in pollinating insects. Seed and fruit morphology may be influenced by changing floral morphology, combined with selection pressures favouring improved dispersal, whereas wood anatomy in the same group can remain unchanged through time.

This pattern of evolutionary change at the organ level, or below, is extremely significant for the palaeobotanist who mostly has to deal with isolated organs. Erroneous concepts regarding the evolution of organisms as a whole can easily arise from the analysis of one or two commonly preserved organs. In spite of differing evolutionary rates between organs, it is the whole plant that functions as a unit, and it is the fitness of the whole plant in relation to its general environment that is the important factor.

For all these reasons phylogenetic studies should be based whenever possible on the reconstructions of whole plants rather than on individual organs.

Extinctions

Extinctions are a common, integral and inevitable consequence of evolution. They are the natural result of competition between species and of the failure to adapt appropriately to changing environments (including the evolution of competing groups). The evolutionary consequences of extinctions are complex but in general terms they are of two types: loss of genetic information and the opening up of niches.

The extinction of any group, whether it be at subspecies, species, genus or higher taxonomic levels, results in loss of genetic information from the total reservoir of life. Millions of years of genetic accumulation and modification, the result of innumerable mutations and genetic innovation, are lost forever. Many advantageous genes, such as those for imparting resistance to particular diseases, which may be unique to the otherwise poorly adapted group, are lost. The unique combination of genes that comprise the group are lost as is the potential for further unique combinations.

The extinction of a group opens up niches for other groups to occupy. Competition from the extinct species, no matter how weak it was prior to extinction, ceases to exist. This can lead to geographic spread of the surviving groups and introduces possibilities for speciation and adaptation. The extent to which this happens is dependent largely on the ecological significance of the extinct group and the suddenness of extinction.

Causes of Extinction

To be successful and survive, a group (for example a species), as well as an individual, must be relatively well adapted to its environment. The environment in this case is not only the physical surroundings but also the competitive pressure of other organisms. However, in the course of developing specific adaptations, particularly those of a morphological nature, the group inevitably becomes canalised along a more restricted evolutionary pathway. The extent of this canalisation varies from situation to situation but in the more extreme cases it can have profound consequences. In general terms the more extreme the adaptation the more dependent the group becomes on the environment's remaining unchanged. However, environments do change and the physical environment in particular changes independent of biological processes. The rate of change of the environment, as well as the magnitude of the change, often determines the number of extinctions and the groups in which they will occur. When environmental change is profound and affects large geographical areas, large numbers of groups may disappear and 'mass extinctions' result. Evolutionary modification may be able to keep pace with a slowly changing environment, but if the changes are rapid, extinctions become likely as the rate of evolutionary change is outstripped. Under conditions of rapid environmental change it is usually the adaptable that survive rather than the adapted.

Some organisms may survive changing environments not because they are adaptable but because a previously neutral character (one not acted upon by selection, therefore not advantageous or disadvantageous) becomes advantageous in the new environment. Such organisms are often said to be preadapted. Organisms may also be preadapted if a character previously advantageous for one reason also proves to be advantageous in the new environment for a quite different reason. An example here might be the evolution of xeromorphic characters in plants frequently exposed to brackish-water conditions. These characters enable the plants to tolerate physiological water stress and they may later prove to be advantageous during and following a climatic change towards more arid conditions. The reverse can be true of course, and a previously neutral or advantageous character may become disadvantageous possibly resulting in extinction.

Extinctions due to extremely rapid environmental change are often linked with the word 'catastrophic'. Catastrophes may be local or global in scale. In a sense the death of any individual could be thought of as a small-scale extinction in that a unique genotype is eliminated and can no longer contribute to the evolutionary process. More realistically the elimination of local populations by, for example, fire is an extinction in that a gene pool has been destroyed. Clearly, the more geographically widespread a taxon is, the less prone it is to catastrophic extinction. However, the fossil record seems to indicate that there were several periods during Earth's history when catastrophic, or at least rapid in geological terms, extinction has occurred on a global scale and across a broad range of taxa. Before trying to hypothesise about the probable causes of these extinctions, we need to examine whether or not they are likely to be real phenomena or artefacts of the fossil record.

As we have seen, the geological column is divided up into a series of units based on the occurrence of fossils. The divisions are usually made on perceived biotic changes that are due to real biological discontinuities (for example extinctions or migrations to and from depositional environments) or geological discontinuities (periods of non-deposition). During periods of non-deposition, evolution continues to take place such that when deposition recommences different organisms are preserved.

Whatever the cause, once a division is established it tends to be self-sustaining. The basis for biostratigraphy is the consistent occurrence over large areas of recognisable fossil forms. These can be single pollen grains or spores or larger megafossils. The stratigraphic range of these forms is the span of the stratigraphic sequence between which the organism first appears and it finally disappears. Now, for ecological or preservational reasons, the organisms may be absent for intervals between the end-points of its stratigraphic range. Under these circumstances the range-through concept is applied, which in essence says that even though the organism was not present in, or close to, the depositional environment, it was alive somewhere on the Earth (but not represented in the fossil record) and therefore should be included in the stratigraphic chart. The same logic is not applied, however, to the end-points of the stratigraphic range. Beyond the first or last occurrences of the organism it is assumed not to exist. This in itself would not unduly prejudice the documentation of extinctions were it not for the resolution of the fossil record.

Because of the difficulties of cross-correlation between diverse groups of organisms, endemic populations and geographic heterogeneity, biostratigraphic resolution is often fairly coarse in comparison to biological time scales. If a taxon becomes extinct within the minimum resolvable unit of geological time, the time of the extinction cannot be plotted with any accuracy. All that can be said is that the taxon under consideration is absent from the succeeding unit. The range chart for that taxon is then plotted up to the end of the unit in which it is perceived to have become extinct. If other taxa become extinct within that unit, they also are plotted up to the unit boundary. This not only reinforces the unit boundary but also results in many taxa apparently becoming extinct simultaneously even though their real extinctions may have been separated by several million years. The limited resolution of the fossil record may therefore produce 'mass' extinctions where none may have occurred in reality, and may invite hypotheses involving single causal factors although the extinctions may have been unrelated.

When we talk about fossil taxa, we have to remember that they are defined solely on morphological criteria which are not necessarily related to breeding limits. Because morphological information is often limited, it is not unusual for quite small morphological changes to be used as the basis for erecting new forms and new names. A change in name of fossil name in the stratigraphic record is often regarded as being synonymous with an extinction and/or speciation event even though in biological terms the gene pool might not have been lost but continued slightly modified. Again by seeking divisions in the continuity of geological time we run the risk of 'creating' extinctions where none may have occurred.

Nevertheless there are periods in Earth's history when a number of diverse and apparently ecologically unrelated organisms appear to have become extinct over, geologically speaking, a short period of time. One such 'extinction event' occurred at the Cretaceous/Tertiary boundary. This event has been investigated in detail from a palaeobotanical point of view in an ecological setting and at high stratigraphic resolution, and we will consider it in detail in Chapter 16.

The Species Concept and Taxonomy in Palaeobotany

The fragmentary nature of plant fossil material, the variety of different preservation states (and therefore the quality and quantity of retained biological information), and the lack of information on breeding limits or genetic composition all mean that there is considerable difficulty in reconciling the biological species concept of neobotanists with 'species' as described by palaeobotanists. The ideal goal of palaeobotanical systematics is to devise a taxonomic scheme that reflects phylogeny. Given the incomplete nature of the fossil record this may be an impossible task and a truly natural taxonomy may never be realised. However, there is no excuse for not attempting to improve our systematic framework in the direction of a system that relates to phylogeny.

In the course of trying to reach the goal of a natural taxonomic system it is inevitable that some groups and suites of specimens will be better understood than others, and there will be a range of degrees of confidence in our taxonomic placement of fossil material. The taxonomic status of many taxa will therefore be different in relation to our perceived goal.

Unfortunately the degree to which a taxon is artificial or natural is not conveyed by the Linnean type of binomial that is applied to isolated organs. Quite clearly the taxonomic status of an isolated smooth-walled spore is quite different from that of a permineralised reproductive unit attached to a vegetative shoot. Similarly the age and morphological complexity of a plant or plant organ will, to a considerable extent, determine the confidence with which taxonomic placement can be made. For example, populations of Tertiary fossil angiosperm leaves may be compared with those of close living relatives in order to determine the morphological criteria necessary for the taxonomic partitioning of the fossil forms (Burnham, 1986). However, for older material, even angiosperm leaves, this type of approach becomes increasingly difficult. It is impossible for extinct forms. For anyone not familiar with palaeobotany it is important that the grades of artificiality of plant fossil taxa be appreciated.

In many respects palaeontologists are at the 'sharp end' of evolutionary studies in that they are constantly exposed to the products of evolutionary processes that have operated over, and are the product of, millions of years of environmental change. The palaeobiologist has to attempt to apply the findings of neobiologists to extinct forms: findings which are severely limited by the infinitesimally short time (in relation to geological time) over which investigations have been conducted and which are restricted to currently extant organisms. Such a body of knowledge is more restricted in some senses than that provided by the fossil record. Speculation with regard to the biology of ancient plants is therefore inevitable. It might be thought that the taxonomy of extant organisms would be relatively clear cut when populations of whole plants are available for study. Unfortunately even here the situation is not as tidy as one might assume. In populations where polyploidy and hybridisation are common (in fact where evolution seems to be taking place), delimiting species except in arbitrary terms is often impossible. It has even led some neobotanists to doubt the value of a single species definition (Levin, 1979; Heywood, 1980). Recognising these limitations Levin (1979) agrees with Cronquist (1978) that the purpose of taxonomy is to impose a mentally pleasing organisation upon natural diversity rather than to impose a framework that reflects phylogeny. This pragmatic view is currently applicable to palaeobotany. The palaeobotanical Linnean-type (binomial) nomenclature should be perceived as being nothing more than a labelling system. Furthermore, palaeobotanical taxonomy should be perceived as providing an artificial but working system that can and should be constantly modified as new data are collected. Strictly speaking a palaeobotanical species is not equatable to a species of a living organism. However, under certain circumstances, notably with more recent forms and where most if not all of a whole plant has been reconstructed, the gap between what is understood as a palaeobotanical 'species' and the concept of a living species becomes narrower. Unfortunately this gap can never be fully closed because fossil forms can never be interpreted as biological species when these are defined in terms of breeding systems. Nevertheless, as our understanding of fossil plants as whole and once-living organisms improves, so will our phylogenetic and taxonomic schemes.

6

SPECIALISM IN FORM AND FUNCTION

Plant evolution in the late Silurian and Lower Devonian had given rise to a variety of plant forms. However, it was in the Middle Devonian that many fundamental tissues and organs such as dissected steles, secondary thickening, leaves and roots made their first appearance. There is also evidence showing that some plants were becoming arborescent in stature, imposing greater complexities on plant communities than had previously existed. During the Emsian, pre-existing groups continued to diversify and form new groups that rapidly expanded in the succeeding Eifelian. By the late Devonian all extant major plant groups, except the Angiosperms, were recognisable.

The Middle and Upper Devonian was an extremely important time for adaptive radiation, although we still have only a limited understanding of the ways in which it must have occurred. One of the problems we have in trying to evaluate evolutionary changes in terms of a series of events is our need to separate plants into taxa. This tends to polarise our thoughts and to limit our comparisons between particular taxa and those which we think are ancestors or descendants of them. This is not inherently bad, but it can restrict our thinking on tissue and organ evolution if it forces us to think in taxonomic terms and in terms of whole plant evolution. One way is to look for the earliest record of tissues and organs irrespective of the taxa involved. This kind of information, portrayed in Figure 6.1, gives us a rather different picture of evolutionary

Figure 6.1 (*page 74*): Diagrammatic Record of the Time of Appearance of Various Features of Vegetative Morphology and Anatomy of Early Land Plants (after Chaloner and Sheerin, 1979). a, Equal dichotomy (e.g. *Cooksonia*);
b, H-branching (e.g. in *Zosterophyllum*); c, annular-thickened tracheids (see Edwards and Davies, 1976); d, stoma with apparently interconnected guard cells (e.g. in *Zosterophyllum*); e, overtopping with 2 : 1 ratio of main axis to laterals; f, stele with exarch, elliptical xylem (e.g. in *Gosslingia*); g, enations (e.g. in *Sawdonia*); h, stoma with paracytic subsidiary cells (e.g. in *Drepanophycus*); i, vascularised microphylls (e.g. in *Drepanophycus*); j, forked microphylls (e.g. in *Sugambrophyton*); k, stele with endarch, cylindrical xylem (e.g. in *Rhynia*); l, spirally thickened tracheids (e.g. in *Asteroxylon*); m, stele with stellate xylem (e.g. in *Asteroxylon*); n, tracheids with mesh-like wall, simulating pits, between scalariform bars (e.g. in *Psilophyton*); o, overtopping with 4 : 1 ratio of main axis to laterals (e.g. in *Pertica*); p, trichotomy of laterals (e.g. in *Trimerophyton*); q, vascularised wedge-shaped megaphylls (e.g. in *Platyphyllum*); r, planated megaphyll (e.g. in *Aneurophyton*); s, polystele (e.g. in *Calamophyton*); t, secondary xylem (e.g. in *Rellimia*); u, trilobed stele related to branch arrangement (e.g. in *Triloboxylon*); v, expanded leaf base (e.g. in *Archaeosigillaria*); w, secondary cortex or periderm (e.g. in *Triloboxylon*); x, tracheids with bordered pits (e.g. in *Leclercqia*); y, leaf abscission (e.g. in *Cyclostigma*)

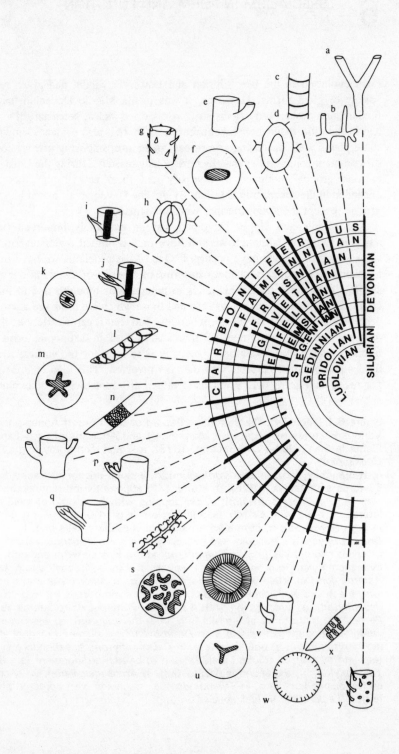

Figure 6.2: Diagrammatic Record of the Early Vascular Land Plants, Showing their Possible Evolutionary Relationships. (After Chaloner and Sheerin, 1979)

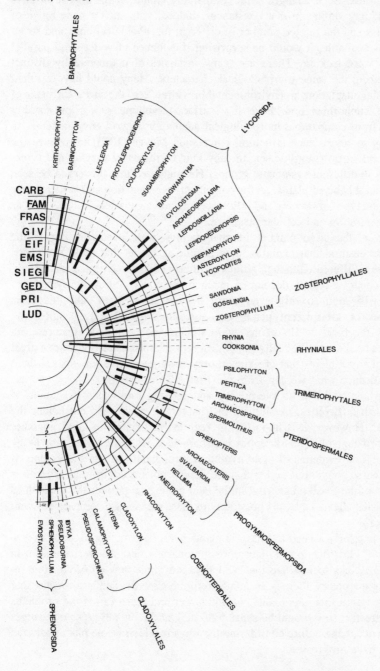

change from that portrayed by comparable taxonomic charts — as in Figure 6.2.

Many plant groups were evolving in a variety of ways during this critical period of time. Such changes from comparably simple beginnings must have been strikingly similar in many instances. Indeed, with such a wide range of possibilities and the limited number of different morphological units and structural parts available, it would be surprising if evidence of widespread parallel evolution were lacking. There are many instances of taxonomically distinct plants sharing the same morphological characters. Many could have resulted from similar adaptations to environmental pressures, e.g. the succulent stems of cacti and euphorbias both have low surface-to-volume ratios, presumably resulting from comparable morphological adaptation to arid environments. In attempting to reveal such structural correlation Meyen (1973) introduced the term 'repeating polymorphic set' for any kind of regularly varying characters appearing in different systematic groups. He suggested that this could be seen in the general habit of plants, leaf arrangement, segmentation and venation, the construction types of flowers and inflorescences and anatomical features such as stelar organisation and cellular structure.

However, the large number of changes that were occurring during the Palaeozoic ensured that evolution was producing plants that must have been recognisably different, although groups obviously still shared central characters. The reproductive organs developed much more widely, allowing systems of plant classification to rely even more heavily upon them. On this basis herbaceous or arborescent plants can sometimes be clearly grouped and classified together, while plants with similar morphological features can sometimes be shown to be really rather different. Unfortunately there are a great number of fossil plants that are known only from vegetative remains, so here we must deduce what we can from our limited information.

This whole problem of interpreting and relating fossil plant fragments is fraught with difficulties, as outlined in Chapter 4, so we need not labour the point here. However, it is important to keep the difficulties in mind, together with the principle of regularly appearing characters, when we try to unravel the ever increasing evidence of plant evolution throughout the late Palaeozioc. It does of course go without saying that the same problems confront us regardless of the age of the fossil plants. But in the later geological periods the difficulties are somewhat alleviated by the possibility of more meaningful comparisons with living taxa.

There is clearly a need to interpret plants before attempting to fit them into a system of classification or into an evolutionary series. As we have seen in Chapter 3, Banks reclassified the Psilophyta into three new divisions after re-appraising the plants; thereby escaping from the classification strait-jacket that had existed until then. As plants become larger and more varied the problems become greater; so it would be sensible to look at the overall types of changes that were occurring before considering the ways in which whole plant evolution seems to have progressed.

Leaves

Leaves appear to have evolved in at least two distinct ways. The lycophyte microphylls were modified enations that became vascularised (Chapter 3). In contrast, those plants that were evolving from the Trimerophytina were developing leaves from modified lateral branching systems. Increasing pseudomonopodial branching patterns can be seen within *Psilophyton* and *Pertica* with a condensation of the fertile lateral branches in *Pertica varia*. Leaves are thought to have developed from the modified vegetative laterals by further suppression of growth, progressive flattening of the branches into a two-dimensional form, and by extension of their cortical tissues to give webbing between the branches. Changes in the branching patterns of the developing leaves would have given variations in venation pattern, and complete extension of the webbing gave rise to the leaf lamina mesophyll. Leaves that evolved in this manner are often referred to as megaphylls.

Although the concept of flattening and webbing in leaf evolution has received general acceptance, it is likely that varying amounts of the lateral branches were involved in their development. There are Devonian plants belonging to the so-called progymnosperm group that suggest leaves were developed only from the ultimate segments. The Aneurophytales are the simplest of the Progymnosperms with none showing any webbing at all. They have monopodial branching systems with a lobed primary xylem, but with ultimate dichotomous appendages having single small vascular traces. There is a similar change in anatomy between stem and leaf in living plants, so we can consider these ultimate divisions as early stages in leaf evolution (Banks, 1970). Within the group there is a gradual change in symmetry and size of the ultimate appendages which correlates with the geological distribution of the genera. *Protopteridium* has no distinct ultimate divisions; those in *Aneurophyton* are once or twice dichotomised and recurved; in *Tetraxylopteris* they are three-dimensional and divided in several different patterns; in *Triloboxylon* and *Proteokalon* they are dichotomously divided and in one plane (Figure 6.3). *Proteokalon* is thought to be the most advanced because its final divisions are flattened and also because they are the largest in the group (15–20 mm as against a maximum of 10 mm in the others) (Scheckler and Banks, 1971a,b). Such a series of changes is thought possibly to have given rise to the widespread group of Carboniferous gymnosperms — the seed-ferns. By the late Devonian the pteridosperms had developed many different forms, but all had broad laminate pinnules with dichotomous venation (see Chapter 9).

A similar succession of changes in branching pattern, leaf arrangement and anatomy can be seen in the second major group of Progymnosperms, the Archaeopteridales. Again, these changes can be correlated with time. There is change in branching pattern from helical in *Actinoxylon* and possibly *Svalbardia* to subopposite, alternate and distichous in *Archaeopteris* and *Siderella*. Phyllotaxy changes from helical and decussate in *Actinoxylon* and

Figure 6.3: A–D: *Tetraxylopteris schmidtii*; A, Reconstruction of a plant showing decussate branching; the rapid tapering is the result of secondary thickening. B–D: Three patterns of branching of the ultimate appendages. E: Reconstruction of *Triloboxylon ashlandicum* showing helical branching and planated ultimate appendages (two are shown complete, the others indicated by cut stumps); F,G: *Proteokalon petryi*; F, reconstruction of a plant showing the decussate branching and the planated ultimate appendages (two are shown complete, the others indicated by cut stumps); G, a planated ultimate appendage with sections at the positions indicated. H–J: Sections of penultimate branches; lined secondary xylem encloses the unshaded metaxylem that contains black protoxylem strands; H, *Tetraxylopteris*; I, *Triloboxylon*. J: *Proteokalon*. (A–E, After Scheckler and Banks, 1971a; F, G, after Scheckler and Banks, 1971b; H–J, after Beck, 1976)

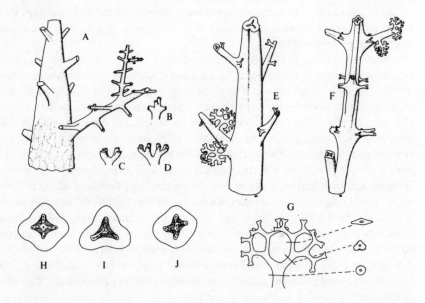

Archaeopteris to alternate and bilateral in *Siderella*. *Archaeopteris* (Figure 6.4) is the best-known genus (see Chapter 8). Originally this arborescent plant was thought to have possessed compound leaves, but we now know that it bore planated lateral branch systems with helically arranged simple leaves very similar to those of some modern conifers (Beck, 1976). The leaves are arranged essentially in one plane which is thought to be the result of developmental orientation in response to their photosynthetic activity (Beck, 1981). Leaf orientation of this nature can be seen in many living conifers such as *Sequoia, Taxus* and *Tsuga*, and in a great number of flowering plants, so we have every reason to believe that it could have happened in these Devonian plants.

The somewhat taxonomically isolated Noeggerathiopsida (characterised by *Noeggerathia foliosa*), which extended through the Carboniferous and into the Permian, been suggested to represent an extension of the progymnosperm

Figure 6.4: *Archaeopteris*. A: Restoration of a lateral branch system of *A. macilenta*. B: Enlargement of a segment of A, showing a portion of the main axis with leaf bases as if the leaves had abscised and a part of a leafy lateral branch. C: Reconstruction of adult plant, about 25 m tall. D: Putative sporeling (*Eddya sullivanensis*) about 30 cm tall. E: Transverse section of B, showing the departing branch traces (br.tr.), protoxylem (pr.x.), leaf traces (l.tr.) and secondary xylem (shaded). (A, B, E, After Beck, 1971; C, after Beck, 1962; D, after Beck, 1967)

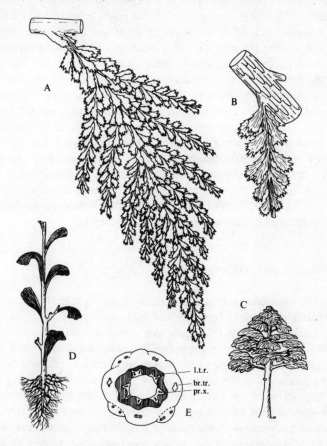

line (Beck, 1981). The decurrent leaves in *Noeggerathia* are alternately arranged in two ranks on the abaxial side of the shoots (Setlik, 1956). Beck proposes that the type of leaf arrangement shown in *Noeggerathia* could have resulted from that of the ultimate branch systems of *Archaeopteris* by the loss of two adjacent ranks of leaves and a 45° rotation of the axis.

It seems likely that this must represent a distinct line of megaphyll evolution that most probably gave rise to the coniferophytic gymnosperms.

It seems equally certain that neither type of series could have given rise to the compound megaphylls (fronds) of the ferns. Ferns evolved leaves much later than the pteridosperms did. The contemporaneous Visean coenopterid ferns *Botryopteris* and *Musatea* possess three-dimensional, slightly webbed fronds with fertile recurved distal units. *Musatea* also possesses a lamina with specialised mesophyll and stomata (Galtier, 1981). The late Upper Carboniferous ferns show a gradual increase of webbing, although the earliest are still quite small with monopodial venation. Ferns comparable in size to the pteridosperms did not evolve until the Westphalian.

Bierhorst (1971, 1973) has suggested that some ferns evolved their fronds rather differently. According to his theory there was a two-stage origin of the megaphyll involving the evolution of pinnae on an axis, as described above, followed by a process of megaphyllisation involving the axis. The theory is based on the fact that certain ferns have non-appendicular fronds, that is leaves which are direct continuations of stems. *Gonocormus* (*Trichomanes*), *Stromatopteris, Psilotum* and *Tmesipteris* have such fronds, and *Gleichenia* and *Actinostachys* (*Schizaea*) are said to have closely analogous morphology.

There is an ever increasing amount of evidence of variation in stem and leaf morphology in Palaeozoic plants. Although the evolutionary relationships between the species are still often difficult to resolve, it seems prudent to accept the idea of a multiple origin of megaphylls. At present the evidence suggests three, or possibly four, lines of origin, but naturally when new evidence comes to light our ideas may have to be modified once again.

Increase in Size and Arborescence

Reference has been made several times to the increasing size of Devonian plants. With increasing competition for space and light, many plants reacted by growing upwards. Limitations upon the height that could be achieved may have been partly overcome by the same crowding that was stimulating them to grow. Mutual support must have been very important, especially if the vertical shoots were being given off at short intervals from a horizontally creeping rhizome. Cortical specialisation must certainly have helped strengthen the stems, but beyond a certain size extra supporting tissues are necessary. The original protosteles of herbaceous plants became dissected in several ways, with each evolutionary development giving greater mechanical strength with increased efficiency of effort.

For any plant to become arborescent a much greater increase in girth is essential. This is usually, although not exclusively, achieved by continued formation of lignified vascular tissue by secondary thickening. There is evidence of secondary thickening in the middle Devonian, and by the late Devonian forest trees had evolved with trunks that were over a metre in diameter. During this time many groups of plants were increasing in size and several methods evolved for attaining the necessary girth increase. Some of these methods are not used by living plants, although they must have been structurally successful for the plants

Figure 6.5: The Probable Trend of Evolution of the Primary Vascular System of Progymnosperms and Gymnosperms. A: Three-ribbed protostele (e.g. *Aneurophyton*). B: Three-stranded vascular system derived from A by longitudinal dissection (e.g. *Stenomyelon musatum*). C: Separation of the strands shown in B with the leaf traces still diverging radially (e.g. *Archaeopteris*). D: A stele similar to C, but the initial divergence of the leaf trace is tangential (e.g. *Lyginopteris*). E: A stele similar to D, but showing the undulating path of the vascular bundles of those living conifers with helical phyllotaxis. (All after Namboodiri and Beck, 1968)

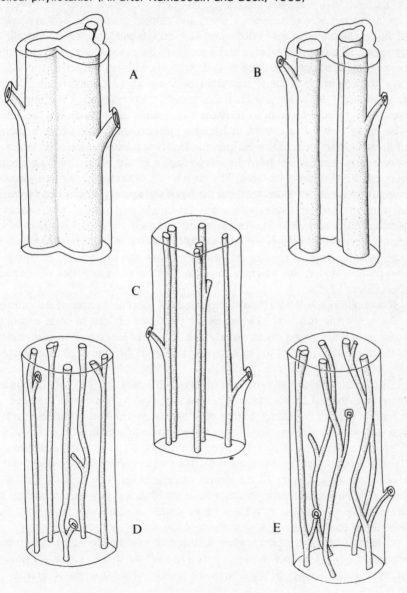

Isolating Mechanisms and Speciation Models

The orthodox view of species formation derived from Darwin's belief in the gradual nature of evolutionary change envisages a continuum of increasing genetic divergence as one moves from populations, to races, to subspecies, and finally species. If this view is valid, we could expect direct correlation between genetic divergence as expressed at the molecular level and changes in the organism as a whole associated with the formation of new species. But this is not so. An explanation for this lies in the role played by geographical isolation.

Geographical isolation not only facilitates the accumulation of genetic differences which may lead to reproductive isolation but it may be the most important factor in speciation. Once geographical isolation is taken into account, the speciation process need no longer be gradual.

The ways in which isolation, and subsequently speciation, may be brought about range from strict geographical isolation to chromosomal changes which may result directly in breeding isolation. Perhaps the most straightforward mode of speciation occurs when a single ancestral species population or hologamodeme becomes separated into two large populations by a geographical barrier that individuals cannot cross, or cross so rarely that the effect of immigration is negligible. The isolation prevents genetic interchange and over a period of time the two geographically separated (or allopatric) populations undergo genetic change such that if at any time they again occupy the same area, that is they become sympatric, they will not be able to interbreed successfully with each other. Because only large populations are involved, genetic drift does not play a major role in bringing about the genetic differentiation. If the two populations remain allopatric, their genetic isolation (and by definition their status as species) can only be tested in the laboratory.

When a very few individuals, or even a single seed, become isolated from a parent population, they serve as founders of a separate population. This isolated group has only a fraction of the genetic variation seen in the parent population, and genetic drift even in the absence of selection will lead to genetic divergence of the two populations. This is known as the founder effect and is important in island speciation.

It is possible to have a ring of populations or races each able to exchange genetic information with its neighbours. Each race has a slightly different genetic composition, however, and these differences may build up along the chain to such a degree that the terminal portions of the ring, even though they may overlap geographically (i.e. they are sympatric), cannot interbreed. Now if the ring becomes broken, two sets of races are produced which already exhibit a high degree of genetic divergence. The extent of the differences between the two sets may be sufficient to prevent interbreeding and the result is two species instead of the previous one.

Many species exhibit variation along environmental gradients. Such gradients, or clines, might include changes in soil pH, temperature, or a that evolved them.

The most lasting type of stelar dissection can be seen in the progymnosperms. It is possible to interpret from them the type of changes that probably gave rise to the vascular systems of living gymnosperms and angiosperms (Figure 6.5). The simplest forms, such as *Aneurophyton*, had a central ribbed stele which initially formed secondary wood between the arms of the primary xylem. In the largest forms, such as *Archaeopteris*, the primary vascular system was a ring of mesarch vascular bundles (a sympodium). Cambial activity formed a continuous cylinder of secondary wood and phloem around the primary xylem of the bundles. Many logs of this wood have been found in widespread localities throughout the Upper Devonian and were given the name *Callixylon*. The dense wood is made of compact thick-walled tracheids with interdispersed medullary rays, thereby being of the pycnoxylic type found in most conifers. Beck, Coy and Schmid (1982) suggest that the Lower Carboniferous *C. arnoldi* had possibly evolved a torus in its tracheid pits, dramatically pre-dating the otherwise earliest record of a torus in Jurassic gymnosperms. The lateral branches of *Archaeopteris* were of limited growth. This type is usually described as determinate growth as distinct from the indeterminate growth shown by the branches of conifers and angiosperm trees. The main length of the branches had a vascular system similar to the main stem, but the basal and apical portions had cylinders of primary vascular tissue with fewer protostele points than in the discrete bundles of the central region. In the ultimate branches a similar developmental change occurred, although here the vascular tissue was even more reduced. In their basal and apical regions the primary xylem was protostelic, changing in the central region to vascular bundles that were never quite separated (Scheckler, 1978).

Namboodiri and Beck (1968) interpreted the vascular system of the conifers to have evolved this way. The ancestral protostele divided to form a ring of vascular bundles giving off an orderly radial spiral of leaf traces. A realignment of the leaf traces, initially to depart tangentially, was then followed by an adaptation to undulating vascular bundles.

The pteridophytes showed some rather different systems of supporting increased size and arborescence. The fern-like cladoxylaleans had developed a complex stele of radiating plates of mesarch xylem that sometimes interconnected by dissection of a highly ribbed protostele (Scheckler, 1975). The Carboniferous marattialean ferns (e.g. *Psaronius*) had stelar patterns rather like those of modern tree ferns (Figure 6.6). The trunks of *Psaronius* have obconical steles consisting of up to 14 concentric vascular rings. This primary stem was formed by an ever increasing apical meristem. Plant stability was maintained by the growth of a massive mantle of adventitious roots over the stem that sometimes reached as much as a metre in diameter (Morgan, 1959).

The arborescent lycophytes grew to about 40 m in height with large crowns of branches. They did not, however, achieve such statures by secondary growth of their xylem, although there was a limited amount formed (Figure 6.7).

Figure 6.6: *Psaronius* sp. A: Reconstruction, about 8 m tall. Leaf scars can be seen on the upper part of the trunk underneath the crown. Most of the trunk is covered with a layer of adventitious roots that thickens towards the base. B–F: Cross-sections of the stele from the base to the apex. Xylem is shown as continuous lines, sclerenchyma as thinner dashed lines. (After Morgan, 1959)

Instead they supported themselves by growth of a thick zone of secondary cortex which even showed evidence of tangential cellular division in the trunk and larger branches. These arborescent lycophytes had a determinate growth pattern with the number of protoxylem points, the pith diameter and the amount of secondary thickening diminishing at every branching (Eggert, 1961).

The largest relatives of the herbaceous living horsetails were the Carboniferous arborescent *Calamites*. They reached about 30 m in height and their main axes were up to 30 cm in diameter. Transverse sections of these main axes showed a large central pith cavity, a ring of primary vascular bundles and large amounts of secondary wood that clearly provided the necessary support. The aerial shoots of *Calamites* exhibited determinate growth like the arborescent lycophytes and *Archaeopteris*, although their branching pattern was rather different as we shall see later in Chapter 7. Growth did, however, continue in *Calamites* via a branching underground rhizome that gave off aerial shoots at intervals.

Figure 6.7: The Anatomy of an Arborescent Lycophyte. A: Reconstructed median longitudinal section of the xylem system of its above-ground axes. Primary xylem is black, secondary xylem is shaded. B–G: Diagrammatic transverse sections of an arborescent lycophyte, from the base to the distal branches of the crown. Primary xylem is black, secondary xylem has radiating lines, outer primary cortex is shaded, periderm has broken radial lines. (After Eggert, 1961)

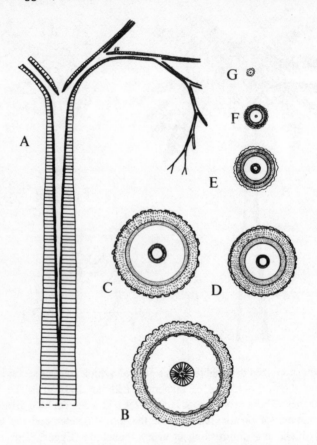

Thus there were several distinct methods of providing the necessary structural support for the increasing size of the plant. Whatever the mechanism employed, the result was similar in permitting the plants to dominate their immediate surroundings. Those which grew in dense patches like *Calamites* must have done so to the virtual exclusion of all other species.

What secondary growth also did for nearly all of these plants was to give them a finite period of life. A growth pattern utilising a creeping rhizome, such as in *Rhynia* and *Calamites*, made the plants virtually immortal. No doubt the vertical aerial shoots died after a relatively short time, but the rhizome could conceivably grow on for ever. In contrast, arborescent plants like *Archaeopteris* and the giant Carboniferous lycophytes were individuals and as such eventually died.

Roots

Increase in size also brought about other changes which affected the plants structurally and physiologically. Probably the most underrecorded structural change in the fossil record was the evolution of rooting organs and true roots. Roots, by their very nature of being underground organs, were less likely to have been transported to areas of sedimentation and therefore were fossilised much less frequently than other plant organs. Occasionally, however, we have evidence of rooting organs preserved where they grew, that is *in situ*, by decay and sediment infilling or by widespread petrifaction of organic remains as in the Rhynie Chert and Carboniferous coal balls.

The earliest land plants, like *Rhynia*, grew by rhizomatous extension and had no need for specialised rooting organs. They absorbed water and mineral salts through the rhizome epidermis and its single and multicellular hair-like outgrowths called rhizoids. *Asteroxylon*, as an early lycophyte, had downward-growing subterranean axes with very similar internal anatomy to that of the horizontal rhizomes. This is very similar to the condition found in the stem-like roots of *Lycopodium*. True roots, however, arise endogenously and not by apical branching as the underground organs of *Asteroxylon*. To evolve true roots from the *Asteroxylon* organs we must invoke a system of delayed differentiation, although it must be made clear that there is no evidence as yet to prove it.

The need for anchorage and support of larger plants, together with their greater requirement for water and mineral salts, resulted in the development of many underground systems differently adapted for the varying environments in which the plants grew.

Archaeopteris presumably had a large rooting system that looked rather like those of modern conifers, but structurally they were very different. Studies of the putative sporeling (*Eddya*) show that lateral roots probably developed by unequal segmentation of the meristem, being therefore very similar to the pseudomonopodial branching pattern of the shoots (Beck, 1967).

The best-understood rooting system of the arborescent lycophytes was a large, dichotomising, shallow-descending axis normally described under the name of *Stigmaria* (see Figure 7.8). Cambial activity formed large amounts of secondary xylem and cortical tissues in the older parts and was no doubt necessary for stability and to support the weight of such large trees. These rooting axes are best called rhizophores as they bear spirally arranged lateral appendages which are rather like endogenously produced roots. Unlike roots of living conifer and angiosperm trees, these appendages became functionally separated from the main axes by a process resembling leaf abscission, thus forming the characteristic circular scars on stigmarian axes (Frankenberg and Eggert, 1969).

Abscission

Perhaps the most obvious change accompanying arborescence is leaf abscission. All plant groups show some form of abscission mechanism, but it is the angiosperms that have utilised the phenomenon to the best advantage. All woody perennial plants shed their leaves, but this is particularly obvious to people living in temperate regions of the world where deciduous angiosperm trees simultaneously shed all of their leaves and stand bare throughout the winter months. In contrast, those trees growing in regions of uniform climate, such as the tropical rainforests, usually shed their leaves steadily in conjunction with the overall growth of the tree. Between the two extremes there is a great variety of abscission timing reflecting the different physiological requirements of the plants.

Leaf abscission can be stimulated by a wide range of external and internal conditions and we can see numerous examples in the fossil record of plants evolving such a capability. The benefits of leaf abscission are many, although it is difficult to correlate beneficial results with the physiological stimuli which must originally have created the necessity for loss of leaves. In other words, plants evolved extra anatomical or physiological features and reproductive strategies that were dependent upon their innate ability to abscise leaves. Such abscission mechanisms have also been utilised by plants for shedding leafy shoots, branches, reproductive organs and, to a much more limited extent, roots.

Organs that are to be shed develop a separation layer in what can be called the abscission zone. It is difficult to identify the physiological factors that are essential to the process, but Addicott (1982) gives three; a lowering of the auxin levels in the abscission zone, a source of energy and the constituents for synthesising the hydrolytic enzymes, and, thirdly, oxygen to provide energy for the synthesis of the enzymes. Protective layers form immediately under the separation layer and their subsequent exposure often reveals characteristic scars upon which species may even be identified.

Abscission seems to have developed very early in the evolutionary history of vascular land plants, if indeed it was not already a feature of their algal ancestors. Edwards (1980) has shown that the Devonian *Rhynia gwynne-vaughanii* shed its terminal sporangia although there was clearly no stem or leaf abscission occurring. It is highly probable that many other early Devonian plants similarly shed their sporangia after spore dispersal, but we have no clear evidence to support this suggestion except for the occasional findings of isolated sporangia. Sporangial abscission was clearly a feature of the vast majority of pteridophytes as there are many examples of isolated reproductive organs in sedimentary rocks of all ages. Gymnosperm seeds and angiosperm seeds and fruits are similarly found in large numbers. Some of these seed propagules have been recovered in an underdeveloped stage showing that plants rapidly developed the capacity to use abscission when they could not bring all their reproductive units to maturity.

Clear evidence of leaf abscission is known from the late Devonian arbores-cent lycophyte *Cyclostigma*, a character that it shared with the many Carboniferous genera of arborescent lycophytes. These plants also shed their reproductive organs either as whole cones or as individual sporophylls and also occasionally lost branches. The calamites shed branches leaving nodal scars arranged singly, in opposite pairs or in whorls. The Carboniferous tree ferns often show leaf scars on their trunks, and generic names have been given for the various configurations. For example, *Megaphyton* has two opposite vertical rows of scars, and *Caulopteris* is covered with helically arranged or whorled scars; other names are given to stems on the basis of different scar characters.

We have no proof that the progymnosperms abscised their developing leafy branch systems, but we suspect that such large plants as *Archaeopteris* must have done so. The large numbers of isolated pteridosperm and cordaitalean (see Chapter 10) leaves in Carboniferous sediments suggest that these plants abscised their leaves. There is little actual evidence, however, except that cordaite bark remains are known that show distinct leaf scars with clear vascular prints (form genus *Cordaicladus*). Similarly the Southern Hemisphere glossopterids (Chapter 10) appear to have been capable of leaf abscission. We do know however that the gymnosperms and subsequently the angiosperms used abscission in much more varied ways than their predecessors, allowing the plants to make use of this extra ability to spread into new habitats and more varied environments.

The initial stimulus causing abscission in vascular plants appears to have been related to loss of function and death of tissues. This must surely have been the case with dehisced sporangia. The subsequent enlargement of plants and their evolution of leaves created whole new problems for them which their ancestors did not face. Leaves typically reach a peak of photosynthetic efficiency on reaching full size. In time this efficiency will decline and eventually the leaf will die and probably be abscised. In many angiosperms the leaves will turn yellowish or red in senescence, accompanied by chlorophyll breakdown. Many factors control leaf fall in living plants so we must assume that a similar reaction would have happened in extinct species. They are low light intensity, alteration of day length, temperature change, mineral and other nutrient (e.g. nitrogen) deficiencies, water stress, and damage or infection.

Early arborescent species presumably brought about the abscission of their own shaded leaves by forming a canopy. This, accompanied by shedding of lower branches and a general thinning of the crown by attrition and abscission of the smaller and weaker branches, would have resulted in a 'modern' tree architectural appearance in such plants as *Archaeopteris*. It is the various combinations of shoot enlargement, orientation and abscission that give trees their differing styles of architecture.

The other important factor controlling early plant growth must have been the availability of minerals. The accumulation of humus and the creation of organic soils must have had a lot to do with changes in plant morphology related to increase in size. Vascular plants must have followed earlier extensive

Figure 6.8: Leaf Abscission in Arborescent Lycophytes. A: Shoot of *Lepidodendron* with detail of single leaf attachment. B: The same after leaf abscission. l.l., Leaf lamina; v.t., vascular trace; p, parichnos print (aerating tissue); l.c., leaf cushion (swollen basal part of leaf); l.s., leaf scar

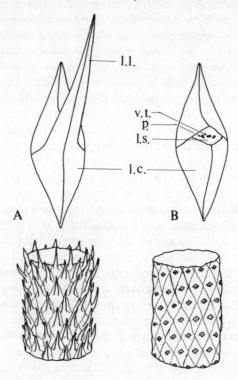

colonisation by algae, fungi, lichens, bacteria and other extinct non-vascular plants. The action of fungi and bacteria as pathogens and decomposers would have continued to build up the organic content of soils as the numbers of vascular plants increased. Invasion of these developing soils by detritus feeders would again have increased the rate of organic accumulation. Soil and humus formation would itself have encouraged further plant and animal life thereby accelerating the cycle of formation. Eventually a stable plant-animal-soil community would have resulted until disturbed by a wave of immigrants, climatic change or geological disturbance. Studies on modern rainforests have shown leaf fall to be very important in maintaining the nutrient level of the soil (Longman and Jenik, 1974). The rapid action of tropical micro-organisms quickly recycles nitrogen and carbon into a form that is suitable for uptake again by vascular plants. Studies of present-day carbon balance in soil detritus show that tropical soils have little detrital accumulation in spite of high productivity and litter fall. Most of the soil carbon is dispersed as light-coloured fulvic acids in the deeper parts of the soil profile (Schlesinger, 1977). The Devonian forest communities were most likely to have been in a similar delicate balance with the soil. Logic therefore tells us that leaf abscission must have developed in conjunction with increasing size, otherwise the organic cycles relying on organic input would seem to have been impossible.

Leaf abscission through senescence and partial shading most probably occurred in the tree ferns, but abscission in the arborescent lycophytes (Figure 6.8) and calamites might have been partially due to another stimulus. Both types of plant grew in waterlogged or flooded environments in the Carboniferous Coal Measures swamps and possessed internal aerating systems to alleviate oxygen deficiency. In spite of their anatomical adaptations it is possible that oxygen deficiency did occur in the rooting organs, thereby impairing the uptake of mineral nutrients. In fact the very formation of peat removes minerals from circulation in the swamp ecosystem. Mineral deficiency often causes leaf fall in living plants (Addicott, 1982), so it is conceivable that it promoted branch and leaf degeneration and abscission in the lycophytes and calamites, keeping their photosynthetic foliage to a minimum. Today the swamp cypress (*Taxodium distichum*) living in present-day bog environments seldom develops great crowns of leaves as larger amounts of foliage might exacerbate the nutrient deficiency already troubling the tree (Schlesinger, 1978).

As we will see later, the subsequent migration and evolution of plants took them into increasingly variable and inhospitable climates. The ability to shed organs was a vital asset for many species, enabling them to survive through periods of adverse conditions. The angiosperms especially put their ability to great use while invading many areas of the world where they would periodically encounter intense heat or cold, times of drought and annual seasons of short day length in high latitudes.

Bark

The formation of secondary cortical tissues (phelloderm) to produce the protective layer, loosely called bark, usually accompanies the development of secondary vascular tissues. As the xylem increases in diameter, the surrounding layers must expand or become ruptured by pressure from within. Development of secondary cortex overcomes the structural and physiological dangers that would result from such a rupture. Usually a continuous series of cortical cambia (phellogen) are initiated in deeper layers of primary cortex and phloem, each forming its own secondary cortex (phelloderm). Cells to the outside of the suberised phelloderm die, resulting in the protective layer that we loosely call bark. Apart from its protective value to the expanding tree, bark can act as a physical barrier to pathogen or predator attack and as an insulating layer against fire.

The earliest reported occurrence of bark in the fossil record was shown by Scheckler and Banks (1971a) to be in the Mid-Givetian progymnosperm

Triloboxylon hallii. Three other members of the Aneurophytales (*T. ashlan-dicum, Tetraxylopteris schmidtii* and *Proteokalon petryii*) also possessed bark. Every subsequent group of plants that underwent secondary thickening formed bark or an equivalent tissue. The arborescent lycopods were probably the most unusual. Their secondary cortex was the main supporting tissue of the plant but it did not immediately produce a dead outer zone. Photosynthetic basal parts of leaves remained active for some time outside the secondary cortex. A certain amount of lateral expansion of these epidermal and sub-epidermal layers did occur (Figure 7.6), but stem enlargement eventually caused them to rupture and then to be sloughed off (Thomas, 1966).

Communities and Floras

The earlier Devonian plants were all quite widespread over the world within the relatively small areas of sedimentation. The later, more worldwide, distribution of plants led to some later Devonian and Carboniferous genera becoming increasingly restricted and often quite localised. This resulted in the development of several different floras with many complex communities. Each community was stimulating or restraining evolution in a variety of interactions between the plants themselves and with their environment. Habitat specialisation was the result, causing some pockets of plants to adapt and evolve in both geographical and genetic isolation. Some of these resultant communities were so dependent upon particular habitats that they perished when the habitats changed. This, for example, led to the extinction of many of the giant pteridophytes that were so successfully adapted to the Coal Measures swamps at the close of the Carboniferous.

Co-evolution of Plants and Early Terrestrial Arthropods

It is highly probable that interrelationships existed between vascular land plants and arthropods from the advent of land colonisation (Figure 6.9). It is equally likely that these interrelationships stimulated co-evolutionary developments to give interdependent forms which together provided an integral part of the expanding ecosystem. Some arthropods were detritus feeders or predators living in, or upon, the soil with other more soft-bodied metazoans. Other arthropods would have had a far greater effect as they fed directly on the plants themselves. There is, however, very little direct evidence of such interactions, so ideas rest upon functional interpretations of fossils, analogous modern phenomena and assessments of proposed phylogenies (Kevan *et al.*, 1975).

Arthropods are known in detail from the Rhynie Chert where apparently predatory forms have been found in empty sporangia of *Aglaophyton major* which they probably occupied following spore-eating. The stems of the Rhynie

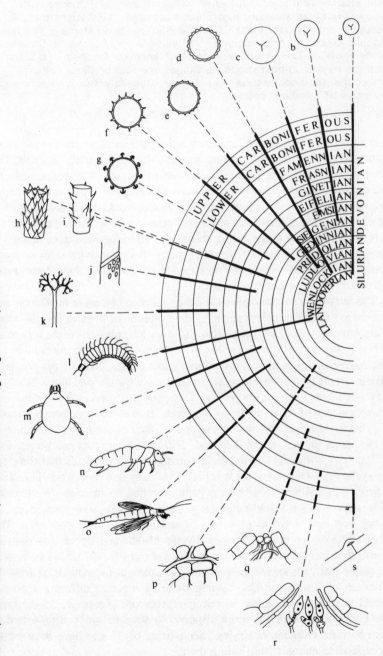

Figure 6.9: The Appearance of Phenomena Indicative of Co-evolutionary Processes in and around the Devonian Period. (After Kevan, Chaloner and Savile, 1975). *See following page for explanation*

Explanation to Figure 6.9 (*see previous page*): a, Triradiate spores
exceeding 50 μm in diameter. b, Spore diameters exceeding 100 μm.
c, Spore diameters exceeding 200 μm. d, Verrcuate spore exines. e, Exines
with small conical spines. f, Exines with long spines. g, Exines with
grapnel-like ornamentation. h, i, Stem structures: spiny stems that
disappear entirely before the Carboniferous, and leafy stems. j, Phloem in
vascular strand. k, Arborescence. l, Myriopoda. m, Acarina. n, Collembola.
o, Pterygota. p, Conjectural phycomycete ancestor (dashed) and branching,
aseptate hyphae in host tissue. q, Conjectural rust uredinial state. r,
Conjectural *Uredinopsis*-type uredinium. s, Mycelia with clamp connections
indicative of Basidiomycetes.

plants also show evidence of attack by arthropods or some other small animals.
Some injuries are extensive, being probably caused by biting animals. Others
show organic plugs in lesions extending into the region of the phloem-like tissue
and can be interpreted as a plant response to attacks by sap-feeding arthropods.
The increasing stem thickness of later plants would have discouraged such
attacks especially if there was any bark formation so arthropod damage would
have been restricted to the young shoots and reproductive organs of such adult
plants. Also, as we saw in Chapter 3, many species evolved stem emergences
but, as was pointed out, it is by no means certain whether these discouraged or
aided arthropods to climb the axes.

The increasing height of plants and the evolution of trees in the Devonian may
have been one of the major causes for the evolution of flight in insects, although
Wigglesworth (1973) suggested that the main advantage of flight was to allow
airborne migration from areas that were drying up to more favourable areas.
The raising of edible shoots and reproductive organs high off the ground may
have given the plants some respite from large-scale arthropod attack but it would
have been brief. By the end of the Carboniferous there were winged insects
clearly capable of reaching any plant organ. However, even though arthropods
must have consumed vast numbers of spores, they must have given some benefit
to the plants through spore dispersal. Different arguments can be put for the
evolutionary advantages of the highly ornamented spore walls that were evolving in the early Devonian, for it is highly unlikely that energy would have been
expended in this way if such an organisation had no function. Wall sculpture
adds support which would be of value during transportation by animals. Long
spines may have acted as processes of attachment to arthropod setae. This is
especially likely in those otherwise diverse Middle and Upper Devonian spore
genera that had grapnel-hooked spines (Kevan *et al.*, 1975). It has even been
suggested that such spore transportation may have had something to do with the
evolution of heterospory by helping to bring together differing spores. We
should not be too confident in our interpretations, however, for sculpturing
could similarly have aided spore dispersal by wind or water or protected them
from harmful radiation or drying. Sculpturing could also have deterred small
spore-feeding animals from eating them.

Another, though less obvious, result of animal-plant interaction is the spreading of disease. Fungi and bacteria-like and virus-like organisms are known from Precambrian rocks, but the beginnings of their relationships with vascular plants are rather unclear. Pratt, Phillips and Dennison's (1978) work on Silurian macerates suggests that septate filamentous fungi had colonised the land during the late Ordovician. Pirozynski and Malloch (1975) have even suggested that the appearance of land plants was a direct result of an endo-symbiosis between an oomycetous fungus and a green alga. Whatever the path of colonisation taken by the fungi, it is clear that they were involved in a variety of ways with the earliest vascular plants. The problem is recognising a parasitic or mycorrhizal fungus as distinct from fungi that have invaded plant tissue after death. *Asteroxylon*, from the Rhynie Chert, and many Carboniferous plants have fungal associations but most relationships cannot be reliably interpreted. Whatever the relationships were, it is extremely likely that arthropods would have spread such fungi, together with bacteria and viruses, through their feeding. Thus although most of our ideas are based upon little direct evidence, it must be obvious that there are many ways in which the evolution of the early vascular land plants and arthropods must have been inextricably linked. As we will see later, other subsequent plant-animal relationships developed that were to stimulate interdependent evolutionary changes that were vital for the success of both.

7 PTERIDOPHYTE SUCCESS AND SPECIALISATION I — LYCOPHYTES AND SPHENOPHYTES

The earliest vascular plants were free-sporing pteridophytes sharing the same basic type of homosporous reproductive life cycle. As we saw in Chapters 3 and 5, the first recognisable groups, the Rhyniophytina, Zosterophyllophytina and Trimerophytina, eventually became extinct. They were superseded by their more successful, and presumably more competitive, descendants the lycophytes, sphenophytes and ferns. Although there was great subsequent diversification in morphology, anatomy and reproductive biology, these pteridophyte groups can be traced through their evolutionary histories to the present day. A vast number of different forms have been described from fossils and from living plants, so it is clearly impossible to mention all the families let alone the innumerable genera that are known to have existed. Instead we can follow the larger scale evolutionary changes that were occurring and attempt to relate these to the ever-changing aspects of the world's vegetation.

A great number of pteridophyte fossils have been found in rocks of all ages, so we have quite a good idea of the major evolutionary changes that have taken place. Comparable variation occurred in the three groups, suggesting that environmental and competitive pressures were stimulating similar adaptive changes. Some forms gained success by achieving dominant status in the vegetation either by increasing in size or by forming dense communities. Others adapted to life as understorey plants, relying on the vegetation as a whole for their survival or restricting their distributions to areas where there was less competition. Many adaptations were evolving, permitting certain species to live in areas of seemingly inhospitable environments. Some species can survive periods of drought whereas others have gone back to an aquatic habitat, developing into emergent, submerged, or floating forms. It is in times of geo-morphological stability that pteridophytes have developed as the dominant plants, thus those pteridophytes in the Carboniferous Coal Measures swamps even evolved into arborescent forms with secondary thickening and quite complex reproductive systems. Pteridophytes generally remained as successful, widespread members of plant communities until the Mid-Cretaceous, when the flowering plants became the dominant element in world floras. It was the competition with angiosperm shrubs and herbs that eventually restricted the numbers of suitable habitats. At this time two important changes occurred. The overall numbers of pteridophytes declined through competition but, conversely, new forms evolved in response to the enormous changes in ecological structure of the world's plant communities.

Specialisations in the Life Cycle

Sporophytes represent only one phase of the pteridophyte life cycle. There is an alternation between the spore-producing, asexually reproducing sporophyte and the gamete-forming, sexually reproducing prothallus. The latter is the much more vulnerable phase and only an infinitesimally small percentage of the millions of liberated spores ever develop to sexual maturity and to form embryos that grow into mature sporophytes.

Some of the earlier pteridophytes had large prothalli that seem to have been able to grow and branch into conspicuous plants, e.g. the Devonian gametophytes *Lyonophyton* and *Sciadophyton*. In contrast, extant homosporous pteridophytes form gametophytes with much more limited growth. The largest belong to the primitive fern genera *Psilotum, Tmesipteris* and *Stromatopteris* where they are extremely similar to the underground rhizomes of the sporophytes. They are subterranean, with well-organised apical meristems, branching dichotomously and by lateral budding in *Stromatopteris*. Their sex organs and rhizoids are scattered over their surfaces, either uniformly or in definite fertile zones. The three-dimensional gametophytes of living homosporous lycophytes are quite variable, being either surface-living and green with a basal pad and a series of short, upright lobes, e.g. *Lycopodiella cernua*; or subterranean, e.g. *Diphasium complanatum* in which they are cone-shaped, *Lycopodium clavatum* in which they are disc-shaped with convoluted margins, and *Phlegmarius phlegmaria* which has a branching axial body. Living ferns similarly have very varied prothalli, ranging from the narrow delicate branching ones of *Trichomanes* to the thallose surface-living forms of the higher filicalean ferns.

Some pteridophytes have evolved a rather different growth pattern for their gametophytes. Two sizes of spore are formed resulting in the condition we call heterospory (Figure 7.1). The larger spores (megaspores) develop prothalli with archegonia; the smaller (microspores) develop prothalli with antheridia. Both types of prothalli form within the spore walls. This condition is said to be endosporic as opposed to the exosporic growth of the larger free-living prothalli of homosporous pteridophytes.

The retention of the prothallus within the spore wall presumably affords some protection against desiccation, although it must also limit the capability to absorb water. However, even if desiccation is less of a problem, the prothalli still cannot fulfil their reproductive role without water because the sperm need to swim from the antheridium of one prothallus to the archegonium of another. Rhizoids are usually formed on the 'female' prothalli which, in addition to acting as water-absorbing structures, might help retain a surface film of water for the free-swimming sperm.

What is probably of much more importance to the overall success of the species is the dramatic reduction in time taken for the prothallus to reach sexual maturity. Exosporic maturation of prothalli could have brought about specialisa-

Figure 7.1: Diagram of a Simplified Heterosporous Plant Life Cycle Showing Alternation of Phases. The sporophyte phase is diploid (2*n*) and the gametophyte phases are haploid (*n*)

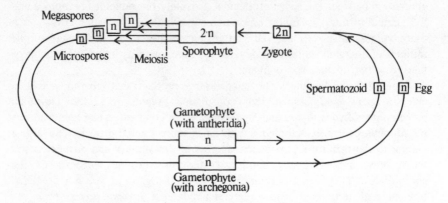

tion resulting in the dimorphic development of prothalli. The much larger size of the megaspores has surely evolved through the need to provide adequate food reserves in an endosporic prothallus for the dependent embryo until it is able to survive on its own. However, such an enlargement of the megaspores should not be thought of as essential for the evolution of sexual dimorphism in prothalli.

Sporangia are completely dimorphic in heterosporous plants except in a very few species, with megasporangia forming smaller numbers of megaspores by reducing the number of tetrads or by aborting members of the tetrads. Developmentally, though, the sporangia are initially similar and the diverging morphological maleness and femaleness are not genetically controlled. Studies of *Selaginella* give us an idea of the essential features of heterospory (Bell, 1979). Microsporogenesis essentially resembles the sporogenesis of homosporous plants where limitations of food supply and an insubstantial amount of cytoplasm restrict spore development to maleness. In contrast, in megasporogenesis there appears to be competition between megasporocytes and a variable number of megaspores are produced. The essential factor in determining the femaleness of the spore is that the crucial nucleocytoplasmic interaction conferring femaleness is much earlier as it occurs at megasporogenesis rather than at oogenesis.

Heterospory is deduced in fossil plants by the recognition of two sizes of spore and in dispersed spore populations by the presence of large spores. But if Bell (1979) is correct in that femaleness is conferred during megasporogenesis, then size enlargement is not an essential feature. Bell's view on the changing size of spores through the Palaeozoic is that homospory was followed by anisospory before heterospory. Anisospory forms spores of only slightly different sizes which gave gametophytes with different growth rates, thereby ensuring the simultaneous provision of both gametes. Duckett and Pang's (1984) observations on growth and sexual behaviour in *Equisetum*

Figure 7.2: Branch of a Bisexual Gametophyte of *Equisetum fluviatile* Showing Antheridial Meristem (a), Dehiscing Antheridia (d), Mature Archegonia (m) and Young Antheridia (y). (After Duckett, 1973)

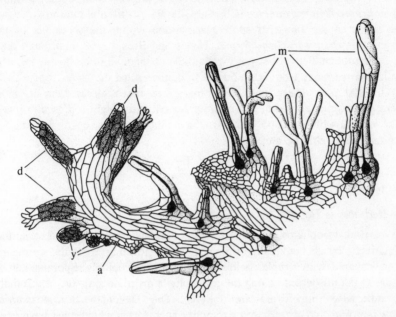

giganteum and in the Australian fern *Platyzoma microphylla* support this idea. Gametophyte morphogenesis and sex organ formation are linked in *Equisetum* although it remains unequivocally homosporous (Figure 7.2). Sex organ formation is determined by environmental conditions after spore germination. *Platyzoma* shows the same association between gametophyte morphogenesis and sex organ formation, but here sexuality is largely fixed during sporogenesis. It is an anisosporous plant with large and small spores within every sporangium. The smaller spores yield filamentous 'male' gametophytes and the larger spores spathulate 'female' gametophytes that subsequently form antheridia.

Isoetes pantii may be also anisosporous as it has spores of three different sizes within its microsporangia (Goswami and Arya, 1968). However, as nothing is known about the germination of the spores the sexuality of their prothalli is a mystery.

The overall suggestion is that mutations forming anisospory might lead to abortion of megasporocytes or meiotic products which could then stimulate the formation of heterospory. The next stage would be increased nucleocytoplasmic activity in the developing spore, leading to extensive growth and differentiation. Chaloner's (1967) records of spore sizes indicate that anisospory gradually increased throughout the Devonian but that heterospory arose suddenly in the Frasnian with a burst of variation and megaspore formation.

There are a few species of fossil plants that provide additional evidence, in that they have spores of more than one size in a sporangium. In most cases the smaller spores have been thought to be abortive megaspores, e.g. *Chaleuria cirrosa* (Andrews, Gensel and Forbes, 1974), *Protopitys scotica* (Smith, 1962) and *Stauropteris burntislandica* (Chaloner, 1958). A different interpretation has been made of the range of spore sizes found within the sporangia of the Devonian *Barinophyta citrulliforme*. Taylor and Brauer (1983), through their ultrastructural studies of the spore walls, believe them all to be functional. The smaller spores range from 30 to 50 μm in diameter and the larger spores from 650 to 900 μm. These two size ranges are much closer than in most heterosporous plants. As these authors point out, this might represent a phase in the evolution of heterospory prior to the evolution of two separate sporangia. It could conceivably be anisospory.

Lycophytes

The Herbaceous Habit

The earliest lycophytes were herbaceous, homosporous forms that spread vegetatively by dichotomising rhizomes while forming erect dichotomising shoots covered with simple, helically arranged microphylls. Sporangia were randomly interdispersed among the microphylls or in the axils of, or adaxially on, otherwise unmodified sporophylls. The Devonian *Baragwanathia, Drepanophycus* and *Asteroxylon* exemplify such forms which must have been very much like the living lycophyte *Huperzia selago*. These ancestral lycophytes must have spread quickly, although they were probably still able to hybridise freely if they intermingled. Variations appeared in the Middle and Upper Devonian which we now recognise as distinct genera, although still loosely grouping them together as the Protolepidodendrales. This grouping is clearly an oversimplification of the situation, and in lumping all the Devonian species we conceal their potentially rather different botanical affinities. For example, *Asteroxylon* has greater similarities with living Lycopodiaceae and is better grouped with these than with other Devonian genera such as *Protolepido-dendron*, which has been used for a variety of Middle Devonian dichotomously branching lycophytes, *Archaeosigillaria*, which is a small dichotomously branched herbaceous lycophyte with helically arranged leaves organised into vertical ranks, and *Leclercqia* which has closely spaced leaves, each dividing to give five slender tips and each possessing a delicate membranous flap (ligule) on its upper surface.

These Devonian lycophytes evolved in several different ways. Some continued relatively unchanged to the present day, some gave rise to further varied herbaceous homosporous and heterosporous genera, and others enlarged to evolve eventually into arborescent forms.

Living heterosporous lycophytes are ligulate, and homosporous lycophytes are eligulate. Until the homosporous Devonian *Leclercqia* was shown to be ligulate (Grierson and Bonamo, 1979) it was generally assumed that ligules would be restricted to heterosporous species. But it now seems likely that ligulate homosporous forms gave rise to ligulate heterosporous ones, although we still have no substantive proof of it.

Although superficial ligules, as found in *Leclercqia*, have persisted into the living *Selaginella*, another adaptation of the ligulate condition rapidly evolved in the Upper Devonian. Most ligules became sunken into cutinised pits in the leaf tissue. Such ligule pits can survive as cuticular cylinders or be filled with sediment and preserved as casts. This can quite often be the only evidence that we have of the ligulate and presumed heterosporous nature of many genera of fossil lycophytes (e.g. Thomas, 1968; Meyen, 1972, 1976).

Devonian lycophytes evolved in several rather different ways. Herbaceous and presumably homosporous forms continued with little change to the present day. *Asteroxylon* lived in the Lower Devonian, but *Drepanophycus* persisted throughout the entire Devonian. *Lycopodites* is a name that is often given to small leafy shoots from the Carboniferous onwards, although it is not a very satisfactory taxonomic unit. Most specimens are sterile and many have been subsequently shown to be terminal twigs of larger lycophytes or even of conifers. Some are undoubtedly herbaceous lycophytes as demonstrated by Harris (1976a) for his *Lycopodites hanahensis* from the Wealden of south-east England. Similar axes are also known from the Yorkshire Jurassic.

Today there are several genera of lycophytes, including the highly reduced and specialised *Phylloglossum* (Figure 7.4A), which is known from only a very few sites in Australia and New Zealand. There are also those genera created from the originally broad concept of *Lycopodium* — *Diphasiastrum, Huperzia, Lycopodiella, Lycopodium* and *Phlegmarius*, which together include about 400 species. These latter genera vary considerably in growth habit and distribution, presumably having evolved differently to suit their individual ecological requirements (Figure 7.3). For example, *Huperzia selago* is an almost arctic-alpine species with erect dichotomising axes, with sporangia in fertile regions up the stem and not in cones. Its spores, however, probably contribute very little towards its reproduction; instead it seems to rely very heavily on wind dispersal and the establishment of bulbils that it forms instead of some sporangia. *Phlegmarius phlegmaria* is a tropical epiphyte with drooping dichotomising axes and terminal cones. *Lycopodium clavatum* has a similar distribution to *Huperzia selago*, but has horizontal pseudomonopodially branching axes and the occasional fertile vertical axis with a pair of cones. *Lycopodiella cernua* is virtually pantropical in distribution, growing in the drier, sandy soils. It propagates vegetatively by each vertical axis dichotomising to give a much branched fertile shoot and a runner-like shoot which roots on touching the ground.

Herbaceous ligulate lycophytes expanded dramatically in number and variety during the Carboniferous. The simplest erect and dichotomising forms had a world-wide distribution in the Lower Carboniferous but became much

Figure 7.3: Extant Herbaceous Lycophytes. A, B: *Huperzia selago*; A, individual erect plant, about 20 cm tall; B, a bulbil. C: *Phlegmarius phlegmaria*, a pendulous plant about 35 cm long. D, E: *Lycopodium clavatum*; D, portion of a creeping plant with a fertile shoot about 10 cm tall; E, sporophyll with its sporangium. F, portion of *Lycopodiella cernua* with a fertile shoot about 40 cm tall. The plant has developed from the end of the runner coming from the right and has formed a new runner to the left. This has subsequently rooted and a new vertical axis is growing

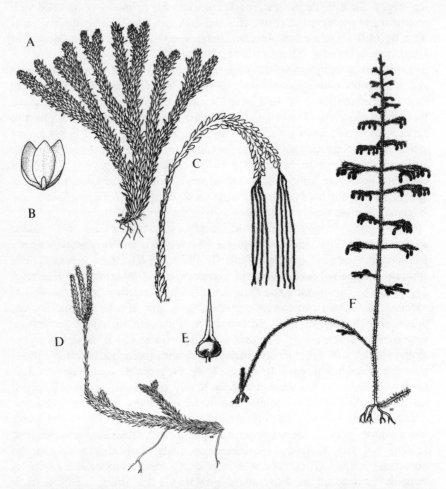

more restricted in the Upper Carboniferous and the Permian. They really only survived successfully in the more inhospitable region of Angaraland where they seem to have escaped competition from the more vigorously growing and actively evolving forms of the other floristic zones (Meyen, 1976).

However, in the Euramerian Upper Carboniferous Coal Measures swamps evolutionary changes were continuing and forms comparable to living

Figure 7.4: A: *Phylloglossum drummondii*. B: *Selaginella vogelii*, an erect
anisophyllous species (the lower branches on one side have been omitted).
C: *Selaginella selaginoides*, an isophyllous species. D, E, F: *Selaginella
kraussiana*; D, a creeping anisophyllous species with rhizophores; E, detail
of the stem; F, leaf showing the single vein and the ligule (l)

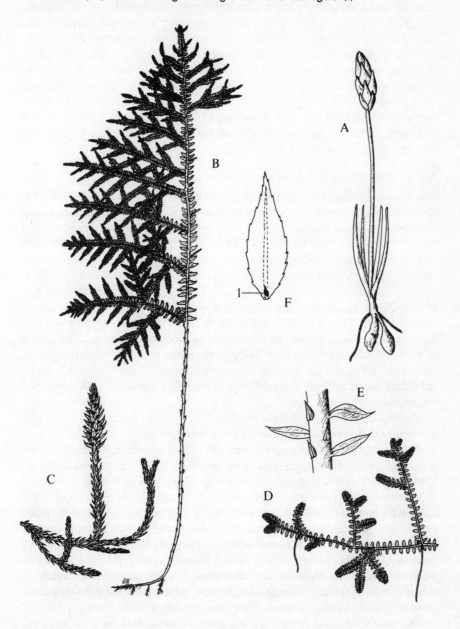

Selaginella were appearing. For example, *S. fraiponti* from the American Pennsylvanian has been pieced together from its constituent parts which were originally described separately (Schlanker and Leisman, 1969). It was sparsely branched, sprawling and possibly of determinate growth. Its cones were terminal and had sporophylls in alternating pairs of verticils. In all respects, except for its having a distinctive basal root-bearing organ, it is virtually identical to some living species of *Selaginella* (e.g. *S. selaginoides* — Figure 7.4). The majority of living species, however, are rather different in being anisophyllous (having two dorsal ranks of smaller leaves and two lateral rows of larger leaves). These anisophyllous species, which are today mostly tropical, are also known from the upper part of the Upper Carboniferous of Europe where they are usually given the name *Selaginellites* (e.g. *S. gutbieri*, *S. primaevus* and *S. suissei*). Some of these are fertile and are known to be heterosporous, just like the living forms of today.

The Development of the Arborescent Habit

The Middle Devonian lycophytes also gave rise to a further group of descendants that was destined to emerge as the most complex and in many ways most successful of all the pteridophytes. These were the arborescent lycophytes with species that would grow up to nearly 50 m tall and form extensive forests. The Devonian and Lower Carboniferous plants called *Lepidodendropsis* were at least shrubby in nature with leaves in low-angled spirals: leaves that had their basal portions expanded into narrow fusiform leaf cushions. Unfortunately only two of the 20 or so species are fertile and they both come from the Lower Carboniferous. They have their sporangia in the axils of seemingly unmodified leaves and one is described as heterosporous (Jongmans, 1954). There were other forms of large lycophytes living at the same time as *Lepidodendropsis*. However, although the generally accepted genera have definable limits, individual specimens are often difficult to identify because of badly preserved vegetative characters.

The evidence suggests that the largest of these plants had a main trunk with a crown of dichotomising branches. Leaves were persistent, although they most probably withered and collapsed in late life. The few fertile specimens known show that some were homosporous whereas others were heterosporous, but we know virtually nothing of their reproductive biology. The rooting organs are thought to be corm-like axes, up to 60 cm long and 23 cm in diameter at the base, with up to 13 basal lobes bearing lateral appendages. These axes, called *Protostigmaria*, may have supported the comparatively larger aerial system by a mechanism similar to that employed by the living lycophyte *Isoetes*. In *Isoetes* the roots in the corm furrows are moved laterally during the corm's growth, thereby 'pulling' the plant into the substratum. If corm growth were similar to this in *Protostigmaria*, it would explain how these plants existed without the more extensive rooting systems of the later lycophytes.

The most specialised arborescent lycophytes developed in the Coal Measures

Figure 7.5: Reconstruction of *Lepidodendron*, about 40 m tall

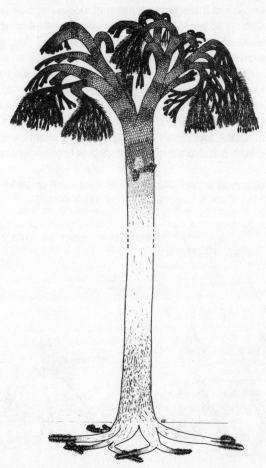

swamps of the Northern Hemisphere Euramerian province. They are grouped together as the Lepidocarpales (Thomas and Brack-Hanes, 1984) although showing extensive and varied modifications in morphology and reproductive systems. Their ability to develop to arborescent proportions relied upon their unusual growth patterns. In *Lepidodendron* the sporeling arose from an endosporic prothallus in a megaspore that may have been only 1 mm in diameter, then grew straight up, expanding in diameter by meristematic enlargement and secondary growth until its apical meristem divided. By this time the plant could have been nearly 40 m tall. The two new meristems formed branches that were only slightly narrower than the terminal part of the main trunk. A successive series of meristematic divisions gave progressively narrower branches until growth finally ceased with slender terminal twigs (Figure 7.5). As we saw earlier (p. 82), this determinate pattern of growth was accompanied by anatomical changes. The original protostele became medullated in the lower part of the trunk and then increased in diameter with a simultaneous increase in the number of protoxylem points. The protostele diameter and the number of protoxylem points then progressively decreased at every branching until they terminated as a narrow unmedullated strand. The amount of secondary xylem and supportive secondary cortex similarly decreased with the branching until the terminal twigs possessed none. Such a growth pattern is completely unlike that of the conifers and angiosperm trees with which we are much more familiar. In these trees apical meristems are always small, forming twigs that steadily enlarge by secondary growth. In the arborescent lycophytes large shoots were formed as such so the terms young and old are meaningless when discussing shoot size.

Figure 7.6: *Leipidodendron* Leaf Cushions. A: Leaf cushion before and after the expansion of the stem's tissues following secondary growth; the intercushion grooves expand and become flattened. B: A transversely cut reconstructed stoma from the leaf-cushion epidermis; the guard cells are sunken in a pit and have internal thickenings of lignin. (A, After Thomas, 1966; B, after Thomas, 1974)

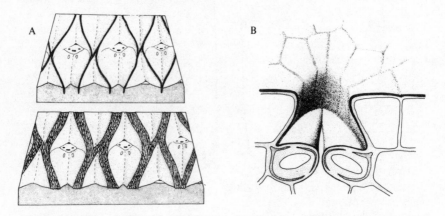

The outward expression of shoot size also reflects the style of growth. The leaves have swollen photosynthetic bases (leaf cushions) that remain when the distal laminae are shed (Figure 6.8). This abscission is the only reliable evidence of ageing. The size of the leaf cushions and the length of the laminae are related to the diameter of the shoots; the smallest twigs having the smallest leaf cushions and the shortest laminae (Chaloner and Meyer-Berthaud, 1983). Secondary growth also affects the outward appearance of the larger shoots by the increase in circumference. Leaf cushions do not enlarge, but there are often signs of limited growth of the superficial tissues between the leaf cushions (Figure 7.6). Secondary growth of the trunk was, however, so extensive that the entire outer tissues were shed, leaving the secondary cortex exposed which itself in time became longitudinally split by the continued internal growth.

Figure 7.7: A: *Stigmaria ficoides*; the Williamson specimen in the
Manchester Museum (about 5 m across). B, C: *Isoetes echinospora*; B, the
whole plant (about 8 cm tall); C, inner surface of the lower portion of a
sporophyll showing the megasporangium (m) partially covered by the
vellum, the ligule (l) and the four air canals (c) in the leaf

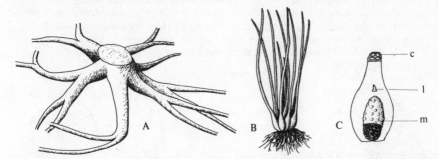

There were several variations on this general *Lepidodendron* type of growth
pattern. Some species often branched pseudomonopodially so it is doubtful if all
these lycophytes were the perfectly dichotomising plants that are so often
portrayed (Thomas, 1978; Dimichele and Phillips, 1985). *Lepidophloios* was
very like *Lepidodendron* in growth but had leaf cushions that enlarged after leaf
fall, thereby increasing the photosynthetic ability of the plants (Thomas, 1977).
Ulodendron in contrast never shed its leaves. *Bothrodendron* only had leaf
cushions on its smaller branches with the larger ones bearing leaves, and subsequently leaf scars, directly on the shoot surface. All four genera also sometimes
show large rounded 'ulodendroid' scars on their larger axes which have been
interpreted both as cone scars and branch scars. The fact that cones are often
found terminally on narrow shoots suggests that they are more likely to have
resulted from branch abscission.

The underground axes of these arborescent lycophytes are called *Stigmaria*
(Figure 7.7A), although we have little idea whether the different stem genera
had different rooting systems. *Stigmaria ficoides* was a large, dichotomising,
shallowly descending axis that spread over an area about 6 m across and was
comparable to a rhizophore. Growth by quite large apices was possible through
the soft substratum in which it lived. It had spirally arranged endogenous lateral
appendages which are best likened to small roots and are very similar to *Isoetes*
roots. The main stigmarian axes had a distinct primary vascular system with
endarch xylem, and vascular cambial activity produced large amounts of secondary xylem. No secondary phloem has been found but diffuse phellogen zones
yielded abundant secondary cortical tissue. The lateral 'rootlets' were functional
for a while, although they were ultimately separated from the main axes by a
process resembling foliar abscission. The circular scars found on stigmarian
casts are the results.

It seems highly likely that it was the invasion of the shallow-watered flats that
brought about the evolution of such specialised rooting organs. Increased overall

Figure 7.8: Fructification and Spores of Arborescent Lycophytes. A–D: *Flemingites*, a bisporangiate cone of *Lepidodendron*; A, a reconstruction of a cone; B, longitudinal section through part of a cone showing sporophylls with adaxially attached megasporangia and microsporangia; C, microspore (about 25 μm in diameter) comparable to the dispersed spore genus *Lycospora*; D, megaspore (about 1 mm in diameter) comparable to the dispersed spore genus *Lagenicula*. E: Reconstruction of *Lepidostrobophyllum alatum*. F: Cross-section through *Lepidocarpon*. G: The single tetrad of megaspores from a *Lepidocarpon* sporangium; the larger fertile spore (about 1 cm long) is comparable to the dispersed spore genus *Cystosporites*. (E, After Boulter, 1968)

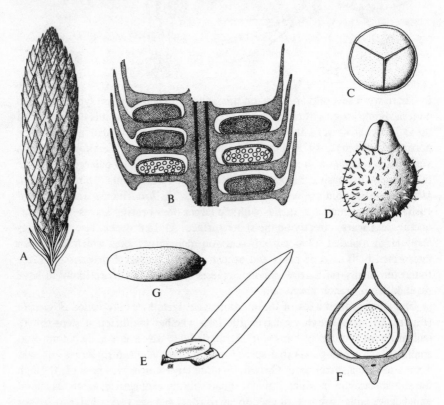

size brought about both greater weight and instability. This, coupled to their living in loosely consolidated soil or peat, necessitated such a shallow, horizontally growing and much branched anchoring system. Those large lycophytes which persisted on the drier and more compact soils of the fringing land retained their smaller corm-like underground organs.

Cones are borne terminally in *Lepidodendron*, *Lepidophloios* and *Ulodendron* but are unknown in *Bothrodendron*. They all have sporophylls helically attached to the central cone axis. Each sporophyll has a large elongated sporangium attached to the upper part of its pedicel. The distal laminae are directed apically to overlap each other and protect the sporangia. The lower surface of the sporophyll also extends downwards to form a protective heel. A sunken ligule is present distal to the sporangium.

Lepidodendron formed bisporangiate cones (*Flemingites* — Figure 7.8) in which the microsporangia usually occupy the apical portion. The microspores are small (about 25 μm in diameter), usually with smooth contact faces and granular distal surfaces. Some have been found with developing prothalli and antheridia, and some even possess inclusions that are very reminiscent of chromosomes (Brack-Hanes and Vaughan, 1978). Megaspores were spheroidal or slightly elongated, sometimes ornamented with spines, and with their three contact faces raised into an apical prominence. The endosporic multicellular prothallus projected between the parted contact faces and developed rhizoids and interdispersed archegonia with one to three tiers of neck cells (Brack, 1970). *Ulodendron* also formed cones that are of the *Flemingites* type.

Lepidophloios formed separate microsporangiate and megasporangiate cones, although we have no idea if the plants were monoecious or dioecious. The microsporangiate cones were structurally identical to *Flemingites* and the two kinds of microspores are superficially very similar, both being called *Lycospora*. The microspores from the microsporangiate cones can, however, be distinguished by their broad equatorial flanges and their smooth distal surfaces (Thomas, 1970). For a long time both cones were called *Lepidostrobus*, but the name is now limited to the microsporangiate cones (Brack-Hanes and Thomas, 1983).

The megasporangiate cones show several evolutionary advantages over their bisporangiate ancestors which can be related to their reproductive biology and their ecology. Only one functional megaspore forms in each sporangium by the reduction of developing tetrads to one and the subsequent abortion of three of its members (Figure 7.8G). These megaspores would have contained a great amount of food reserves, giving the developing sporeling an advantage over its competitors and a greater ability to survive as an understorey plantlet. For this reason it is likely that *Lepidophloios* was a plant of mixed established communities rather than a pioneer tree like *Lepidodendron*. The megaspores were also so large that they could not have been lost from the sporangia and so were shed together with the sporophyll as one unit (Figure 7.8E). The large wing-like lamina would have acted as a 'sail' to help disperse the spore rather like the seeds of *Fraxinus, Acer* and *Pinus* (Thomas, 1981). This again is suggestive of a plant that lived in relatively open communities.

There are two main genera of these monosporangiate cones. *Achlamydocarpon* has its megaspore trilete suture directed towards the cone axis and covered by a cap (massa) that probably helped retain moisture. *Lepidocarpon* had its megaspore trilete suture directed away from the cone axis and the spores had no massa. Instead the sides of the pedicel were extended to form lateral laminae, called integuments, that almost completely enveloped the sporangia (Figure 7.8F).

Fertilisation must have occurred within the sporangia, although it is not certain how and when this occurred. The massa in *Achlamydocarpon* and the integuments in *Lepidocarpon* have been variously described as water- or moisture-retaining structures or as protective adaptations. Why they should need these adaptations any more than the free-sporing bisporangiate cones is not clear unless it is something to do with fertilisation. Both adaptations could hold water for the free-swimming sperm, but they could also trap microspores as do the pollen drops in gymnosperms. The counterpart microspores have equatorial flanges which would have made them more buoyant for wind dispersal, so perhaps 'pollination' or even fertilisation occurred on the parent tree (Thomas, 1981).

There is a further type of ligulate megasporangiate sporophyll with one functional megaspore in the sporangium. This is *Miadesmia* from the European Carboniferous. Lateral laminae with elongated tentacle-like extensions enclose the distally opening sporangium. The affinity of such cones is, however, open to question, for the sporophylls are much smaller than those of *Achlamydocarpon* and *Lepidocarpon*, being only 3 mm long. For this reason it is quite possible that they represent a specialised development of the herbaceous rather than arborescent lycophytes. Therefore they would seem better included with the *Selaginella*-like plants in the Selaginellales.

Sigillaria was rather different in morphological and reproductive characters. The largest plants were probably just over 30 m tall, although many must have been considerably smaller. One of the most conspicuous differences from the other large lycophytes is their virtual lack of branching; some probably did not branch at all. Another difference is the way in which the helical leaf scars were secondarily arranged into vertical ribbed files. This seems to have allowed, and been accentuated by, secondary expansion of the outer tissues without an accompanying abscission of the superficial tissues. Wrinkles in these ribs also indicate expansion in this superficial tissue by cell division. There are no true leaf cushions in *Sigillaria*, although in some species the scars are on raised mounds. The stem epidermis, like that of *Bothrodendron*, has a great many stomata, being presumably highly photosynthetic. This increased photosynthetic ability must have helped counteract the relatively few leaves that would have been present on such unbranched plants. It also explains the need to retain the photosynthetic outer stem tissues attained by the vertical ribbing. The underground organs of the sigillarians were very like the *Stigmaria* of the other arborescent lycophytes, although there were a few minor anatomical differences (Eggert, 1972).

Sigillaria produced megasporangiate and microsporangiate cones that were very much alike (both being called *Sigillariostrobus*), although again we do not know if individual plants were monoecious or dioecious. The pedunculate cones were formed in zones between the leaves, presumably with the spores being released after the surrounding leaves had been shed. The sporophylls were in low helices and were most probably shed from the apex to the base.

Habitat Changes Leading to the Extinction of the Arborescent Form

The arborescent lycophytes had evolved and specialised to live in the Coal Measures swamps where they dominated the vegetation by their great size. Their widespread rooting organs were shallow and supportive for a swamp tree. Aerating tissue throughout the aerial shoots and subterranean axes enabled them to live in waterlogged habitats. However, their great success made them vulnerable to climatic change and dependent upon the swamps for their survival. At times when the swamp basins subsided too fast for peat formation to keep pace, the swamp plants were restricted to the edges of the basins. From here they reinvaded the basins when subsidence ceased or slowed sufficiently for the accumulating sediments to permit recolonisation.

But at the close of the Carboniferous the conditions necessary for swamp formation completely disappeared and those plants which were dependent on the habitat dried out. The extinction of the arborescent lycophytes was therefore the result of geomorphological change, coupled with the inability of the plants to live outside the waterlogged substrates of the swamps and mud flats. The morphological adaptations of their underground organs were the most likely cause of their extinction rather than failure of their reproductive systems. Given the opportunity, the more adapted and advanced forms with *Achlamydocarpon* and *Lepidocarpon* seed-like fructifications might have competed quite well with other land plants. However, growth in the drier and firmer soils would seem to have been very difficult, if not impossible, for the large and presumably delicate apices of *Stigmaria* that have been described by Rothwell (1984).

The loss of the Euramerian swamps certainly did not kill all but the herbaceous lycophytes, for there were medium-sized forms living during the subsequent Mesozoic period. These pleuromeialans have often been described as transitional forms in an evolutionary reduction series running from *Lepidodendron* to the living *Isoetes*, via *Pleuromeia*-like and *Nathorstiana*-like plants. *Isoetes* is a relatively inconspicuous perennial with a small swollen axis bearing quill-like sporophylls in a dense helix and helically arranged monarch roots (Figure 7.7B, C). The plants are heterosporous and monoecious with all sporophylls fertile. Each sporophyll has one large sporangium which is sunk into the swollen basal tissue. Much early stress was laid upon the similarity of the roots of *Isoetes* and *Stigmaria*. Both possess a single, monarch stele in a narrow cortical sheath that is free in a cavity, or connected to the outer cortex by a strand. However, our increased knowledge of the many other forms of lycophytes now suggests the evolutionary picture to be much more diffuse with several possible different lines of change.

The idea of a reduction series, of course, has its attractions. The Triassic *Pleuromeia* (Figure 7.9A) has an unbranched trunk on a basal four-lobed rooting rhizophore very suggestive of a diminished rhizophore of *Lepidodendropsis* (i.e. *Protostigmaria* — Figure 7.9C). The plants were about 2 m tall and shed their leaves to form superficial leaf scars. Some bore solitary terminal cones,

Figure 7.9: Reconstructions of Lycophytes with Corm-like Bases. A:
Pleuromeia longicaulis, about 45 cm tall although a fully mature plant may
have been considerably larger. B: *Nathorstiana arborea*, about 12 cm tall. C:
Protostigmaria eggertiana, about 16 cm across. D: *Chaloneria cormosa*,
about 2 m tall. (A, After Retallack, 1975; B, after Mägdefrau, 1956; C,
after Jennings, 1975; D, after Pigg and Rothwell, 1983)

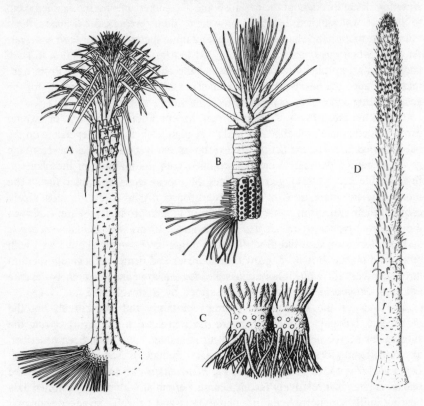

although there is evidence that others produced several cones of the form that
been previously called *Cyclostrobus* (Retallack, 1975). The Lower Cretaceous
Nathorstianella and *Nathorstiana* (Figure 7.9B) can easily be thought of as
reduction forms. Reduced height to less than 30 cm in *Nathorstiana*, smaller
basal lobes in *Nathorstianella* and their loss in *Nathorstiana*, and the reduced
size of the leaf scars and their closer proximity support the idea.

There is, however, other evidence that seems to provide a better line of
evolution. The Upper Carboniferous woody axes, with corm-like bases,
described by Pigg and Rothwell (1983) as *Chaloneria* (Figure 7.9D), could well
be taken for an intermediate between *Protostigmaria* and the *Isoetes* form. This
is supported by the recognition of Triassic plants that are closely related to, if
not identical with, *Isoetes*.

Survival of all these plants must surely have depended upon the ability of certain of the Carboniferous forms to survive the disappearance of the swamps. As we have seen, those plants with corm-like bases most probably grew in the drier habitats with firmer soils, and were thus able to migrate along the coastal plains or to invade the wetter inland habitats. *Pleuromeia* has been described from coastal deposits in Europe, Asia and Australia, suggesting that it migrated along the Triassic shorelines. *Nathorstiana* probably grew in coastal areas, and *Nathorstianella* inhabited the sandy banks of streams or lakes. *Isoetes* and its close relatives could also have spread this way resulting in its present world-wide distribution. *Isoetes*, by invading the more inland areas, seems to have secured its survival whereas those others more committed to coastal areas or river-banks became extinct. *Isoetes* has also continued to evolve and to invade new habitats. There are terrestrial species, shallow-water species (of which some have emergent sporophylls and others are totally submerged) and species living on the bottom of deep-water lakes. This migration into water has resulted in several anatomical changes. The most noticeable and probably the most significant are the loss of stomatal function in shallow-water species (e.g. *I. malinverniana*) and the total loss of stomata in the deeper-water species (e.g. *I. lacustris*). The plants have compensated for the loss of available atmospheric carbon dioxide for photosynthesis by evolving the ability to utilise crassulacean acid metabolism (CAM).

Isoetes has no method of sporophyll abscission, nor of spore dispersal, other than through the destruction of the sporophyll and the sporangial wall. Limited growth in the truly aquatic species results in their sporophylls being gradually sloughed off. In contrast, the terrestrial species survive the dry seasons through the complete death of their sporophylls. In both instances the spores are eventually released, although there is clearly very little chance of the terrestrial species spreading very rapidly. Many populations of *Isoetes* are therefore likely to be very ancient ones.

Sphenophytes

The evolutionary history of the sphenophytes parallels that of the lycophytes. The earliest recognisable form appeared in the Devonian; the group reached its maximum development in size and diversity in the Carboniferous, then it went into decline, leaving only *Equisetum* living today with about 20 species but a virtually world-wide distribution.

The main characteristic feature of the plants included in this group is that the axes are radially symmetrical, with nodes from which the subsidiary organs arise in whorls. There is thus a distinctive whorled pattern to its branching, and to the leaves and loosely arranged cones. These cones had whorled modified branches called sporangiophores, which formed sporangia that were directed back towards the cone axes. Whorls of sterile bracts usually alternated with the sporangiophore whorls.

The Herbaceous Habit

The Devonian *Pseudobornia ursina* showed the typical nodal pattern of branching which enables us to include it in this group. It was a rather large plant and could be up to 20 m tall and 60 cm in basal diameter. Unfortunately we know nothing of its anatomy nor of its underground organs. The fertile branches at the ends of the primary branches had whorls of bracts and sporangiophores, each of the latter bearing about 30 sporangia. Such large plants were, however, probably not the ancestors of all the later sphenophytes especially the Lower Carboniferous trailing herbaceous forms. The Hyeniales were generally thought to be primitive sphenophytes but current thinking includes them with the primitive ferns (Taylor, 1981). Similarly, the Middle Devonian *Ibyka* has been described as a member of the complex from which Carboniferous sphenophytes arose (Skog and Banks, 1973). It had spirally arranged branches with dichotomously divided, three-dimensional, ultimate appendages that were sometimes fertile. Its close similarity to the Hyeniales, as indeed was argued by its authors, does, however, suggest that it may equally have been a link between the Trimerophytina and the Hyeniales leading to the ferns.

The Sphenophyllales were a group of herbaceous sphenophytes that most probably grew as trailing and scrambling plants on the floor of the wetter Carboniferous forests. There is one common genus of stems, *Sphenophyllum*, that grew both by the formation of derivative cells from the tetrahedron-shaped apical cell and from nodal intercalary meristems. They were sometimes secondarily thickened and give us the best example of a functional bifacial vascular cambium in any spore-forming group of vascular plants (Eggert and Gaunt, 1973). The resultant stems branched several times and were up to 2 cm in diameter. The leaves were fan-shaped, up to several centimetres in length, and usually in whorls of between six and nine. The size, shape, venation and degree of lobing of the leaves provide the generally accepted method of distinguishing species. However, as variation is known to occur even within a single specimen, it is doubtful if many of the species recognised are really valid (Figure 7.10). Adventitious roots were formed by many stems and branching of these occurred only very rarely.

There are several kinds of cone that have been described as belonging to *Sphenophyllum*. They are *Bowmanites, Litostrobus, Sphenostrobus, Cheirostrobus, Tristachya* and *Eviostachyd*; others found attached to *Sphenophyllum* shoots are given no name of their own. *Mesidiophyton* was described as a fertile herbaceous shoot different from *Sphenophyllum* (Leisman, 1964). The cones all differ in the number and relative positions of sterile bracts and sporangiophores, the number of sporangia and their mode of attachment to the sporangiophore. Some spores are trilete whereas others are monolete, there being a difference even between species in *Bowmanites* (Storch, 1980). All the cones were apparently homosporous, although the wide range of spore sizes in some species (e.g. 95–162 μm in *B. dawsoni*; 58–70 μm and 75–140 μm in *Sphenophyllum tenerrimum*, where the smaller ones were described as abortive)

Figure 7.10: *Sphenophyllum emarginatum*. A: Partial reconstruction of the vegetative parts of a plant. B–I: The variations in its leaf shape. (After Battenburg, 1977)

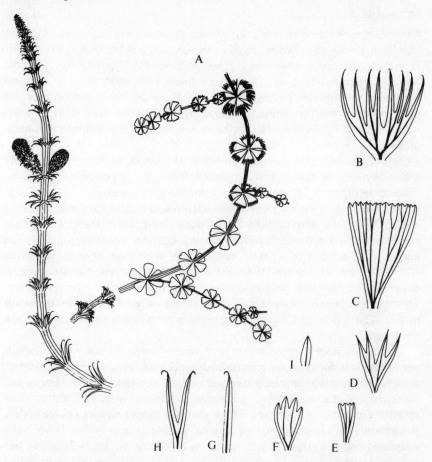

suggests that anisospory was probably evolving within the group.

The wide variation in reproductive organs could be taken to mean that we have only a limited idea of their corresponding vegetative shoots. We may be trying to compare fructifications of several different sphenophyte evolutionary sequences. On the other hand it may be that the vegetative characters were much more conservative in evolutionary changes, simply because they were successfully adapted to their mode of life as understorey plants. If this was the case, it is likely that this conservatism led to their reduction in numbers at the close of the Carboniferous and their ultimate extinction during the Triassic through competition with other more adaptable herbaceous plants. Asama (1970) believes that the Permian Asian *Parasphenophyllum*, *Trizygia* and *Paratrizygia* arose from the Carboniferous Euramerian *Sphenophyllum*. Two main evolutionary trends are suggested to have occurred, one involving the increase and the other the reduction of leaf segmentation.

Increase in Size

The other major line of evolutionary development that commenced in the Devonian quickly gave rise to arborescent plants. The oldest recognisable genus is *Archaeocalamites* whose range extended from the Upper Devonian to the Lower Carboniferous. Erect stems that were several centimetres in diameter grew from underground rooting rhizomes. The whorled leaves were slender and dichotomising with a similarly branching vascular strand. Cones called *Pothocites* and *Protocalamostachys* have been described as belonging to these plants.

The Carboniferous swamps provided the ideal environment for the sphenophytes as it did for the lycophytes. There the largest sphenophytes — *Calamites* (Figure 7.11) flourished, sometimes growing up to nearly 20 m in height. The majority grew from large underground rooting rhizomes that were able to force their way through the unconsolidated peaty swamp soils. They probably did not grow indiscriminately through the lycophyte forests as an understorey but rather by forming dense thickets in clearings or on the waterside edges of the forests. Barthel (1980) has described one species, *Calamites gigas*, as growing in pure plant communities in the shallow waters of lakes and rivers. This species does, however, differ from the majority of species in its growth pattern in having solitary and erect stems, with roots emerging directly from the basal parts.

An unusual mode of preservation gave some of the fossils that we call *Calamites*. These sphenophytes were secondarily thickened, resulting in a cylinder of secondary xylem with inwardly directed wedges of primary xylem. Decay and breakdown of the internal pith parenchyma, together with an infiltration of sediments, could give a pith cast. These show the former positions of the nodes, and longitudinal furrows mark the original primary xylem points. Other scars associated with the nodes sometimes reveal branching and the positions of leaf scars and vascular rays. Anatomically preserved stems are properly called *Arthropytis, Arthroxylon* or *Calamodendron*, being distinguished from each other on selected features of the secondary xylem. None of these has any secondary phloem so there is no evidence for a bifacial cambium as in *Sphenophyllum*.

We seem to face a rather complex taxonomic problem here, for although the anatomical differences are consistent, these genera cannot be correlated with the differing types of cone that are known to have been formed by the plants. Good (1975) has suggested that the differences might be partially explained as developmental stages. The continued use of such different generic names can, however, be defended on the grounds of descriptive anatomical convenience. Similarly, genera such as *Calamites* and *Arthropytis* describe the mode of preservation as well as providing the basis for binomial taxonomic descriptions. In practice the whole plant is referred to as *Calamites* and the group is

Figure 7.11: Reconstruction of *Calamites carinatus*; about 10 m tall. (After Hirmer, 1927)

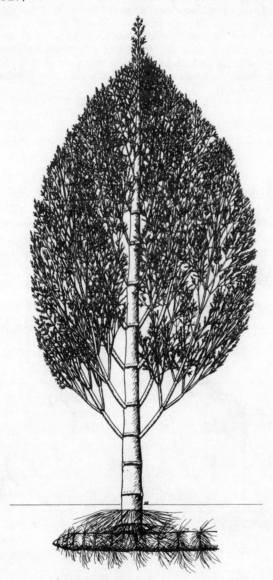

commonly called the calamites.

The aerial stems grew from large apical cells with four internal triangular faces that gave rise to cell derivatives. The vascular strands first started to mature at about the third node, and at about the fifth node the primary xylem broke down to give spaces — the carinal canals. The underground rhizome must

Figure 7.12: Details of Carboniferous Calamites. A: Portion of fertile shoot with *Calamostachys paniculata*; each cone about 3 cm long. The bracts have been omitted from the lower fertile lateral shoot. B: Diagrammatic longitudinally cut section of *Palaeostachya andrewsii*, about 1.5 cm in diameter. C: Spore from *Calamostachys binneyana*; spore body about 50 μm in diameter with three elaters. D: Portion of a terminal leafy shoot of *Annularia sphenophylloides*; the largest leaves being about 1 cm long. The shoot is drawn as though viewed from above, showing the orientation of the leaves. (A, After Weiss, 1876; B, after Baxter, 1955; C, based on Good, 1975)

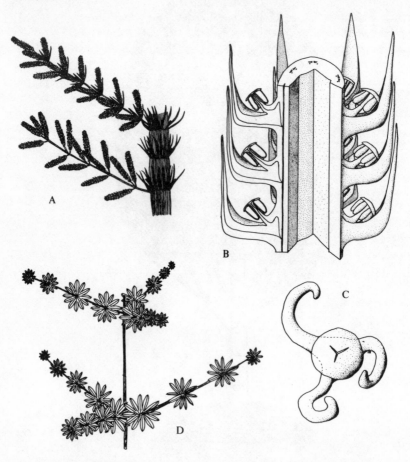

have expanded quickly from its original size with a corresponding increase in aerial root size. Thereafter the rhizome continued to grow and to branch occasionally, and to form big aerial shoots. These primary shoots had large numbers of vascular strands and formed abundant secondary xylem in their basal regions. Good (1975) has described a large stem of *Arthropytis communis* that shows how the basal regions developed. By the 13th internode the pith diameter had enlarged from 0.5 mm to 2.2 cm and the number of vascular strands increased from 11 to a constant 48. The amount of secondary xylem decreased with the increase in primary tissue size. Branching resulted in similar periods of primary expansion but gave rise to smaller apices, narrower shoots, fewer vascular strands and less secondary xylem. This reduction continued with every branching order, ultimately giving rise to narrow branches with fewer vascular strands and no secondary thickening. Their growth was predetermined like that of the arborescent lycophytes.

The roots, like the shoots, have been given several generic names. They branched irregularly and formed smaller steles with fewer primary strands as they did so. Secondary xylem developed, although in much smaller amounts than in the shoots.

The leaves of the calamites can be called *Asterophyllites* or *Annularia* (Figure 7.12D). Those of the former are narrow and needle-like, while those of the latter are lanceolate to spatulate and basally fused into a disc. *Annularia* has evolved the ability to orientate its developing leaves to maximise their photosynthetic capabilities. This has resulted in the majority of shoots being compressed with leaves in a series of flattened stars rather than being embedded in the sediment. This configuration of leaf orientation must have been determined during the early stages of shoot maturation, for anatomical studies show us that the strengthening nodal diaphragms were formed in the same plane as the leaves.

Calamite cones showed variations on the general theme of alternating whorls of sterile leaf-like appendages (bracts) and sporangiophores (Figure 7.12A,B). Each sporangiophore was a peltate structure bearing a ring of elongated sporangia directed back towards the cone axis. Much has been written on the subject of the possible evolutionary development of the sporangiophore. Page (1972a) and Good (1975) review the alternative views but agree that fossil evidence and studies on *Equisetum* cones suggest that sporangiophores represent reduced three-dimensional branching systems. Page also found a continuous range of morphological variation between vegetative leaves and sporophylls in abnormal *Equisetum* cones. These were suggested to be evidence for homology between the leaf and the sporangiophore stalk plus the upper half of its peltate head, indicating different evolutionary specialisation from an ancestral unit (Figure 7.13).

The two most important features used in distinguishing different forms of cone are the number of sporangia per sporangiophore and the position of sporangiophore attachment. Gastaldo (1981) rightly argues for the retention of generic names on this basis. The earliest recorded cones, from the Lower Carboniferous of Europe, have their sporangiophore whorls midway between the bract whorls. Such cones, called *Calamostachys*, seem to have been the ancestral forms giving rise to the other Upper Carboniferous forms. *Palaeostachya* and *Weissistachys* have their sporangiophores originating immediately above the bract whorls, resulting in their being obliquely aligned. *Mazostachys* has its sporangiophores slightly below the bract whorls, whereas in *Cingularia,*

Figure 7.13: *Equisetum* and the Evolution of its Sporangiophore. A: Portion of the underground rhizome with a fertile aerial shoot terminating in a cone. Roots are shown at the nodes of the rhizome and underground portion of the vertical shoot. B–E: Suggested evolutionary stages of the *Equisetum* sporangiophore (E) from the terminally fertile primitive appendage (B). Broken lines indicate the appendage outline and solid lines the vascular supply. (B–E, After Page, 1972a)

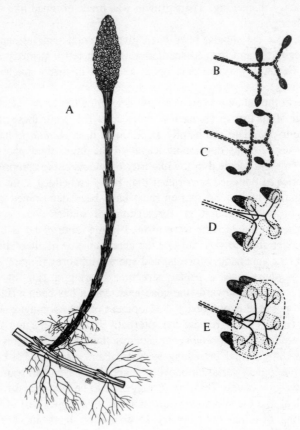

Metacalamostachys and *Pendulostachys* they are partially fused to the underside of the bract whorls.

Most of the cones included within these genera are homosporous, although a few heterosporous species have been described. The spores probably all had elaters (Figure 7.12C), although many have been thought to have had none. This seems to be the result of the failure of observers to recognise spores with unusual wrinkled or furrowed surfaces, or a poor state of preservation, or because of different stages of ontogenetic development (Good, 1975). They have therefore been called *Vestispora* when they have tightly coiled elaters and *Calamospora* when they are apparently without elaters. The morphological features of calamitalean spores do not permit clear distinctions within the broad-based genus *Calamospora*, where the taxonomy of dispersed spores is now very confused.

The most highly evolved cone is *Calamocarpon* in which only one functional megaspore developed in each sporangium. It also seems highly likely that the spores were retained within the sporangia, being therefore very similar to the condition known in *Achlamydocarpon* and *Lepidocarpon*. It is probable that *Calamocarpon* cones were borne on those vegetative stems assignable to *Calamodendron* rather than on *Arthropitys* as were the other types of cone.

Habitat Changes Leading to the Extinction of the Larger Forms

Although they were undoubtedly very successful in the Euramerian Coal Measures swamps, the largest of the sphenophytes failed to survive the changing conditions at the end of the Carboniferous. As in the case of the arborescent lycophytes, it was probably their inability to grow in the more compacted soils of the surrounding areas that caused their extinction. Their massive underground rhizomes were just not capable of growing except in the unconsolidated ground of the swamps (Thomas, 1985a). Other smaller sphenophytes survived of course, being more capable of growing in a wider range of habitats. There were relatively large Mesozoic forms which are described under the name of *Neocalamites* and some associated cones have been called *Neocalamostachys*. Unfortunately we have little knowledge of their overall morphology or of their reproductive biology.

Sphenophytes also lived in other parts of the Palaeozoic world. Some are known from the Upper Palaeozoic of both Angaraland and Gondwanaland. *Phyllotheca* has its leaves fused into sheaths and cones corresponding to the fructification genus *Tchernovia* (Meyen, 1971). Meyen and Menshikova (1983) suggest that these Angaran sphenophytes show archaic features in both their fructification structures and nodal anatomy. The Angaran *Tchernovia* and *Gondwanastachys* may be paralleled with the Euramerian Lower Carboniferous genera *Pothacites* and *Protocalamostachys*. Meyen (1969) has also reviewed some other sphenophytes called *Barkeria*, suggesting that their dichotomising leaves are like those of the earlier Euramerian forms. The smaller overall size of the Angaran sphenophytes, like that of the lycophytes, can also be related to their growing in a colder climate.

The only surviving sphenophyte is *Equisetum* which contains about 20 species. It is widely distributed in diverse habitats throughout most of Eurasia, America, Africa and east to north-eastern Asia, the Philippine Islands and New Guinea, New Caledonia and Fiji (Tryon and Tryon, 1982). The various species grow in ponds, ditches and marshes, riverside and lakeside banks, wet woodlands and dense forests, and some also in the drier soils of banks, loose mountain soils and arctic/alpine turf. The uniform chromosome number of $n = 108$ seems to be a highly derived level by allopolyploidy, suggesting an extinction of lower chromosome levels and a great antiquity for the genus.

All species are herbaceous perennials, with some dying back to the ground each year. All grow by a horizontal rhizome that forms aerial shoots. Most species branch and the largest, *E. giganteum*, has stems up to 13 m tall. None has vascular cambia so the tallest stand by virtue of mutual support. Their leaves are non-photosynthetic, reduced and fused into sheaths at the nodes of the photosynthetic shoots. Cones are generally terminal on the main aerial axis, although some species also form them on the tips of their lateral branches.

Equisetum can be split into two subgenera, *Equisetum* and *Hippochaete*, on morphological characters of both the sporophytes and gametophytes (Page, 1972b; Duckett, 1973). The interrelationships of the taxa are very difficult to assess and it appears that the living species are remnants of a much larger group. Hybridisation and polyploidy must both have been important in stimulating evolutionary change, although as yet we are still without much direct evidence. Ideas have differed on the relative antiquity of the two subgenera, but on the fossil evidence it seems that *Equisetum* is the older. This subgenus is known from the Jurassic whereas the earliest record of *Hippochaete* is *E. clarnoi* from the Eocene of Oregon (Brown, 1975).

The fossil evidence suggests that *Equisetum*-like plants, if not *Equisetum* itself, had evolved by at least the Triassic. There are earlier, Carboniferous, putative records, but these are really too doubtful to be included with any degree of certainty. There are three major differences distinguishing *Equisetum* from the earlier calamite forms. It has no vascular cambium, its leaves are fused into sheaths and its cones have no bracts between the whorls of sporangiophores, although there is evidence to suggest that these latter two differences are only apparent. The leaves of *Annularia* and *Asterophyllites* do show basal fusion, albeit only to a very small extent. Abnormal *Equisetum* cones, as we have already seen, do sometimes have bracts interspersed between the sporangiophores, although even in these there is not the regular alternation seen in calamite cones. Good (1975) therefore considered that the absence of secondary thickening alone was not sufficient for distinction at the family level and included the calamites within the Equisetaceae.

It seems reasonably certain that the calamites gave rise to *Equisetum*, although it may not have been the largest of them that were the ancestral forms. Reduction in size with fusion of leaves and a compaction of cone units seems feasible in a series of evolutionary changes even though we have no direct evidence from fossils. Whatever the cause for evolutionary change or the direction of the trends, it seems certain that *Equisetum* is a vigorous plant that will survive into the future. Its rapid growth capability through its rhizomes makes it an ideal coloniser of open ground. Fragmentation of its rhizome, followed by water transportation, can help to spread those species living in or near streams and rivers. Similarly, those *Hippochaete* species that form rhizome tubers can be very easily dispersed by water transportation of these organs. Vegetative growth and dispersion, together with the antiquity of the genus, obviously help to explain the widespread distribution of many of the species.

8 PTERIDOPHYTE SUCCESS AND SPECIALISATION
II — FERNS

Early 'Ferns'

The early fern-like ancestors of the plants that we like to call true ferns seem to have evolved in many different ways. Extant ferns show such a range of vegetative and reproductive characters that they must surely represent different end-products of evolutionary change. Some show what are regarded to be primitive characters and usually have long fossil records, whereas others with more advanced characters have less of a fossil record or seemingly none at all.

Living ferns generally have large megaphyllous leaves, called fronds, with well-developed venation systems. The young fronds of most ferns start to grow coiled up and later characteristically unroll, a developmental pattern which is termed circinate vernation. Mature fronds may be undivided or compound, often with many subdivisions. The divisions are called pinnae and the ultimate segments pinnules. As these fronds have evolved from three-dimensional branching systems their vascular tissues are closely integrated with those of their stems. Fern stems often have complex vascular systems of dissected cylinders. These have evolved through portions of the cylinder effectively departing as leaf traces, forming what are called leaf gaps in front of their departure points. Sometimes the cylinders can be further broken by holes not associated with leaf traces. These gaps are called perforations.

Fern sporangia are varied in shape, size and method of dehiscence. The presumed ancestral type of sporangium was similar to that of the Trimerophytes in being elongated and dehiscing by a lateral slit. The sporangial wall cells generally had a greater amount of thickening on their inner tangential and radial walls than on their surface walls. Loss of water and shrinkage would therefore have caused stress in the unequally thickened cells leading to sporangial wall rupture and dehiscence.

Fern-like plants evolved in the Devonian from the Trimerophytina, and some persisted until the Permian. Three major groups (usually given at Class level) are generally thought to have existed, although many fossils show barely enough evidence to be assigned with any certainty. All were without planated leaves and their stelar systems were often rather unusual. The two main groups were the Cladoxylopsida and the Coenopteridopsida, with the Rhacophytopsida being the third, smaller, group.

Cladoxylopsida

The Cladoxylopsida is not so much a biological group as an assemblage of plants with similar highly dissected vascular systems. These vascular systems may have evolved from protosteles by dissection to give a greater surface area to the xylem, rather in the way that the star-shaped xylem in *Asteroxylon* must have done. Some systems divide further at higher levels in the plant, presumably in response to the evolutionary development of the leaves.

Plants assigned to this group appear to have been very varied in appearance. The Middle Devonian *Cladoxylon scoparium* from Germany was a 35 cm-tall plant with a branched main axis that bore helically arranged segmented leaves and fan-shaped organs with terminal sporangia. *Pseudosporochnus* was much larger, being about 3 m tall with terminal branches on a trunk. Some ultimate segments of these branches were replaced by ellipsoidal sporangia. *Calamophyton* and *Hyenia* grew rhizomatously producing dichotomising aerial branches that bore sterile and fertile branchlets in low helices or pseudowhorls. This apparent whorled arrangement and the fact that the sporangia are sometimes borne on recurved branches were the reasons why the genera were orginally thought to have association with the sphenophytes. Recent anatomical studies of petrified axes show them to be more fern-like, so they are better included here.

Coenopteridopsida

The Coenopteridopsida again probably represents an artificial group of plants that share anatomical similarities. In this case it is more the anatomy and morphology of the fronds and leaf traces that are diagnostic rather than the features of the main axis. The group includes many diverse members, and several attempts have been made to establish meaningful families. An accumulation of knowledge of their reproductive organs has allowed recent judgements to be made in a manner similar to those employed for living plants. As a result three families, the Botryopteridaceae, Tedelaceae and Anachoropteridaceae, have been removed from this group as they now seem to be better included with the ferns. This leaves the Stauropteridales and the Zygopteridales.

The three known species of *Stauropteris* alone constitute the Stauropteridales. The plants were small and bushy with pairs of unusual pinnae-like leaflets, called aphlebia, at each branching. The vascular system changed with each order of branching, decreasing in size and altering from a four-lobed to a terete protostele. The plants therefore appeared to have had a limited and determinate pattern of growth. Thin-walled sporangia were borne terminally. The Upper Carboniferous *Stauropteris oldhamia* was homosporous with sporangia that dehisced apically. In contrast, the Lower Carboniferous *S. burntislandica* is described as heterosporous. However, Chaloner's (1958) description of comparable dispersed megaspores as tetrads of two large spores (580 μm) and two small spores (45 μm) suggests that the plants might more correctly be thought of as anisosporous.

The Zygopteridales existed from the Devonian to the Permian and includes plants of diverse shape from rhizomatous forms to erect tree-like ones. The group is generally subdivided and the plants are named on the highly artificial basis of anatomical evidence. The plants have lateral branching systems that have dorsiventrally organised anatomy and their distal units are non-planated and non-laminated. There does, however, appear to be an overall trend within the group indicative of leaf evolution. The plants bear lateral organs on their stems which can be thought of as primitive petiolate leaves, but because of the overall dissimilarity of these organs to leaves the 'petioles' are often called phyllophores. Galtier (1981) has suggested that *Musatea* and *Corynepteris* arose by evolving some degree of bilateral anatomy and morphology before the development of webbing. The Lower Carboniferous *Musatea globata* shows the beginning of webbing and had specialised mesophyll, rather like palisade tissue, and stomata on its lower epidermis. The steles were generally stellate in the main axes, but the bundles appear H-shaped in transverse section in the petioles. For this reason such isolated petioles are called *Etapteris*. The petioles were subtended by pairs of aphlebia and themselves bore four ranks of pinnae that sometimes ended in laminar pinnules. The Lower Permian *Nemejcopteris feminaeformis* had such laminar pinnules and also modified pinnules that bore sporangia. There are other forms of petiole with different-shaped vascular traces. One of the better known forms has hourglass-shaped traces; for which reason they are called *Clepsydropsis*. Some of these had four ranks of primary pinnae as in *Etapteris* (e.g. *C. leclercqii*), although the majority had two ranks (Galtier, 1966). This apparently fundamental difference is not as unusual as it may seem because *Clepsydropsis* petioles have been found attached to more than one type of stem. It merely emphasises the difficulties encountered in attempting to understand the biological affinities of these early fern-like plants while producing a workable taxonomy at the same time.

The Upper Carboniferous Angaran fern *Geperapteris* may be affiliated to the zygopterids as it has unilayered sporangial walls with a wide annulus-like band covering a large portion of the rounded sporangium. It is very similar in this respect to *Discopteris* and *Norwoodia* which are most probably zygopterid. *Geperapteris* may therefore be a descendant of zygopterids which had emigrated from the earlier Euramerian Lower Carboniferous floras.

Rhacophytopsida

The third group of fern-like plants is the Rhacophytopsida which presently contains only two genera of Devonian plants of rather disputed affinity. The best-known species, *Rhacophyton ceratangium*, has been described in some detail by Andrews and Phillips (1968). It was an erect plant that must have been at least 1 m tall. The fronds were borne in a close spiral, although there is no evidence of how the fertile and sterile fronds were spatially arranged on the plant. The vegetative fronds were alternately and bipinnately branched but although generally two-dimensional in appearance still showed some tendency towards being three-dimensional. Fertile fronds were obviously three-dimensionally branched and appear to have been quite variable in morphology. Each had between 12 and 18 fertile nodal units that consisted of a pair of divergent sterile pinnae and a fertile pair that dichotomised several times to form dense ellipsoidal structures. The sterile pinnae also bore much branched three-dimensional basal pinnules which are best thought of as aphlebia. The sporangia were ovoidal, with terminal slender tips, and dehisced laterally. *Protocephalopteris* similarly had large bipinnate vegetative fronds that were apparently two-dimensional. There is, however, a major difference in the fertile fronds as the ultimate sporangia-bearing appendages in these plants were quadriseriately branched.

The pinnate branching system of the vegetative fronds of these plants suggests an intermediate stage in the evolution of flattened, webbed leaves similar to that outlined for the Progymnosperms in Chapter 5. Indeed, it is obviously intermediate characters such as these that again make it so difficult for us to decide on their exact botanical affinities with any certain degree of finality.

True Ferns

The rest of the ferns can be thought of as much more advanced than the preceding groups. For that reason they constitute what we like to think of as the 'true ferns'.

Marattiaceae

The oldest extant family is the Marattiaceae which is generally raised to Order level as the Marattiales. These ferns are known from the Carboniferous period onwards. They appear to have been present in quite large numbers in some Upper Carboniferous floras and must have been important members of these plant communities. In the Coal Measures swamps lived the largest of all these ferns which, in being called *Psaronius*, is given the same name as the genus for its petrified stems. As we saw earlier, the stems were polystelic but supported by an adventitious root mantle. Sporangia were large, sessile and dehisced through simple lines of weakness. Small numbers of sporangia were united into what we term synangia on the lower surfaces of the much divided fronds called *Pecopteris*. The most common petrified fertile fronds are called *Scolecopteris*, although there are others (Millay, 1979).

The Palaeozoic marattiaceous ferns differed from the living species in two important and interrelated respects. Their fronds were much more divided with smaller ultimate segments and their synangia were rounded, not elongated. However, the Carboniferous petrified fertile frond *Acaulangium* has slightly elongated synangia, and Millay (1977) suggested that if the pinnules had expanded to give additional space on the crowded lower surface, then the synangia would probably have elongated to fill this extra area. Such a phase of evolution involving the fusion of smaller pinnules into larger units was occurring in the late Carboniferous. Asama (1960) suggested that this was due to changing

Figure 8.1: Marattialean Ferns. A: *Scolecopteris altissimus*. Reconstruction
of a fertile pinna segment with distal synangia. The decurrent pinnules are
drawn slightly wider spaced than in life. B, C: *Zeilleria avoldensis*; B,
reconstruction of a fertile pinna with marginal synangia; C, a monolete
spore from a sporangium. D–F: *Angiopteris evecta*; D, fertile pinnule with
submarginal synangia; E, synangium with eight dehisced sporangia; F, trilete
spore. (A, After Millay, 1982)

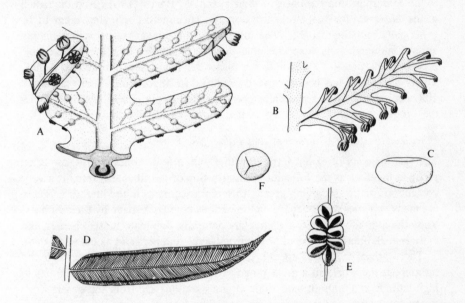

climatic conditions, basing his ideas on the fact that the foliage of living
Angiopteris lygodiifolia becomes less complex when the plants are grown at
cooler temperatures.

The synangial arrangement of the sporangia also gives some evidence of their
coenopterid ancestry. *Chorionopteris* and *Zeilleria* (Figure 8.1B,C) have
synangia that are borne on the edges of the pinnules and not on their under-
surfaces. As such they might easily represent intermediates between the non-
laminated fronds of the coenopterids with their terminal sporangia and the
laminated marattialean fronds. A 'phyletic slide' of the sort proposed by Bower
(1926) and Mamay (1950) could have transferred the *Chorionopteris* and
Zeilleria form to that of the marattialeans.

There are also two kinds of synangial arrangement in the Carboniferous
forms matching those of the extant genera. *Eoangiopteris* had laterally free
sporangia that were only held together by a common mound of tissue. This can
be taken as an early representation of the 'free-sporangiate' living genera such
as *Angiopteris* (Figure 8.1D–F) and *Macroglossum*. In contrast, *Acaulangium*
and *Scolecopteris* (Figure 8.1A) seem to be of the more completely synangiate
type as *Marattia* and show similar methods of sporangial dehiscence (Millay,

1976). In both *Scolecopteris* and *Marattia* dehiscence is preceded by a lateral separation of the sporangia before they split longitudinally along their inner walls.

The evolution of the modern forms seems to have been initiated in the Palaeozoic but probably from the smaller, less numerous forms rather than from the tree ferns. There are several Mesozoic and Tertiary species that are referable to the Marattiales on the basis of their synangia. Harris (1961) used the extant genus *Marattia* for his Yorkshire Jurassic specimens, believing them to be generically indistinguishable. The full size of these later Mesozoic ferns is not known, although it is suspected that they were smaller than their Palaeozoic predecessors, possibly like the living forms that have short unbranched trunks. Today the family has become greatly reduced in importance. There are about 100 species in what has generally been accepted as six genera. All live within the Tropics and most have very limited ranges of distribution.

Filicalean Ferns — Characters and Palaeozoic Forms

The vast majority of extant ferns, together with a few extinct groups, are often grouped together as the Filicales. Classification of the filicalean ferns is a very contentious matter receiving greatly different treatments in the literature. There are many schemes dependent upon the various emphases made by their authors. Reproductive and vegetative characters naturally dominate most schemes, but there is also often some use of base chromosome numbers and biochemical data.

There are over 12 000 species in at least 400 genera of living ferns found throughout the world in a great range of habitats; from tropical rain forests to high-latitude and high-altitude arctic-alpine situations and from sub-desert scrub to sea-coast cliffs and mangrove swamps. The ferns also show a wide range of size and form. There are tree ferns to very small floating aquatic species. Some ferns are solitary with fronds arising from small erect or suberect axes and can be either terrestrial, epiphytic or lithophytic; others spread by horizontal rhizomes and can again be terrestrial, scrambling or climbing. Page (1979) has given an excellent review of the diversity of ferns in an ecological perspective, relating fern morphology to environment and describing the characteristic morphological composition of a cross-section of different fern communities worldwide, grouped within the habitats and environments in which they occur. It is apparent that many structural adaptations are associated with their ecological role and some ferns have adapted to produce very different morphologies in different habitats. Conversely, ferns from very different families can show very similar morphologies in similar habitats as the result of morphological evolutionary convergence. Similar environmental pressures must have acted on ferns throughout their evolutionary history, so we must be extremely careful in making morphological comparisons and environmental deductions.

Reproductive characters that are used in distinguishing plant groups can also be given as supportive evidence for ideas of evolution. These characters incorporate the structure of the sporangia and their patterns of distribution on the plants. Filicalean sporangia can be isolated but more often are grouped together into clusters that we call sori. A very few ferns have their sporangia on specialised portions of the fronds and some even produce separate fertile spikes. Some genera have all their sporangia maturing simultaneously, e.g. *Gleichenia*; this we regard as primitive. Others have a basal meristematic region to their sori and form new sporangial initials, resulting in a gradual maturation of sporangia towards the centre of the sorus; this type of sorus we call gradate, e.g. *Hymenophyllum, Dicksonia* and the Carboniferous *Psalixochlaena* (Holmes, 1981). The majority of extant ferns instead have mixed maturation of sori, e.g. *Dryopteris, Pteridium* and *Polypodium*. This enables a greater number of sporangia to develop and spreads the period of spore liberation over a greater period of time. Mixed development is more efficient than the other two and we regard it as an evolutionarily advanced adaptation.

Those ferns with marginal sori usually protect them by an infolding of the frond margin. Many of those with superficial sori protect their developing sporangia with a membranous flap called an indusium. On maturation of the sporangia these protective structures are either forced aside by the expanding sporangia or may shrivel. Some of the Carboniferous ferns show some evidence of laminal infolding, but there is no evidence of indusium-like structures. These are first seen in the Jurassic *Aspidistes*.

As we saw earlier, the primitive fern sporangium had a simple wall structure that was only slightly modified to stimulate a vertical line of dehiscence. Gradual localisation of surface cells with thickened walls into discrete areas would have led to the sporangia having both more efficient growth and development and better dehiscence mechanisms. Such a discrete area of thickened cells is known as an annulus. There are several ways in which the area of thickened cells has been reduced to an annulus (Figure 8.2). In the Carboniferous *Tedelea* and *Sermaya* the annulus is limited to one side of the sporangium. In the extant *Schizaea* and *Lygodium* and the Carboniferous *Senftenbergia* it is a terminal cap. In the extant *Osmunda* it is an equatorial patch. Similarly, in the Carboniferous *Botryopteris* it is an equatorial patch, but in this genus it also extends to near the base of the sporangium. There is, however, much variation in the sizes of the annuli in these genera and in the number of their constituent cells. The differences between genera are often slight, posing problems to anyone suggesting evolutionarily sequential changes.

The variation in reproductive characters shown in the early ferns is to be expected if the evolutionary changes of sporangial wall structure were progressing in different ways at the same time. The Coenopteridales were most probably simultaneously giving rise to the various Carboniferous fern families that are currently recognised; the Botryopteridaceae, Anachoropteridaceae, Sermayaceae, Tedeleaceae (Taylor, 1981) and Psalixochlaenaceae (Holmes, 1981). These Carboniferous ferns must then have continued to evolve into forms which ultimately are recognised as belonging to modern families. Several lines

Figure 8.2: Fern Sporangia Showing a Range of Different Annular Structures. A: Carboniferous *Botryopteris* with no distinct annulus. B: Carboniferous *Senftenbergia* with a terminal multilayered annulus. C: Extant *Schizaea* with a terminal single-layered annulus. D: Extant *Osmunda* with a lateral patch of annular cells. E: Extant *Gleichenia* with a lateral annulus. F: Jurassic *Aspidistes* with a linear annulus

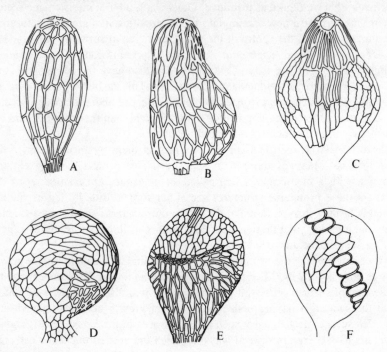

of evolution of the filicalean ferns seem to be the most likely way of arriving at the modern extant fern families.

The Carboniferous *Botryopteris* sporangial structure is very reminiscent of those of the Osmundaceae which are relatively large and sessile and have an equatorial annulus consisting of a patch of cells. Sporangia may be randomly dispersed in the Osmundaceae, in clusters over the lateral veins, or on specialised non-laminated portions of fronds. In *Botryopteris*, however, they are simply aggregated into clusters.

The Osmundaceae has a wide range of form that can be traced back to Upper Permian petrified genera. Some species have creeping rhizomes, others have erect stems covered with densely packed fronds, and some Permian forms reached arborescent proportions. As in the Marattiaceae the rate of evolutionary change in the Osmundaceae seems to have been greatest in the Palaeozoic with subsequent changes being relatively slow and slight. The wide range of Permian material shows how the typical dictyoxylic protostele of the extant *Osmunda* probably evolved. The development of pith parenchyma, as seen in *Thamnop-*

Figure 8.3: *Todites princeps* (Gothan) Presl from the Jurassic of Iran. The main plant has a trunk about 50 cm tall giving off basal runners that form accessory plants. (After Schweitzer, 1978)

teris, seems to have preceded the formation of leaf gaps through the xylem cylinder. Intermediate forms like *Chasmatopteris*, with rudimentary stelar invaginations, must have preceded true leaf gaps as in *Palaeosmunda*.

Miller (1971), using multiple character correlation, suggested that the group originated from the Upper Carboniferous and Permian genus *Grammopteris* (Anachoropteridaceae). This naturally conflicts with ideas based upon sporangial features, once again highlighting the difficulties encountered in proposing evolutionary relationships. However, all of this early evolutionary evidence is of petrified stems and it is not until the Triassic that indisputable foliage is known in the form of *Todites fragilis* described by Ash (1969). By then the family was in decline and although there were many Mesozoic species, e.g. *T. princeps* (Figure 8.3), further extinctions reduced the numbers of genera in the family. Today it is a mere remnant, having only 16 species in three genera.

Mesozoic Diversification

Two other extant families have been generally accepted as having lived in the Palaeozoic but are now thought to have evolved later than this. The Upper Carboniferous *Senftenbergia* was included in the Schizaeaceae on the evidence of terminal annuli on its sporangia. The annulus does, however, have more than the one row of cells found in the more acceptable members of the family. Jennings and Eggert (1977) have also shown its petiole anatomy and its sporangial arrangement in double rows to indicate a closer affinity with *Tedelea*. Similarly, the Carboniferous *Oligocarpia* is now thought to be the compression of *Sermaya* rather than a member of the Gleicheniaceae. There are Triassic ferns that bear a resemblance to both these families (Ash, 1969). *Cynepteris lasiphora* has sporangia with apical annuli formed almost entirely of a single row of elongated cells. However, Ash noted that several characters of these ferns are not found in the Schizaeaceae, especially the scattered superficial arrangement of the sporangia and the reticulate venation of the pinnules. He therefore preferred to create a new family, the Cynepteridaceae. In contrast, *Wingatea plumosa* has enough characters closely comparable to the Gleicheniaceae for it to be included in the same family, even though it does not show the pseudodichotomous type of branching typical of all living members. The problems encountered in studying early ferns should be very clear from these two families. Evolutionary changes of different characters at different rates must have given rise to ferns that do not fit all the characters of extant families. We can only try to assess the available evidence, knowing that new discoveries can easily make us modify our ideas.

Ferns diversified in the Mesozoic and apparently increased in numbers. They were so abundant in the Yorkshire Jurassic flora that Harris (1961b) thought they might well have formed the dominant herbs on land. This seems highly likely as there were few smaller gymnosperms and the lycophytes and sphenophyes were definitely rare and seemingly in decline. Angiosperms, of course, had not yet made their appearance.

Ferns referable to the Marattiaceae and Osmundaceae still persisted in the Mesozoic, although by now other families were present in greater numbers. *Klukia* (Schizaeaceae) was also a common fern in the Jurassic and Cretaceous. The Matoniaceae was widespread throughout the Mesozoic, even though it apparently contained relatively few taxa. *Phlebopteris, Matonidium* and *Weichselia* were all common members of this family. Ash, Litwin and Traverse (1982) have described *Phlebopteris smithii* (Figure 8.4B) from the Upper Triassic of southwestern USA as one of the oldest members of this family. It has most of the characters of the living members of the Matoniaceae except that it has no indusium. Alvin (1974) in his studies of the Wealden of southern England has shown *Weichselia reticulata* (Figure 8.4A) to be a most successful plant. It possessed a number of xeromorphic features (e.g. a cuticle, fibrous

Figure 8.4: Mesozoic Matoniaceous Ferns. A: *Weichselia reticulata* from the European Wealden. B. *Phlebopteris smithii* from the Triassic of North America. (A, After Alvin, 1971; B, after Ash *et al.*, 1982)

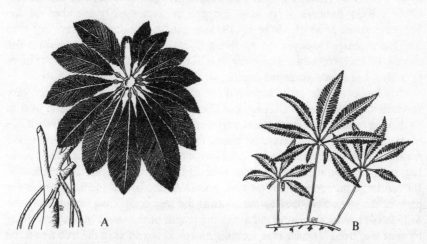

tissues, and sunken stomata) that permitted it to dominate communities growing in habitats that were subjected to periods of extreme drought. In contrast, the Gleicheniaceae were much rarer, being represented in the Jurassic by few and fragmentary specimens of *Gleicheniopsis*. The Dipteridaceae made their appearance in the Upper Triassic with *Clathropteris walkeri* from the USA (Ash, 1969) and *C. meniscoides* from Greenland (Harris, 1931). In the Jurassic, *Clathropteris* is found in Yorkshire, Japan, Korea and Central Asia. *Dictyophyllum* and *Hausmannia* are two other widespread Mesozoic genera referable to this family.

Living Cyatheaceae and Dicksoniaceae form tree ferns that have evolved the same supporting root mantle as in the extinct Carboniferous *Psaronius*. Members of the Cyatheaceae are the larger, with some species of *Cyathea* being up to 20 m tall with fronds of 6 m length. Both families are known from the Mesozoic. The Dicksoniaceae were especially widespread in the Jurassic in the form of the fertile fronds called *Coniopteris*. These early dicksoniaceous ferns diversified and the thyrsopteridoidean line, which includes the living *Culcita*, gave rise to the extant dryopteroid ferns (Harris, 1973b). The Cyatheaceae appeared in the Jurassic, although the extant members seem to be a close group of genera that have only recently evolved. Tryon and Gastony (1975) in studying the biogeography of the Cyatheaceae came to the conclusion that they show a strong development of local endemism brought about by the establishment of small peripheral populations. Ecological specialisation then isolated them in small unique environments.

The most intriguing Mesozoic fossil fern is perhaps *Aspidistes thomasii* which was described by Harris from the Yorkshire Jurassic. Harris (1973b) thought it to represent the only Mesozoic polypodiaceous fern except perhaps for *Gleicheniopsis* (included here in the Gleicheniaceae). *Aspidistes* has sori with persistent indusia and sporangia with vertical annuli of typical polypodiaceous form. The evidence also suggests that the sporangial development was mixed. Lovis (1975) believes it is best thought of as an early member of the Thelypteridaceae, which Holttum (1973) argued had a common ancestry with the Cyatheaceae. However, as the living *Cyathea* has more primitive soral and sporangial characters than *Aspidistes*, it is likely that the Cyatheaceae is the more primitive and hence ancestral family.

One family of Mesozoic ferns did not survive, possibly because of its very specialised growth pattern. *Tempskya* (Tempskyaceae) grew to heights of 6 m by having many entwined stems and petioles enmeshed within adventitious roots. This would have given the plants the appearance of having a trunk like the tree ferns of the Marattiaceae, Dicksoniaceae and Cyatheaceae. The sporeling probably commenced life with a vertical stem that began to branch as it formed large numbers of adventitious roots. As the height increased the older parts of the stem died, leaving the roots as the sole conducting organs from the soil. Small leaves were attached along the length of the stems thereby covering most of the 'trunk', rather than forming an apical crown as in the true tree ferns (Andrews and Kern, 1947).

Polypodiaceous Ferns

The evolution of ferns through the Mesozoic period seems to have been a gradual process with some species appearing that were comparable to modern species at the family level. Some, like *Weichselia*, were extremely successful in their local environments, being comparable to bracken in their domination of the community (Alvin, 1974). However, they did not face very much in the way of competition as there appear to have been few other plants of comparable size. The appearance of the angiosperms in the late Mesozoic and their rapid expansion in the late Cretaceous and early Tertiary altered this. They created completely new communities, and the evolution of the angiosperm herbs gave direct competition to the ferns. It was this rise of the angiosperms that stimulated the ferns to evolve new and presumably more competitive species in the Tertiary. Unfortunately records of Tertiary ferns are very scanty, and there is very little indisputed evidence that they belonged to the Polypodiaceae. The problem encountered with evaluating records of Tertiary ferns is that authors tend to equate them readily with living plants. The vast majority have been identified on gross morphology with little knowledge of soral structure and none of sporangial structure. Although Andrews and Boureau (1970) quite rightly point out the futility of referring sterile fragments to extant genera such as *Dennstaedtia, Asplenium, Dryopteris*, etc., they list 15 species in 13 genera of Tertiary Polypodiaceae, with only *Acrostichum preaureum* having recognisable sporangia. A few anatomically preserved remains of rhizomes and rachises have also revealed the presence of Tertiary members of the Dennstaedtiaceae. Arnold and Daugherty (1964) have described some specimens from Oregon as

Dennstaedtiopsis, and Ribbins and Collinson (1978) closely compared others from the London Clay with living *Histiopteris*.

The paucity of fossil remains does not necessarily mean that our ideas of evolutionary timing are wrong and that polypodiaceous ferns were virtually absent in the Tertiary. The problem here is that most polypodiaceous ferns have fronds that not only die on the plants but they collapse and wilt as they do so. In contrast, the dominant flowering plants shed their leaves in vast numbers. Tertiary leaf floras must therefore give an over-representation of flowering plants and a corresponding under-representation of ferns. Collinson's (1978) description of dispersed sporangial masses and individual sporangia from the British Tertiary clearly illustrates the importance of looking for these smaller remains. In this instance their assignment to *Achrosticum anglicum* confirmed the presence of the Polypodiaceae in the British Tertiary. Similar studies should reveal the family's presence throughout the Tertiary deposits of the world.

Reproductive and Habitat Factors Affecting Fern Evolution

The lines of evolution encompassing the majority of extant ferns must represent a series of interlocking sequences rather than a single continuously diverging sequence of events. Much morphological, cytological and geographical study has been carried out on these ferns and the reader would do well to consult *The Phylogeny and Classification of Ferns* (Jermy, Crabbe and Thomas, 1973) for further information on these topics. This work, together with evidence obtained from breeding experiments, has enabled us to understand the evolutionary relationships of some taxa.

The life cycles of the ferns are seemingly very simple at first, but a close examination is essential to an investigation of their evolution. The diploid fern liberates haploid spores which grow into bisexual haploid prothalli. The motile sperm swim from the antheridia to an archegonium, being attracted by the liberation of bimalate ions (Brokaw, 1958). Fusion of sperm and egg cell results in a diploid zygote which grows into a new fern sporophyte. Mature prothalli are of varied shape and size, but all develop initially into a plate or filamentous stage which can form large numbers of antheridia (Atkinson, 1973). The haphazard dispersal of spores, the production of large numbers of antheridia on the younger prothalli and the simple archegonial process for attracting sperm can result in sperm finding their way to archegonia of other prothalli. Many interspecific hybrids result from such crossing over of sperm from one prothallus to another, for there seems often to be very little incompatibility at this stage of the life cycle. Some hybrids are sterile and survive a normal lifespan without producing viable spores. Others are fertile, effectively becoming new species.

Cytological evidence does, however, reveal that the distinction between sterility and fertility in hybrids is not as clear cut as it may seem. Some species can hybridise quite naturally, forming discrete fertile plants, but in at least one instance, between *Pteris quadriaurita* and *P. multiaurita* in Ceylon, hybridisation forms fully fertile sexual hybrid swarms (Walker, 1958). Some hybrids,

Figure 8.5: The Evolution of the Polypody Fern (*Polypodium vulgareagg*). Double lines indicate chromosome doubling and the production of a normal sexually reproducing species. The chromosome base number (*X*) is 37

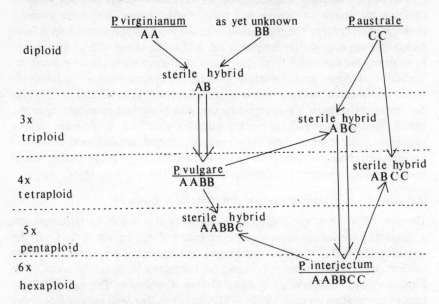

which are effectively sterile with their mixed chromosome content, initiate a doubling of their chromosome numbers during sporogenesis to form viable offspring that then breed true. This is alloploidy and forms the majority of fern polyploids whose cytological origins have been studied. For example, the complex of similar species of *Polypodium* has been interpreted as in Figure 8.5. Here a knowledge of their hybridisation and polyploidy has been obtained through cytological study and experimental hybridisation (Shivas, 1958). There are of course known occurrences of autoploidy as well in ferns.

Sterile hybrids may survive for many generations by vegetatively reproducing before becoming fertile. Compensating processes can sometimes occur which permit the production of viable spores. There are three different processes, but all result in viable spores being formed with no reduction in chromosome number. The gametophytes then form baby sporophytes autonomously from prothallial cells without involving any sexual organs. Ferns that have a morphological alternation between a conspicuous sporophyte and an inconspicuous gametophyte with no cytological alternation in chromosome numbers are said to be apomicts and the process is known as apomixis. For these apomicts to survive, the gametophytes must suppress their archegonia, otherwise sexual reproduction would cause a doubling of chromosome number with every cycle. This would effectively polyploid the species out of existence in a very few generations. Apomictic ferns can therefore never be facultative apomicts as in some flowering plants; they must always be obligate apomicts.

Apomictic prothalli do, however, form antheridia, and the diploid sperm that they liberate are usually more vigorous than usual. Hybridisation between an apomict and a sexual fern usually results in an apomictic triploid. As most apomicts are triploid it appears that these hybrids are common in the wild.

All of these departures from the 'normal' life cycle can result in altered progeny, especially in chromosome content and number. Polyploidy occurs, at a conservative estimate, in 50–60 per cent of the ferns. High base numbers are also an outstanding feature of most ferns that are assumed to have arisen by multiplication of an ancestral lower number (Chiarugi, 1960). For a more detailed account of the cytogenetics of ferns the reader should consult Walker, whose paper is one of many useful surveys in *The Experimental Biology of Ferns* edited by Dyer (1979).

How does this information help us interpret the evolution of the filicalean ferns? We know that polyploidy has been occurring for a long time, but it cannot be related to the longevity of a group for it is apparently absent in the Osmundaceae. The distribution of polyploids shows that the percentages of them in fern floras vary considerably, but these differences cannot be related to any one geographical or climatological factor. There are, however, a few interesting and relevant facts that do emerge. For instance, the Canary Islands ferns have a low percentage of polyploids (25 per cent) and are thought to be part of the relict European Tertiary flora surviving in the high mountains of the islands (Page, 1973). In contrast, areas of the world that have suffered geological or climatological disturbances seem to have relatively high levels of polyploidy.

The Pleistocene Ice Age climatic fluctuations caused large-scale floral migrations. Such migrations of ferns would have caused a mixing of hitherto separate species stimulating hybridisation and in some cases alloploidy. This is especially noticeable in Japan. Ferns migrating southwards from the Asian mainland during the Pleistocene glacial stages have stayed in the islands by moving into the cooler mountainous regions. Since the end of the last glacial period the return of species that had migrated southwards into warmer areas brought two formerly separated fern populations together. This has resulted in large numbers of new hybrid species endemic to Japan (Sleep, 1970).

However, what we are seeing as a result of the Pleistocene is only part of the immense series of changes brought about by the mixing of fern populations. They would have commenced as a result of the devastating effect that the evolution of the angiosperms had on the world's vegetation. New plant communities and competition from herbaceous angiosperms would have caused many ferns to change habitats or to migrate. This, coupled with the large-scale floral migrations that were occurring throughout the Tertiary, must have caused widescale and continuous mixing of fern species, thereby stimulating hybridisation and enormously increasing their rates of evolutionary change.

Fern evolution is clearly proceeding and it is probably best seen in the Aspleniaceae. The family has no definitely acceptable fossil representatives, and the living species (about 700) are unusually alike for such a large group. There is one large genus, *Asplenium*, and a debatable number of smaller genera. Many of these other genera have species that hybridise with *Asplenium* species, suggesting that their recognisable morphological differences have not yet been genetically isolated. The family contains a profusion of evolutionary lines that are still active. It can be thought of as a young family similar to the angiosperm Orchidaceae, Poaceae and Asteraceae, which also contain many hybridising genera.

Ophioglossales

This group includes about 70 species in the three genera *Ophioglossum*, *Botrychium* and *Helminthostachys*. The virtual circumpolar distribution of the group indicates some degree of antiquity, but the only fossil record is of some early Jurassic and Cretaceous spores from the USSR. Their evolutionary history can therefore only be deduced from comparative morphology. All the plants appear very simple, with each aerial system consisting of a sterile lamina portion and a fertile spike. This is generally regarded as a primitive morphology, although Bierhorst (1971) suggests that the filicalean ferns may be more primitive in many respects. He proposes that they are most likely to be derivatives of the Aneurophytales, or some form in a lineage between the Trimerophytina and the Aneurophytales, citing several anatomical similarities as evidence. This idea, however, is unacceptable if the Aneurophytales are thought of as belonging to the line of plants that were to give rise to the gymnosperms (see page 140). The other possibility is that the Ophioglossales are derivatives of the coenopterids or the botryopterid ferns, but although this is possible, there is no real evidence to substantiate the idea. Clearly we are waiting for macrofossil evidence to help us interpret the evolutionary history of the group.

Salviniales

There are two genera of floating heterosporous ferns, *Salvinia* and *Azolla*, that have fossil records extending back to the Cretaceous (Figure 8.6). *Salvinia* has whorls of three leaves on a rootless rhizome. Two of each whorl are floating and covered on their upper surfaces with air-trapping hairs; the other is highly branched and submerged. *Azolla* rhizomes are thin and branching and bear small alternated leaves in two rows. Each leaf is two-lobed and contains a colony of the nitrogen-fixing blue-green alga *Anabaena*. Both genera have their sporangia in highly modified indusia which form hard, resistant cases called sporocarps. Each sporocarp therefore represents a single sorus and develops either microsporangia or megasporangia.

Although there are vegetative remains, the fossil record consists mainly of very distinctive dispersed spores. The most obvious of all are the megaspores

Figure 8.6: Salviniaceae Megaspores. A, C: *Azolla prisca* from the Oligocene of the Isle of Wight, UK. A, Intact megaspore apparatus with three groups of three vacuolate floats beneath an apical cap. The perine is partly removed to reveal the megaspore. C, Section of megaspore wall showing its stratification. B, *Salvinia aureovallis* from the Tertiary of North Dakota, USA. f, Float; t, trilete mark; m, megaspore; p, perine; h, homogeneous layer; g, granular layer; e, exine; s, striated layer; b, basal layer. (A, C, After Fowler, 1975; B, after Jain and Hall, 1969)

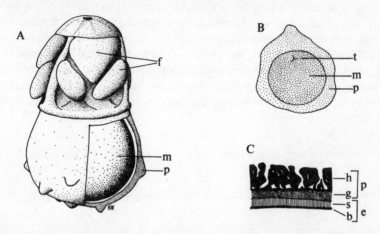

of *Azolla*, which have floats derived from other aborted megaspores. Nearly 90 per cent of *Azolla* species are fossils extending back to the Lower Cretaceous. The ancestral forms had 24 floats which evolved into other forms by many different pathways of float fusion. It appears likely that numbers 21, 18, 15, 12, 9, 6, 3 and 1 are also known in fossil species and 9 and 3 in living species. However, even with 60 described species we still have no clear idea of the evolutionary development of the genus other than that there seem to have been many different lineages involved (Collinson, 1980).

Salvinia megaspores do not have floats, although they are covered by a foamy perine. Such megaspores are known from the Upper Cretaceous of North America, which supports earlier evidence of leaf impressions from the same region (Jain, 1971). Unfortunately none of these fossils gives us any idea of the evolutionary history of the genus.

The group has been suggested to be related to the filmy ferns, the Hymenophyllaceae, as both have simple vegetative characters and indusiate sori and exist in hydrophilic habitats. However, we suggest that no direct evolutionary connection should be made on these characters as they are more likely to represent similar evolutionary adaptations to the environment rather than some ancestral relationship.

Marsileales

This second group of heterosporous ferns contains about 70 species of extant members of *Marsilea, Regnellidium* and *Pilularia*. They are rhizomatous, rooting plants that often grow in dense mats. Like *Salvinia* and *Azolla* they have evolved sporocarps, but these are of foliar rather than indusial origin. The fossil record of this group is sparse, although a few Cretaceous and Tertiary megaspores probably belong here. However, they tell us nothing of the evolutionary history of the group. Comparative morphology suggests a relationship with the primitive 'leptosporangiate' ferns, especially the Schizaeaceae. Morphological change must, however, have been stimulated by adaptation to living in an aquatic environment so their apparent simple vegetative characters may be derived rather than ancestral. We can only wait for more fossil evidence to give us clues of how heterospory and sporocarps arose in this fascinatingly different group of ferns.

9 PROGYMNOSPERMS AND OVULE EVOLUTION

As we saw in Chapter 4, many pteridophytes evolved specialised reproductive biologies; changing from an originally homosporous life cycle to a heterosporous one. A few took this process further to form single spore tetrads in each megasporangium from which only one megaspore developed to functional maturity. Such large megaspores were of necessity retained within their sporangia even through fertilisation and the subsequent germination of the embryo. The most specialised of all pteridophyte megasporangia was *Lepidocarpon* (page 107). Its single-spored megasporangium was both protected by sporophyll outgrowths and aided in dispersal by the enlarged wing-like lamina. Plants bearing *Lepidocarpon* were certainly successful in their own habitat, but when the Coal Measures swamps disappeared they became extinct. Even though their reproductive organs might have functioned in drier areas, the total biology of the plants was too unadaptable.

The evolution of a single-spored megasporangium by species already adapted for life in the drier habitats would produce much more successful plants. Endowed with both physiological and reproductive advantages, they would be set to dominate the world's vegetation.

Progymnospermophyta

A group of plants arose in the Devonian that was to have a profound effect on this course of plant evolution. It was the Progymnospermophyta whose members are generally accepted as anatomical, and most probably phylogenetic, intermediates between the earlier vascular pteridophytes and the later seed plants.

These plants developed large, secondary-thickened, gymnosperm-like axes, permitting many of them to attain arborescent proportions. They were the first true trees. Their free-sporing reproductive habit was, however, essentially pteridophytic, which concealed the true nature of such fertile specimens for many years. Until the anatomy of this fern-like foliage was shown to be gymnospermous by Beck (1960), we were totally unaware of the existence of such botanically interesting and evolutionarily fascinating plants.

We now accept between 15 and 20 genera of Progymnospermophyta, ranging from the Devonian into the Lower Carboniferous. They can be divided into three orders on morphological and reproductive characters: the Aneurophytales, Archaeopteridales and Protopityales; although there are naturally the usual problems encountered in assigning fragmentary plants to any particular group. The combination of gymnospermous secondary tissues and free-sporing sporangia distinguishes these plants from others, but what can we do if there is evidence of only one character? Gymnospermous secondary tissue in stems of the right geological age, like *Stauroxylon* and *Cairoa*, seems acceptable evidence if we assume that it evolved only once and in a single group of plants; but free-sporing sporangia on their own cannot possibly be included here. However, even without either character it seems sometimes justifiable to identify such plants as progymnosperms; for example Stein (1982) included *Reimannia* in the Aneurophytales solely on the evidence of its primary vascular architecture.

Within the group there are some recognisable morphological series that can be thought of as evolutionary trends. The Devonian Aneurophytales had at least three orders of helically arranged branches, with three-dimensional, non-laminated, ultimate appendages. The Middle Devonian to Lower Carboniferous Archaeopteridales had flattened, laminated leaves which were opposite and decussate on the ultimate branches although helically arranged on the penultimate ones. Anatomically the Aneurophytales were simplest with lobed, mesarch, primary xylem. The Archaeopteridales had a ring of mesarch primary bundles surrounding a pith, whereas that of the Protopityales had endarch bundles. These can be thought of as stages in the evolution of the gymnospermous stele (Figure 6.5). Sporangia were pinnately arranged in the Aneurophytales, with *Tetraxylopteris* having quite complex fertile regions of sporangial complexes. The ultimate shoots of *Archaeopteris* have the characteristics of strobili with fertile appendages as sporophylls (Figure 9.1). These stroboli have basal vegetative leaves integrating through a short transition zone to the central sporophylls. The sporophylls are apically dissected with sporangia attached to their adaxial surfaces. The strobili may also have a few apical vegetative leaves that progressively diminish in size and are less dissected (Beck, 1981).

The Aneurophytales were probably homosporous, although the range of spore size in *Tetraxylopteris*, 73–176 μm diameter, suggests that some might have been anisosporous or even heterosporous. Similarly, the spores from the few known fertile specimens of the Protopityales ranged from 82 to 163 μm. In contrast, *Archaeopteris* contained heterosporous species with microspores, 30–70 μm, and megaspores, 150–500 μm, in separate sporangia. There were also apparent homosporous species, but as their spores were very similar to the microspores of the heterosporous species it could be that we have not yet found their corresponding megaspores.

In his discussion of the Aneurophytales, Stein (1982) points out how two rather separate lines of evidence have been accumulating overall morphology and reproductive biology from compressions and a diversity of internal structure from anatomically preserved specimens. This is to some extent similarly true in the Archaeopteridales, although here we have a much fuller picture of the overall plants. *Archaeopteris* is the best known of all the Progymnospermophyta as we have detailed knowledge of its morpology and anatomy, its juvenile stage

Figure 9.1: *Archaeopteris macilenta*. Reconstruction of a strobilus borne on a penultimate axis of a lateral branch system. (After Beck, 1981)

(*Eddya* — Figure 6.4D) and its reproductive organs. It was cosmopolitan in distribution and some species most certainly formed extensive forests. Such a widespread distribution must surely have resulted in geographical isolation that stimulated speciation and evolutionary diversification. Indeed, the amount of variation known suggests that *Archaeopteris* may encompass several genera that we are not yet capable of distinguishing with confidence (Beck, 1981).

The Progymnospermophyta are generally accepted as the ancestors of the gymnosperms, although there is still not enough evidence for us to be certain of the various suggestions that have been made. The Aneurophytales might be the progenitors of some of the Lower Carboniferous seed-ferns, and *Archaeopteris* has been suggested many times to be the ancestor of the conifers. Beck (1981) has, however, outlined some very interesting homologies in reproductive structures between those of *Archaeopteris*, *Noeggerathiostrobus* and *Archaeosperma*, the Lebachiaceae and *Cordaites*. Individual sporophylls of *Archaeopteris* could have formed the primitive cupule-like bract of *Archaeosperma*, and their compaction in different ways would have resulted in strobili like those of the other groups (Figure 9.2).

Neoggerathiopsida

This group of plants, extending through the Carboniferous and Permian, has been at various times assigned to nearly every major group. The foliage of *Noeggerathia* has two opposite rows of obovate leaves that have numerous dichotomising leaflets. Its cone, *Noeggerathiostrobus*, has bract-like units in semicircular discs on a central axis. Sporangia appear in rows on the adaxial surface of the discs, although we do not know how they were attached. *Lacoea* also has semicircular discs, but they are more dissected than those of *Noeggerathiopsis* and have several rows of sporangia on their upper surfaces. *Discites* is similar to *Lacoea* but has completely whorled disc bracts. Such a series could

Figure 9.2: Suggested Homologies in Reproductive Structures of
Archaeopteris (A), *Noeggerathiostrobus* (B), Lebachiaceae (C), *Cordaites*
(D), *Archaeosperma* (E). The white arrows indicate organs of specialised
branch systems (the lebachiacean female cone and both male and female
fertile branch systems of *Cordaites*) which are thought to be homologous
with the entire fertile branch system of *Archaeopteris*. Black arrows indicate
organs thought to be homologous with the strobili (ultimate fertile shoots)
of *Archaeopteris*. The drawings of *Archaeopteris, Cordaites* and the
Lebachiaceae are stylised for clarity. (After Beck, 1981)

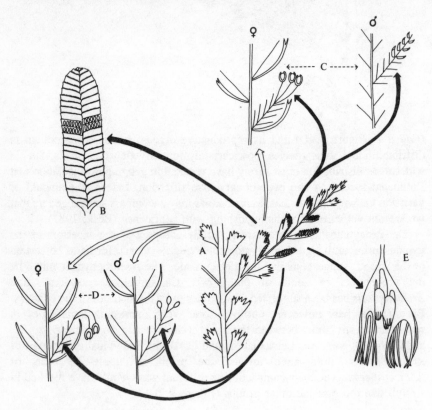

represent the gradual consolidation of the noeggerathian cone from the more lax
Archaeopteris strobilus (Figure 9.2B). There is also some supporting evidence
for this idea from the spores. Although the megaspores are rather different, the
microspores are so similar that both would be called *Cyclogranisporites* if found
isolated. Indeed, the evidence for a close evolutionary relationship between
Archaeopteris and the Noeggerathiopsida seems so strong that Beck (1981) has
suggested that the Noeggerathiopsida might represent an extension of the
progymnosperm line through the Carboniferous. The arguments for this are not
conclusive but seem sound enough for us to accept them, thereby including the
Noeggerathiopsida in the Progymnospermopsida.

The Seed Habit

Probably the most significant event in the evolutionary history of land plants was the evolution of the ovule. This process would endow many plants with a more efficient reproductive biology, thus making them more competitive and hence successful in spreading and adapting to many diverse ecological conditions. The earliest changes towards this type of reproductive biology can be traced to those plant groups that were exploiting the advantages gained through heterospory. A reduction in the numbers of spores developing to maturity led to one functional megaspore per sporangium and ultimately to the complete loss of the other three aborted megaspores. Sporangium rather than spore dispersal was then the result. But of course this modification alone does not make an ovule, for there are other important and necessary changes. The megaspore must remain embedded in sporangial cells rather than becoming isolated from the parental tissue. An independent system of terminology developed before we had any idea of their morphological identities, so in ovules we call the sporangium the nucellus and the megaspore wall the embryo sac. The archegonia, usually three in number (Taylor, 1982a), were embedded in the distal end of the megagametophyte. This necessitated a method for allowing the male generative nuclei to reach them. Receptive systems were evolved for trapping wind-borne pollen, and protective outer tissues gave rise to seed coats.

Our ideas on the development of the seed habit have some factual basis. There are seed-like structures that help us to understand how these evolutionary changes occurred (Figure 9.3). The most primitive, although not the oldest, ovules known are the Lower Carboniferous species of *Genomosperma*. These are radially symmetrical, about 1.5 cm long, with a nucellar tip elongated into a trumpet-shaped structure, called a salpinx, that probably functioned as a pollen trapping device. Small triradiate spores have even been found in the salpinx of some specimens (Rothwell, 1972). Between eight and eleven vascularised processes (integuments) are attached immediately proximal to the nucellus and extended over it in a protective role. This morphological organisation suggests that the integuments arose from a dichotomous branching system by a series of axes reductions. A terminal sporangium would thus gradually become enveloped in a ring of sterile branches. Such a hypothesis is essentially the telome concept of Zimmermann (1930, 1952). In *G. kidstoni* these integument lobes are free whereas in *G. latens* they are basally fused. There are a number of contemporaneous ovules that show further stages of integumental fusion; for example, in *Eurystoma* the integuments are fused except for four apical lobes whereas in *Stamnostoma* they are completely fused, leaving only a small apical aperture.

Wind-borne pollen could have been trapped by these ovules in a number of ways. Some had integuments covered in hairs, as in *Salpingostoma*, or sticky projections, as in *Physostoma*. Others probably exuded a fluid that projected from the salpinx as a pollen drop. The fusion of the integumental lobes with each other and with the nucellus led to the complete encasement of the nucellus and

Figure 9.3: Early Pteridosperm Ovules. A: *Genomosperma kidstoni* with eight free integuments. G: Median longitudinal section of its distal end. B: *Genomosperma latens* with partially fused and adpressed integuments. C: *Salpingostoma dasu* with integuments fused to the nucellus. H: Median longitudinal section of its distal end. D: *Physostoma elegans* with more fusion of the integuments. E: *Eurystoma angulare* with virtually complete integumental fusion. I: Median longitudinal section of its distal end. F: *Stamnostoma huttonense* with complete fusion of its integuments. (After Andrews, 1963)

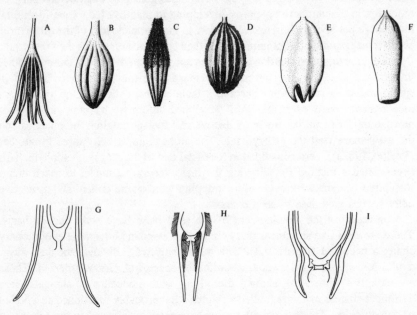

the embryo sac except for the terminal pore — the micropyle. The salpinx was redundant and lost, for now the pollen was captured by the integuments and not the nucellus. The pollen drop exuded out through the micropyle and hung from the tip of the integument.

The early ovules also often show additional sterile processes that were either partially or completely enveloping them. These structures, which we call cupules, once again show stages in fusion from the earlier more divided and bilaterally symmetrical, bract-like kinds to the later enveloping and radially symmetrical, bell-like forms. The cupule units themselves can be thought of as coalesced portions of a branching system. The earliest Devonian seed-like structures such as *Archaeosperma arnoldii* (Pettitt and Beck, 1968) and the similar but older forms from the Famenian of West Virginia, USA (Gillespie, Rothwell and Scheckler, 1981) have flared, rather open, bifurcating cupules (Figure 9.4). Some Lower Carboniferous ovules are known to have been enclosed in similar cupules whereas others show a range of further coalescence (Matten and Lacey, 1981). All the cupules are stalked and, except for *Calathospermum scoticum*,

Figure 9.4: *Archaeosperma arnoldii*. A: Reconstruction of cupule complex. Each cupule contains two ovules. B: Semi-diagrammatic representation of an ovule showing the megaspore tetrad inside the ovule. The dashed line represents the probable position of the nucellar apex. (After Pettitt and Beck, 1968)

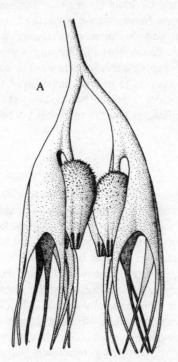
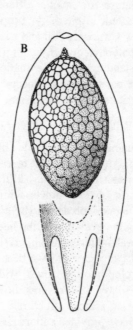

show a basal dichotomy. *Stamnostoma huttonense* and *Genomotheca scotica* have dichotomously branching cupules like *Archaeosperma*. *Eurystoma angulare*, *Hydrasperma tenuis* and *Calathospermum fimbriatum* have unequal branching above the basal dichotomy. *Calaspermum scoticum*, *Gnetopsis ellip-tica* and *Hydrasperma longii* have basally fused cupules. As we will see later, this gradual fusion of the cupule continues and becomes more pronounced in Upper Carboniferous and later ovules.

There are other quite marked differences between these cupulate ovules. The number of ovules in each cupule varies quite considerably from 48 to two, although the reduction to one only occurred in later derived forms. Most ovules were borne on straight stalks, but a few were not. In *Camptosperma* the ovules were attached at right angles, i.e. campylotropous, and in *Anasperma* they were recurved, i.e. anatropous (Long, 1977).

Meyen (1984), in recognising trends in the evolutionary development of seeds and cupules, believes there to be nine major seed types in gymnosperms derived from two initial ones. These he believes cannot yet be reduced to one ancestral type. It is the symmetry of the seeds that Meyen stresses. His *Hydrasperma* type contains seeds with radially symmetrical (radiospermic) vascularised integuments in cupules, whereas the *Lyrasperma* type has bilaterally symmetrical seeds (platyspermic) not in cupules. The presence of only two vascular bundles extending along the main plane of the *Lyrasperma*-type seed to the top of the two integument horns and the absence of nucellar vasculature are major differences that can be taken as evidence for the polyphyletic origin of gymnosperms. The discovery of a Devonian platyspermic seed by Chaloner, Hill and Lacey (1977) supports this hypothesis. The concept of a polyphyletic origin for the gymnosperms is of vital importance to our understanding of their evolutionary histories. It will be returned to later when we are considering gymnosperm diversification and the relationships between the various extinct and extant groups.

Reproductive Biology of Early Seed Plants

The function of the cupule, like that of the integuments, is still a matter for debate. Any theory must take into account the evolutionary changes that were producing such great variability in ovule and cupule morphology. The simplest and most obvious function for the early development of both was protective either against predators (by concealment) or the environment (by enclosure). During the evolution of the integument some of the cells became elongated and their walls thickened. This resulted in the discrete zone we call the sclerotesta. There also seems to have been a tendency in the later forms for the sclerotesta to develop into two zones, with the outer having radially aligned cells (Taylor and Millay, 1981). This probably conferred many advantages apart from producing a more structurally sound integument. A heterogeneous sclerotesta could have provided a more elastic seed coat, permitting some limited expansion of the gametophyte and embryo. This would also have been better at discouraging predators and possibly permitted seeds to develop thinner walls. Some cupules are covered in gland-like hairs that could have attracted arthropods or insects, thereby stimulating an early entomophilous pollination association, or equally may have deterred animals from eating them.

Both integuments and cupules have been suggested to assist in directing wind-borne pollen towards the salpinx or micropyle. Experiments have even been carried out in an attempt to support this hypothesis (Niklas, 1981, 1983), although the methodology of using exposed and rigidly fixed models is very much open to question. Ovules are usually illustrated in an upright position, with the salpinx, micropyle or cupule opening directed upwards. Sometimes they are described as erect, although often no mention is made in the text of how they might have been attached to their parent plants. This is of course a major problem as we have little or no idea of the parent plants of most ovules. A little evidence does, however, show that some ovules were far from erect. Specimens of *Lagenospermum imparirameum* (Figure 9.5) from the Lower Carboniferous of Virginia had at least 22 culpulate ovules borne on pendulously fertile

Figure 9.5: Reconstructions of the Lower Carboniferous Ovule
Lagenospermum imparirameum. A: Portion of an ovuliferous branch. B:
Partially dissected cupule showing the single ovule. (After Gensel and Skog,
1977)

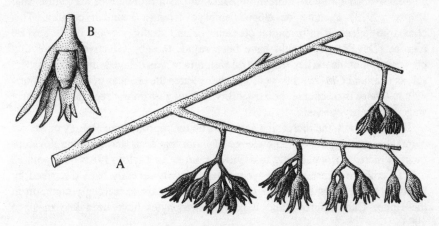

branching systems (Gensel and Skog, 1977). Lax fertile shoots such as these
would have moved in the wind, accentuating wind turbulence that would be
carrying pollen in all directions. In such a fertile system the cupule could have
assisted only marginally in pollen reception, but it would have been very efficient in protecting pollen drops from rapid desiccation or from rain splash. In
an erect position cupules could very easily have acted as rainwater traps, thereby
hindering rather than helping pollination.

Fertilisation in the majority of ovules was mostly achieved by the pollen
liberating sperm that swam through the remnants of the pollen drop to the
archegonia, although there is evidence that some pollen formed pollen tubes
(Rothwell, 1972). Microgametophytes have been found in several pollen grains.
The best were of *Callistophyton* where the preservation is so good that an axial
row of prothallial cells is clearly visible, together with ovoid bodies that can be
interpreted as the generative cell and the tube cell protoplast.

The breakdown of the nucellus apex gave a pollen chamber in which sperm
could swim, but the archegonia still needed to be exposed if the sperm were to
gain access to them. Some of the most primitive ovules retained the triradiate
mark at the apex of the embryo sac which could have split open to reveal the
prothallus. Later forms had lost this mark so the embryo sac was ruptured by
an overall increase in size of the megagametophyte or by apical growth to form
a conical extension. This so-called tent-pole forced its way through the embryo
sac into the pollen chamber and sometimes even enlarged to block the
micropyle, thereby trapping any enclosed pollen grains. Other ovules sometimes
closed their micropyles by cellular proliferations from the integuments or by the
coagulation of the remnants of their pollen drops.

Although we have quite a large amount of information about the earlier forms of ovules, we know virtually nothing about their developmental biology as seeds after fertilisation. Long (1975) has described one Lower Carboniferous embryo with a hypocotyl and two vascularised cotyledons; Stidd and Cosentino (1976) a cellular mass in a cordaitalean archegonium as a proembryo; and Miller and Brown (1973) a more developed embryo from a Permian conifer. This comparative lack of information on embryos and seedlings could be for several reasons. Development could have been rapid, thereby effectively preventing preservation at the earlier stages. If development was delayed until after dispersal, as in extant *Gingko*, this would further reduce the chances of embryo preservation. There is of course the possibility that we just do not recognise the early stages of embryogenesis.

However, with the increasing emphasis on the reproductive biology of extinct plants, we can expect our knowledge of embryogenesis and seedling development to increase dramatically in the near future. As Taylor (1982a) has pointed out, an embryo of arborescent lycophytes has only recently been described by Phillips (1979), even though megasporangial units are exceedingly common in the Upper Carboniferous and female gametophytes have been known since 1904.

PART III: THE SECOND PHASE OF DIVERSIFICATION

10 EARLY EVOLUTION IN THE GYMNOSPERMS

Seeds first appeared in the Devonian and, as we have seen, rapidly diversified during the Lower Carboniferous. These naked seeds, so called because they were not enclosed in ovaries as in the flowering plants, are taken as the basis for grouping large numbers of varied plants together as gymnosperms. For many years they had been thought to be a natural group, the Gymnospermae, but this idea is now generally abandoned in the light of our increasing knowledge of fossil plants. Gymnosperms are instead regarded as a heterogeneous group of seed plants that once constituted much of the world's dominant vegetation throughout the late Palaeozoic and Mesozoic. Since then gymnosperms have steadily declined in number and distribution as the angiosperms have spread throughout the world. There are about 70 genera and 730 species of gymnosperms living today.

Palaeobotanical evidence suggests that seeds evolved independently in more than one group of Palaeozoic plants. Although it is difficult, and often impossible, to trace the evolutionary history of any group of plants within a restricted period of geological time, deductions can be made on the basis of the overall variation of fossil and living gymnosperms. A great deal of information is known about the morphology and internal anatomy of vegetative organs, but emphasis has been placed, quite correctly, on the seeds (or ovules) and seed-bearing organs. It is on these organs that the major groups of living gymnosperms are distinguished. Comparatively little was known until recently about the pollen-producing organs, but a resurgence of interest in these organs and new studies of the ultrastructure of pollen walls have now given further bases for interpreting evolutionary sequences in the gymnosperms. However, in spite of an ever-increasing number of well-known fossil plants, a number of these extinct gymnosperms remain enigmatic. Even so, they are both interesting and perplexing because of their morphological and reproductive diversity, persisting as potential links between better understood fossils and living plants.

A line of approach that is gaining popularity is to consider interpretations of plant phylogenies in terms of the phytogeographical and ecological conditions in which the taxa occurred. Such stress has been laid on interpretations of flowering plant phylogeny (e.g. Takhtajan, 1969; Stebbins, 1974), but it has been comparatively ignored by gymnosperm workers. Florin recognised northern and southern groups of conifers, but did not attempt to analyse all the data available from fossil plants. A difference in distribution between the two hemispheres was also pointed out by Schopf (1976) in his acceptance of two major groups of gymnosperms. Meyen (1984) further emphasised the effects of geographical separation in his review of gymnosperm systematics and phylogeny. In so doing he effectively illustrates how plant migration and dispersal can both stimulate evolutionary divergence and permit the survival of more primitive groups away from the main areas of evolutionary change.

Pteridospermophyta — the Seed-ferns

The recognition that some plants possessed gymnosperm anatomy and pteridophytic reproduction organs was one vital step in our understanding of the progymnosperms in the context of gymnosperm evolution. An equally important contribution was the realisation that some Palaeozoic plants had fern-like foliage but reproduced by means of seeds. A belief that there once existed a group of plants intermediate between the ferns and the gymnosperms led R. Potonié to propose his ideas of plants that he called the Cycadofilices (Potonié, 1899). The first proof of such plants came a few years later from the work of Oliver and Scott (1904). They established their group of seed-ferns, or pteridosperms, on the basis of organ association and the possession of distinct epidermal features on them. Since then our knowledge of these plants has dramatically increased. The Pteridospermophyta, still commonly referred to as pteridosperms, are now generally subdivided and grouped into eight orders: the Calamopityales (Upper Devonian to Lower Carboniferous), the Lyginopteridales (Carboniferous), the Medullosales (Carboniferous and Permian), the Callistophytales (Upper Carboniferous and possibly Permian), the Glossopteridales (Upper Carboniferous to Triassic), the Caytoniales (Triassic to Cretaceous), the Corystospermales (Triassic), and the Peltaspermales (Triassic).

The Pteridospermophyta, in being so central to our understanding of gymnosperm evolution, have been studied intensively and large numbers of both sterile and fertile taxa have been named. Pteridosperms have been shown to be more diverse in their reproductive biology than any other group of vascular plants. In attempting to present an overview of the evolutionary significance of the group it is impossible to mention more than a small fraction of these taxa. The reader can pursue the others through the quoted references or through the very detailed account given by Taylor (1981).

Calamopityales

There are four recognisable orders of Palaeozoic pteridosperms, but of these the Calamopityales should only be thought of as being tentative members. The order consists entirely of stems and petioles as no foliage or reproductive organs have ever been found attached. It must therefore be accepted as a heterogeneous group of Upper Devonian and lowermost Carboniferous plant fragments rather than as a natural botanical group.

The Lower Carboniferous *Stenomyelon tuedianum* was probably the most primitive of the stems, forming a solid triangular protostele dissected almost into three by thin plates of parenchyma and a broad band of secondary xylem with radially aligned tracheids and numerous medullary rays. *S. heterangioides* illustrates how the broadening of the parenchyma plates could have brought about the dissection of the primary xylem into discrete bundles. As suggested in Chapter 6 (see Figure 6.5), further stelar dissection could have resulted in the forms known from the medullosan pteridosperm *Lyginopteris* and the callistophytan pteridosperm *Callistophyton*.

There are some early Lower Carboniferous forms of microsporangiate organs that might belong to the stems making up this family, but none has been found organically attached. Many appear very similar to some progymnosperms and pre-fern types but anatomical evidence suggests a pteridosperm affinity (Millay and Eggert, 1979). The sporangia are generally not synangiate, but are borne terminally in pairs. These sporangia are thin-walled and some have no apparent dehiscence mechanisms. Variation in the sporangial arrangement and in the structure of their pre-pollen (a term used when we believe that germination was still proximal and not through the distal wall as in true pollen) suggests that they were ancestral to the forms found in other groups of seed-ferns. For example, *Paracalathiops* and *Schuetzia* had large numbers of clustered sporangia which could have led to the many and varied forms of fused medullosan synangia, while *Zimmermannitheca* with fewer sporangia that were arranged in uniseriate whorls could have given the *Telangium* kind of lyginopterid synangia.

Lyginopteridales

The lyginopterid seed-ferns include the plants upon which Oliver and Scott (1904) established the pteridosperms. The stem *Lyginopteris* was shown to have leaves of the *Sphenopteris* type which bore cupulate seeds that we call *Lagenostoma*. Even so, this group of seed-ferns is still poorly understood and very few complete plants are known.

The plants were probably less than 2 m tall, or were climbers in habit. The stems underwent secondary thickening producing large amounts of secondary xylem and a characteristic outer cortex with longitudinally running bands of sclerenchyma. Taylor and Millay (1981) in reviewing the morphological variation of this group recognised a trend in stelar organisation involving both stelar dissection and an increase of parenchyma. A series can be seen from *Schopfiastrum* in having a protostele with little parenchyma, *Microspermopteris* with radially aligned plates of parenchyma, *Heterangium* and *Retinangium* with large clusters of tracheids isolated by radial and transverse plates of parenchyma, to *Lyginopteris* with its large pith. There also seems to be an increasing amount of secondary xylem and ray parenchyma in the geologically younger taxa.

The leaves are widely spaced on the stems (Figure 10.1A). This, together with the presence of axillary buds, supports the anatomical evidence suggesting many species to be lianas or vines. Lyginopterid leaves are quite variable although generally they are relatively small. The rachis characteristically dichotomises a short distance above its base before repeatedly dichotomising to give a highly dissected and many times pinnate leaf. Perhaps the best known

Figure 10.1: Lyginopteridaceae. A: Reconstruction of a basal portion of the leaf of *Heterangium* to illustrate the vascular system and pinnule morphology. B: Reconstruction of the cupulate seed *Tyliospermum orbiculatum*. C–E: *Telangium*; C, reconstruction of a portion of a fertile branch system; D, pair of synangia; E, longitudinal section through a synangium (the stippled area is xylem). (A, B, After Taylor and Millay, 1980; C–E, after Millay and Taylor, 1979)

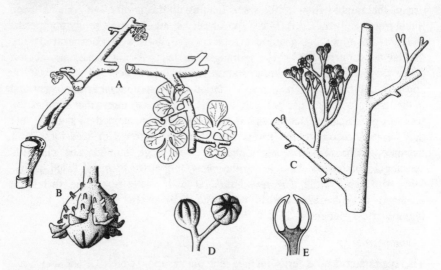

lyginopterid foliage is *Sphenopteris hoeninghausii*, belonging to *Lyginopteris oldhamia*, and the many species of *Mariopteris* that probably belonged to *Schopfiastrum*.

Lyginopterid ovules were quite variable but as yet we can interpret no overall evolutionary trend. All had a relatively thin integument which was attached to the nucellus for most of its length. In the later stages of ovule maturation the nucellus epidermis separated to form an apical pollen chamber leaving a central core of parenchyma tissue. In *Lagenostoma* this separation was minimal producing a narrow pollen chamber, whereas in all the other lyginopterid ovules the chamber became much larger by the displacement of the central parenchyma. After pollination an apical portion of gametophyte tissue meristematically divided to form a column of cells that we call the tent-pole. The vertical expansion of gametophytic tissue both ruptured the embryo sac to expose the archegonia and pushed the central parenchyma upwards to seal the opening of the pollen chamber.

The earliest lyginopterid ovules were loosely aggregated in rather open cupules (e.g. *Eurystoma*) whereas the later ones were single in closed cupules. In *Tyliospermum* the vascularised cupule lobes were basally fused covering a large amount of the seed (Figure 10.1B). This cupular envelopment is even more pronounced in *Lagenostoma* where the lobes completely cover the immature ovule. The cupule lobes are thought to have separated to expose the apex of the

Figure 10.2: *Medullosa noei*. Reconstruction of a plant about 10 m tall. (After Stewart and Delevoryas, 1956)

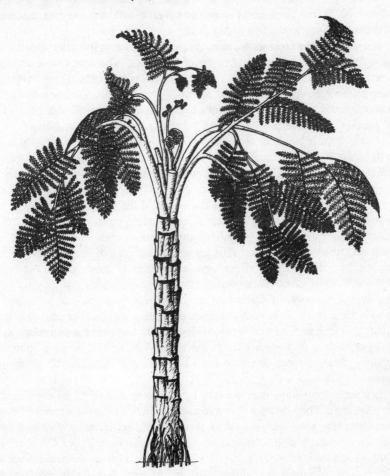

developing ovule.

The earliest pollen organs were terminal, upright synangia on totally fertile branches (e.g. *Telangium*). Some later forms were on the tips of otherwise sterile and planated branches, or pendent on rather more morphologically mixed branches (e.g. *Crossotheca*). The pre-pollen produced by all the pollen organs was spore-like, presumably rupturing by their trilete sutures to release motile gametes.

Medullosales

The medullosan seed-ferns extend from the Lower Carboniferous into the Permian and include some very common genera of stems, leaves and reproductive organs. They were probably very variable in size, and the largest were at least 10 m tall. They may have looked superficially like tree ferns at a distance (Figure 10.2), but their stems were not covered with a dense root mantle and of course their reproductive organs were very different.

Medullosa is the commonest stem genus, being known from petrified material throughout the geological time range of the order. The stems were originally described as polystelic and until recently this view was only rarely challenged. They are now thought to have been essentially monostelic, constructed of axially interconnected bundles that make up a highly dissected stele (Stidd, 1981). Over 40 species of *Medullosa* have been named, some only distinguished by the numbers of their bundles, so there can be little doubt that many really belong to the same biological species. Individual bundles are elongated tangentially or radially; their protoxylem is exarch, with much parenchyma, and is generally completely surrounded by secondary xylem. The rather unusual secondary xylem is typically thickest towards the centre of the stem and consists of very large tracheids that are 250 μm across and possess up to 12 rows of bordered pits on their radial walls. Medullary rays, two to eight cells wide, are very common throughout the wood. The cortex has many elongated secretory canals, patches of sclerenchyma fibres and sometimes a band of thicker walled phelloderm cells in its inner zone.

Adventitious roots of *Medullosa* are commonly found in coal balls, and are up to 2.5 cm in diameter with abundant secondary tissues. The central primary xylem is exarch with up to five protoxylem points, whereas the secondary xylem is either limited to wedges outside the metaxylem or formed as a continuous cylinder. Lateral roots originate from the pericycle opposite the protoxylem points.

Leaves of the medullosans are very large with many leaf traces entering them from the stem. They originated in a regular phyllotaxis but the number of bundle traces entering them varies as does their source; some leaves received traces from two or more stem bundles. Many of the common Upper Carboniferous compression leaves that were once thought of as fern fronds are medullosan with *Alethopteris, Neuropteris, Odontopteris* and *Linopteris* probably being the best known. Though very large and pinnately divided in life, they fragmented so much before fossilisation that only small fragments or single pinnules are usually found. Even so, they can still be distinguished on their venation and the mode of attachment of the ultimate segments. Occasionally, larger pieces of leaf are discovered and there is evidence that some at least had basal stipules of the form normally described as *Cyclopteris*.

Medullosan ovules are quite common and preserved as petrifactions, compressions and casts. They are radially symmetrical although some are slightly flattened. *Pachytesta* is the generic name we give to the best known of the petrified ovules, and there are currently fourteen species which range throughout the Upper Carboniferous (Taylor, 1965). *Hexapleurosperma, Stephanospermum, Rhynchosperma* and *Albertlongia* are closely related genera of similar age. *Pachytesta* ovules range from less than a centimetre to about 11 cm in length. The nucellus is attached basally to the integument although it tightly fills the available space. The integument is distinctly three-layered like those of living cycads and *Ginkgo*. It consists of an inner single-celled layer (endotesta or inner sarcotesta), a middle fibrous layer (sclerotesta) and an outer parenchymatous layer (sarcotesta). The sclerotesta in all of the species except *P. gigantea* has three longitudinal ridges. Both the nucellus and the integument are profusely vascularised, having about 25 and 42 strands respectively. Pollination was effected through the integumental micropyle and a simple pollen chamber at the distal end of the nucellus. Cellular megagametophytes have been found in three species of *Pachytesta*. *P. hexangulata*, has three archegonia at the distal end of its megagametophyte.

The genus *Trigonocarpus* is used for seed (or ovule) casts and compressions of the *Pachytesta* type. Many are essentially internal casts of the integumental sarcotesta layer showing three ridges at their distal ends. They often show a small pimple somewhere on their surface which is evidence of how they were formed. The seeds must have been a valuable source of food for many animals if they could get through the protective sclerotesta. An empty seed if filled with fine sediment could result in a *Trigonocarpus* cast, with a pimple formed from a partial cast of the entry channel made by the animal through the sclerotesta.

Very few ovules have been found attached to medullosan foliage, and several of these reports are rather unconvincing and probably erroneous. It is sometimes actually rather difficult to distinguish between ovules and pollen organs when they are preserved as compressions or impressions. The limited number of attached ovules does, however, show their distribution on the leaves to be variable (Stidd, 1981). Some are attached terminally as in *Neuropteris heterophylla* and *Alethopteris davreuxii*, whereas others are attached adaxially as in some other species of *Alethopteris*.

Medullosan pollen organs are varied, being grouped together and assigned to the group principally on their pollen. Virtually all of them produced bilateral, monolete pollen that would be called *Monoletes* if found dispersed. These distinctive grains are 100–600 μm long with a bent monolete proximal suture and two longitudinal distal grooves. The walls consist of an inner homogeneous layer and an outer mesh-like sexine zone of sporopollenin units that expands during wall development (Figure 10.5). Taylor (1978) has suggested that the wall may have been adapted in this way to contain substances recognisable to insects, possibly promoting insect pollination in the medullosan pteridosperms.

The pollen organs are large synangiate structures, occasionally up to several centimetres in diameter (Figure 10.3). Millay and Taylor (1979) list 15 genera, some based on petrified material and others on compressions. Nevertheless they all consisted basically of a ring of radially or bilaterally symmetrical sporangia that are usually curved around a central hollow. They were either solitary, aggregated or compound in being composed of developmentally fused synangia. *Codonotheca* was one of the simplest, having a ring of partially fused sporangia.

Figure 10.3: Medullosan Pollen Organs. A: *Codonotheca cauduca* with a simple ring of six basally fused pollen sacs. B: *Parasporotheca leismanii* with a pinnate branching system bearing aggregates of curved synangia. The cut section shows the pollen sacs (black) alternating with lacunae. C: *Halletheca reticulatus* with five pollen sacs fused to each other and basally to a central fibrous zone. D: *Whittleseya* sp. with a stack of flattened synangia. E: *Potoniea* sp. synangium with a cut-away exposing radial clusters of elongated pollen sacs arranged concentrically. (A, After Sellards, 1903; B, after Dennis and Eggert, 1978; C, after Taylor, 1971; D, after Millay and Taylor, 1979; E, after Stidd, 1978)

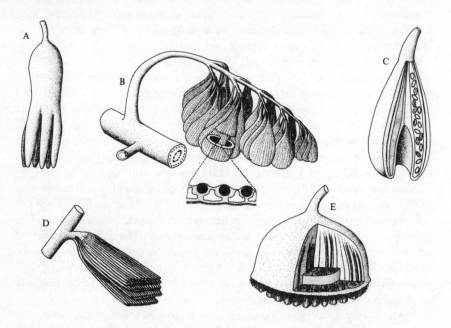

Halletheca synangia were more completely fused surrounding a basal fibrous zone and a distal hollow; the sporangial cavities alternated with large lacunae and all were embedded in a thick ground tissue. *Parasporotheca* had clusters of curved scoop-like synangia. *Whittleseya* had flattened synangia aggregated into stacks.

Dolerotheca was perhaps the most complex. It has been described as a campanulum characterised by radiating pairs of sporangia. This arrangement was most probably formed by the reduction of a small planated branching system accompanied by the folding and fusion of tubular synangia (Figure 10.4).

Both *Parasporotheca* and the bell-shaped *Potoniea* contain pollen grains that are very different from the *Monoletes* type of the other fructifications. Those of *Parasporotheca* are monolete but have two vestigial sacci, one at each end of the grain. *Potoniea* unlike all the others has radially symmetrical grains with trilete sutures. The wall development of these grains was also very different

Figure 10.4: The Evolution of the Medullosan Pollen Organ *Dolerotheca*.
A: The presumed ancestral condition with fertile bifurcating pinnae.
B: Back-to-back fusion of sporangia depicts the origin of the paired rows
prominent in *Dolerotheca*. The rachises and their vascular bundles
correspond to the radiating septa and the vascular system in *Dolerotheca*.
C. Restoration of *Dolerotheca* with cut-away portion showing the internal
structure of sporangial rows embedded in a solid mass of parenchyma and
sclerenchyma. Dehiscence grooves at the distal face correspond in position
to vertical plates of sclerenchyma. (After Dufek and Stidd, 1981)

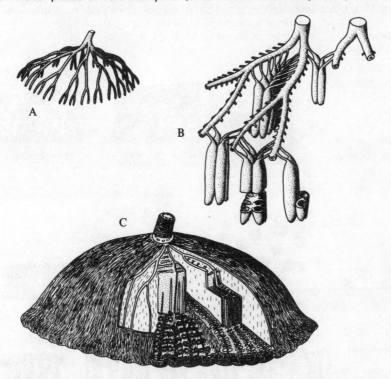

from that of *Monoletes*. Although both types received some sporopollenin from
tapetal secretion, that of *Potoniea* received a great deal more (Figure 10.5). The
outer sexine layer therefore resulted from a combination of lamella exfoliation
and tapetal activity (Taylor, 1982a).

Pollen organs are usually found as isolated structures. Thus, as with the
ovules, most of their parent plants are unknown. Stidd (1981) suggests that
Alethopteris and *Neuralethopteris* are the foliage types of most monolete pollen-
producing organs. For example *Dolerotheca* has been found attached to a
Myeloxylon frond with *Alethopteris* pinnules, taking the place of a penultimate
pinna on the frond (Ramanujan, Rothwell and Stewart, 1974). Others were
borne along a fertile axis either alternately as in *Schuetzia* or in pairs as in
Codonotheca. In contrast to the *Monoletes*-producing pollen organs,

Figure 10.5: Stages in the Development of Pteridosperm and Cycad Pollen. A–D: *Monoletes* from *Dolerotheca*; A, homogeneous layer probably constructed of lamellae; B, early differentiation between sexine and nexine (wavy lines); C, expansion of sexine layer; D, mature sporoderm with alveolate sexine and lamellate nexine. E–H: *Potoniea* pollen; E, homogeneous lamellated layer; F, suggested apposition of tapetal sporopollenin; G, accumulation of tapetal sporopollenin; H, mature sporoderm with lamellated nexine and sculptured sexine. I–M: Pollen from the living cycad *Ceratozamia*; I, initial ectosexine component; J, appearance of radial tubules representing initial sites of sporopollenin accumulation; K, continued development of sporoderm showing expanded alveolate organisation; L, thickened ectosexine; M, initial appearance of lamellated nexine. (After Taylor, 1982a)

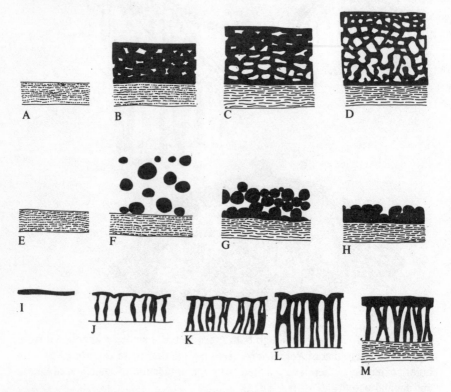

Potoniea may have been borne on *Sutcliffia* leaves with *Linopteris* foliage. If this arrangement is firmly established, it will probably provide the basis for establishing a separate taxonomic unit. The evidence provided from studies of spore wall development certainly supports this argument. Stidd (1981) has in fact suggested that the Medullosales might eventually contain three families: the Medullosaceae containing all the *Monoletes*-producing plants, the Sutcliffiaceae and a family based upon *Parasporotheca*. However, taxonomic grouping on the basis of pollen-producing organs does not seem the soundest way of proceeding,

Figure 10.6: *Callistophyton*. A, B: Fertile pinnule segments showing the mode of attachment of the pollen organs (at A) and ovules (at B); the right-hand ovule is drawn in mid-longitudinal section with details of the megagametophyte omitted. C: Reconstruction showing the general growth habit. (A, B, After Rothwell, 1981; C, after Rothwell, 1975)

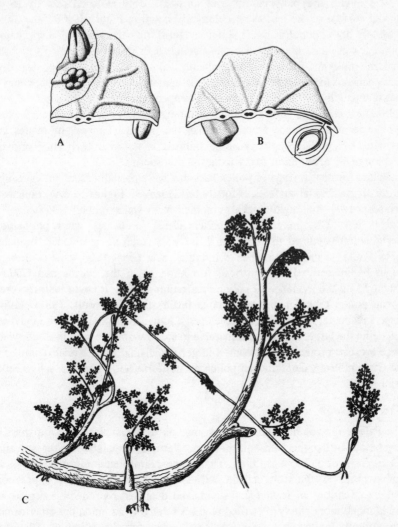

and we should wait until we have a much clearer idea of the total biology of the medullosan plants.

Callistophytales

As proposed by Rothwell (1981), this is a comparatively small group of shrubby, scrambling plants (Figure 10.6). They branched profusely, were eustelic, rooted adventitiously and had very divided fern-like leaves with sphenopterid pinnules that developed circinately. Their simple morphology is, however, coupled with a reproductive biology unlike that of the Lyginopteridales and Medullosales.

The ovules are platyspermic and most probably attached directly to the abaxial surface of the leaf as speculatively shown in Figure 10.6B. They were definitely not in cupules, but it is not known if the parent plant had specialised fertile leaves. The nucellus was only attached to the integument by its basal chalaza (stalk) and formed a pollen chamber at its distal end. Free nuclear division occurred in the megaspores of small immature ovules forming coenocytic megagametophytes. Two archegonia developed on either side of a central tent-pole. The archegonia quite probably had tubular necks as in those of many extant gymnosperms. Evidence from pollen in the pollen chambers of ovules and remnants of a pollen drop suggest that pollination was at an early stage of ovule development; again as in many extant gymnosperms.

Pollen formed in rings of pollen sacs that were partially fused into synangia borne on the abaxial surfaces of totally fertile leaves (Figure 10.6A). Pollen sac dehiscence was by longitudinal slits on their inner walls to liberate *Vesicaspora*-like pollen. This monosaccate pollen clearly indicates wind pollination. Germination occurred through the distal wall resulting in slender, branched tubes similar to the pollen tubes of living cycads, *Ginkgo* and some conifers. It might be interpreted as haustorial formation as in the cycads and *Ginkgo*; although, on the evidence of pollen maturation studies, it could well represent a true pollen tube for siphonogamous fertilisation (Rothwell, 1981). Before dispersal occurs, the pollen had developed a four-celled axial row with the distal cell being the largest. This is comparable to the row of three seen in *Pinus* where there are two prothallial cells and a larger antheridial cell. It is also similar to the Carboniferous Cordaitalean pollen grain *Florinites*, which has a five-celled axial row.

Glossopteridales

The Palaeozoic southern land mass known as Gondwanaland had a distinctive vegetation during the Carboniferous and Permian populated by the so-called *Glossopteris* or glossopterid flora. The gymnosperms that dominated the vegetation are still very far from satisfactorily known. Most remains are impressions and compressions of leaves and wood, and detached reproductive organs are found much more rarely. Petrified wood is relatively common but other organs are virtually unknown as petrifactions except for the silicified Australian material described by Gould and Delevoryas (1977).

The flora derives its name from the commonest fossils; lanceolate leaves that we call *Glossopteris* with their distinct midribs and reticulate venation. *Gangemopteris* is another common leaf distinguished from *Glossopteris* by its larger size, lack of a midrib and different epidermal cell patterns. Both can be separated from the other similar but less common leaves by the anatomy of the

Figure 10.7: Reconstructions of *Glossopteris* and its Reproductive Organs.
A: *Eretmonia* with its two clusters of pollen sacs. B: *Lidgettonia* with two
rows of ovule-bearing capitula. C: The cone-like *Kendrostrobus cylindricus*.
D–G: *Dictyopteridium*; D, leaf and capitulum; E, capitulum from the
underside showing enclosed ovules; F, TS of capitulum showing the ovules
and the multicellular hairs. G: LS of ovule showing the integument clothed
with multicellular hairs and separated from the enclosed nucellus. H:
Ottokaria bengalensis with the capitulum stalk fused to the upper surface of
the leaf. I: *Denkania indica* with its cupule-like capitula. J: Reconstruction of
a tree (other reconstructions show the tree with many more branches and
leaves). (A, After Surange and Maheshwari, 1970; B, C, after Surange and
Chandra, 1974; D–G, after Gould and Delevoryas, 1977; H, after Schopf,
1976; I, after Surange and Chandra, 1973; J, after Pant and Singh, 1974)

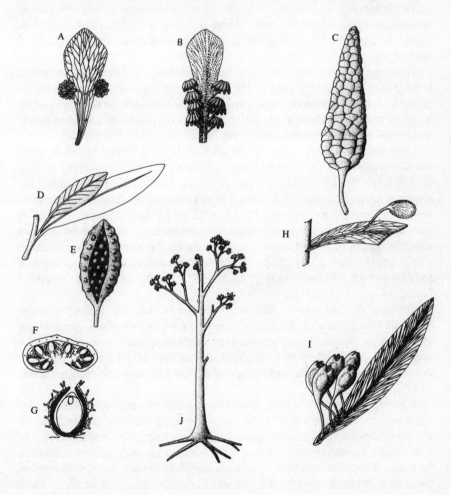

bundles in their midribs. There are probably a great number of species of these leaf genera; at least 80 in *Glossopteris* (Lacey, Van Dijk and Gordon-Gray, 1974).

The *Glossopteris* plant is commonly accepted to be a tree (Figure 10.7J) as its roots, called *Vertebraria*, are known to have been up to 150 cm in diameter (Schopf, 1976). Regular growth rings in the wood of its roots and of its stems suggest a seasonal periodicity of growth. The large numbers of isolated leaves and to a much lesser extent reproductive organs also suggest that the plants were deciduous. Studies on the fossil floras of varved sediments support this hypothesis as numerous leaves are found in the autumn/winter sediments whereas they are absent in the spring/summer sediments.

The leaves were probably arranged either alternately or in tight helices thereby appearing almost in whorls. There is also a possibility that their growth pattern resulted in a long and short shoot arrangement as in *Ginkgo* (Gould and Delevoryas, 1977).

There are several different types of both ovuliferous and staminate structures belonging to this group of plants. They were borne separately, although we do not know if they were on different parts of the same plant or on different plants. Many of these reproductive organs are also poorly preserved and imperfectly understood so some reconstructions are of necessity very interpretive.

More than 30 genera of fructifications have been attributed to *Glossopteris* and *Gangemopteris*. The ovulate organs all have a basic dorsiventrality in their stalked fertile head (capitulum). The capitulum may be divided or entire but always bears its ovules on the lower surface, which may be infolded. Some have been found attached to fertile, leaf-like bracts, and Gould and Delevoryas (1977) have shown at least one to be attached to true leaves. They may be attached either in the axil of a bract or to its upper surface. The whole unit can be called a fertiliger (Schopf, 1971). Fertile bracts may sometimes bear several capitula as in *Lidgettonia* or *Denkania* (Figures 10.7B,I); Schopf called these compound fertiligers.

The more divided capitula, like *Arberia*, seem to have been axillary whereas the entire forms, such as *Dictyopteridium*, were surface attached. These plants are therefore unique among gymnosperms in having leaf-borne reproductive organs. It is highly likely that they originated in the axillary position and were later incorporated on to the leaf surface as in *Tilia* and other angiosperms (Pant, 1982).

Not surprisingly the capitula show a wide range of variation in the number and arrangement of their ovules (Figure 10.7D–I). Some, like *Scutum*, bear up to 75 randomly spaced ovules. Others, like *Senotheca*, have reduced capitula with closely pressed ovules in two rows. *Denkania* has only one ovule on each capitulum which through folding appears very cupule-like. In *Dictyopteridium* the capitulum edges curve over to enclose the ovules in a mass of hairs, while in *Ottokaria* a filamentous structure fills the space between the ovules. Gould and Delevoryas (1977) described similar cellular filaments in their petrified

Australian material. The ovules themselves were sessile and directed towards the centre of the enclosing capitulum. Not only were the spaces between the ovules filled with filaments but conical webs of filaments extended out from each micropyle. This organisation suggests that pollen-drop liquid may have completely filled the structure directing the pollen towards the micropyles as its meniscus withdrew. Pollen was occasionally found in the pollen chambers but no embryos have been observed.

Staminate structures of the glossopterids are not so clearly understood as the ovuliferous scales and it seems probable that many have been confused with these scale leaves. Isolated pollen sacs, given the names *Arberiella, Lithangium* and *Polytheca*, are all uniloculate and borne on slender stalks. The pollen grains are bisaccate in *Arberiella*, monolete in *Lithangium* and trilete in *Polytheca*; monocolpate pollen has also been described from the latter two.

Pollen sacs are borne on branched axes arising from the surfaces of leafy bracts. The commonest are called *Glossotheca* or *Eretmonia* (Figure 10.7A). Some cone-like structures have also been described. *Squamella* seems to be a cone of *Eretmonia*-like scales bearing pollen sacs of the *Arberiella* type (White, 1978). *Kendostrobus* (Figure 10.7C) is a rather different type of cone that has been suggested to belong to the glossopterids. It has groups of pollen sacs helically arranged on an axis with no associated bracts. The pollen sacs are covered with minute pores but show no evidence of a dehiscence mechanism. Its monolete spores are like those of the *Lithangium* type of dispersed pollen sac.

The rather varied nature of the ovules and their modes of attachment make it difficult to recognise the ancestors of the glossopterids with any precision. Indeed it is by no means certain that it is a single group of plants. They most probably evolved from the earlier Northern Hemisphere pteridosperms (Gould and Delevoryas, 1977; Pant, 1977; Rigby, 1978), although Schopf (1976) has suggested that they might have arisen from the Cordaitales.

Much more has been written about the relationships of the glossopterids to their descendants. They have been linked to every major group of gymnosperms and to the angiosperms. Unfortunately there is no clear evidence for them to be the ancestors of any geologically younger group. The claim of them to be ancestors of the angiosperms can certainly be rejected as it was based upon the erroneous supposition that the fructifications were bisexual and on their possession of reticulately veined leaves. It seems that they are best thought of as a highly successful and diversifying group of gymnosperms that dominated large areas of vegetation in its time and environment, but one that subsequently became extinct either through climatic change or the migration of more competitive plants into its habitats.

Peltaspermales

This is a small group of Triassic plants known from Greenland and South Africa. Their general morphology is unknown except for fragments of stem and bipinnate leaves with broadly based, open-veined pinnules (Thomas, 1955).

The pollen organs, *Antvesia*, were pinnately branched to give short dichotomously dividing laterals which bore two rows of pendulous pollen sacs on their ultimate branches (Townrow, 1960).

The ovuliferous structures that these plants bore are very distinctive and suggest their name, *Peltaspermum*, and the name of the group. Between 10 and 20 ovules were attached in a ring on the underside of peltate, disc-like organs that were helically arranged on a central stalk. Unfortunately we have no idea how they were borne on the parent plants. Each ovule was about 7 mm long with an obviously elongated and curved integumental tip (Harris, 1937).

Meyen (1984) has recently included some Upper Permian plants with simple and once pinnate *Tatarina* leaves on the basis of their compact heads of several peltate organs. He also suggested that the group evolved from the Carboniferous Corystophytales and gave rise to both the Ginkgoales and the Leptospermales and possibly even the Caytoniales.

Corystospermales

These plants are also Triassic, but only known from Gondwanaland. Most were probably small plants with pinnately divided and dichotomously veined leaves, although some large trunks (*Rhexoxylon*) were up to 50 cm in diameter. No parts of the plants have ever been found attached, although they are reliably linked together on the evidence of both association and similar epidermal characters.

The pollen-producing organ, *Pteruchus*, has alternately arranged microsporophylls on an axis. Each terminates in a flattened, rounded head that bears clusters of elongated pollen sacs on its underside. The pollen grains are bisaccate and identical to those that have been found in the pollen chambers of *Pilophorosperma* (Townrow, 1962).

The ovuliferous organs were planated with lateral branches in the axils of bracts. Stalked, cupulate ovules were borne in opposite pairs. The cupules were either recurved and helmet-shaped as in *Umkomasia*, or hemispherical as in *Pilophorosperma*, and are borne on branched axes. Each contains a single ovule whose elongated integumental tip projects out to expose the micropyle clearly (Pant and Basu, 1979).

The similarity of these ovules to those of the peltaspermaleans suggests that the corystospermalean cupule might have evolved through the overfolding of a peltate head to give a cupule; in this case with the development of only one ovule. Unfortunately no details are known of the internal structures of the ovules so there is no further evolutionary evidence here. The ovules might simply reflect similar adaptations from Palaeozoic pteridosperms rather than an evolutionary change.

Caytoniales

These were Mesozoic Northern Hemisphere pteridosperms whose dispersed organs are found in Triassic to Cretaceous rocks of North America, Greenland and northern Europe. Their overall growth habit is unknown, but the plants

Figure 10.8: Reconstructions of Caytonialean Organs. A–C: *Caytonia*. A, Cupulate-bearing axis; B, recurved cupule showing its lip-like projection; C, LS through a cupule showing the ovules and the canals connecting them to the exterior. D: *Sagenopteris*: palmate leaf with an inset of the venation of the basal portion of a leaflet. E–G: *Caytonanthus*. E, Branched fertile organ with pollen sacs; F, single structure of four fused pollen sacs. G, Pollen grain. (A, After Thomas, 1925; B, after Harris, 1940; C, after Harris, 1933; D, G, after Harris, 1964; E, F, after Harris, 1937)

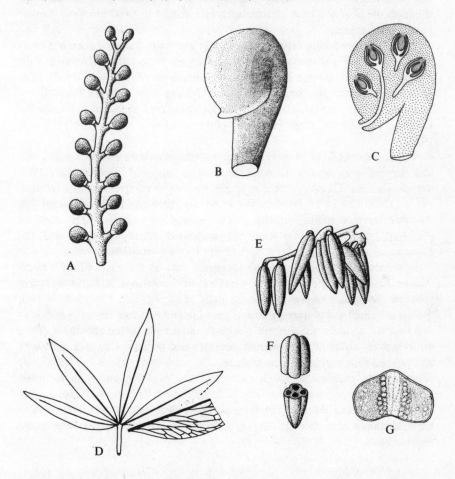

cannot have been herbaceous as they had palmate leaves (*Sagenopteris*) with three to six leaflets (Figure 10.8D). Each leaflet was up to 7 cm long, lanceolate, with a prominent midrib and reticulate venation.

Caytonia is the name given to the ovule-bearing cupulate structures that were borne in subopposite pairs on an axis about 5 cm long (Figure 10.8A). Each was nearly circular in outline, although a lip-like projection near its attachment to a stalk reveals it to be a recurved structure. Between eight and thirty platyspermic seeds developed inside and along the midvein of the different species of cupules, each borne orthotropously on a short narrow stalk. The integument has an epidermis, a distinct sclerenchymatous layer and an inner parenchymatous layer with two vascular strands. The nucellus is free from the integuments and its tip extends into a small and simple pollen chamber (Figures 10.8B,C). Large numbers of isolated ovules or seeds are often found in association with *Caytonia*, suggesting that they were dispersed from the cupules, although the absence of a fleshy sarcotesta seems to preclude their further dispersal by animals.

The pollen-producing organs have slender axes with flattened branches bearing up to three oval structures about 1 cm long, each consisting of three or four pollen sacs fused around some central parenchyma (Figures 10.8E,F). The pollen grains are quite small, being only 30 μm long, but are distinctively bisaccate (Figure 10.8G). These pollen grains have been found on and inside the *Caytonia* cupules and often inside the pollen chambers of the ovules (Harris, 1958).

The origins of all the Mesozoic gymnosperms have been debated many times and varying evolutionary schemes have been proposed. Delevoryas (1962) thought that the *Caytonia* cupules might have evolved from peltaspermalean ovulate structures by recurvature and enclosure. Stewart (1983) suggested that the two groups might equally have evolved separately from cupulate pteridosperms, and in his chart of suggested relationships showed the Caytoniales as probable descendants of the Lyginopteridales.

In contrast, Reymanova (1974) suggested that on the basis of anatomical studies these plants appear to be descendants of the Palaeozoic forms that have subsequently been grouped together as the Callistophytales. The seeds of both groups are small and platyspermic and have integuments free from the nucellus, two vascular strands, scalariform tracheids and simple pollen chambers. They do, however, differ in integumental structure and because *Caytonia* unusually has no resistant megaspore membrane. The callistophytalean ovules are of course not in cupules, although the acceptance of *Caytonia* as an incurved megasporophyll segment resembling a *Sagenopteris* leaflet could explain this apparent difference. Meyen (1984) supports the idea, linking the Caytoniales to the Callistophytales through the probability that the Peltaspermales were intermediates.

11 THE ORIGIN AND EVOLUTION OF CONIFERS

Conifers have the longest fossil record of all gymnosperms, the earliest record being in the Upper Carboniferous (Scott and Chaloner, 1983). From the Permian onwards the group seemed to diversify consistently and achieved world-wide dominance by the Mesozoic era. Only the later evolution of the angiosperms was to oust the conifers from many of their habitats. Even so, they still constitute a very important group that dominates quite large areas of the Earth's vegetation.

The living conifers represent rather diverse products of the very variable evolutionary changes that have occurred since the Permian. They have certain overall characters, but there always seem to be exceptions to them. Generally conifers are large, much branched shrubs and trees, forming pycnoxylic wood with bordered pitted tracheids. Their leaves are needle-like, scale-like or occasionally broader with many veins.

Family and generic recognition both involve a knowledge of vegetative and reproductive characters. Therefore it can be very difficult to classify many fragmentary remains. Conifers show a great deal of morphological variation, with some growing to great heights as in the 110 m-tall redwoods (*Sequoia sempervirens*), whereas others such as *Pinus mugo* form small shrubs.

Conifers seem to have had a great capacity for responding to environmental change. They share certain xeromorphic characters with the early ancestral conifers that probably evolved during the comparatively dry Permian. They have leaves with low surface-volume ratios, usually being narrow and rigid or scale-like and decussate, and with sunken stomata and thick cuticles.

Most living conifers drop their foliage in one of two natural ways and similar mechanisms can be inferred in many fossils. Either individual leaves or intact leafy shoots may be abscised. Species may be either evergreen or deciduous. Reproductive organs are similarly shed after fulfilling their role. However, as Harris (1976a) has emphasised, leaves or shoots may be prematurely lost by violent action. Therefore any reproductive organs still attached to fossil leafy shoots are likely to be immature.

Phyllotaxy is another variable character, although it has seldom been examined closely in fossil conifers. Harris has demonstrated there to be three kinds. The leaves may be attached singly in a regular helix, in pairs forming double helices, or in alternating pairs (or more) at nodes. These arrangements may be further subdivided according to the numbers of oblique rows (parastiches) arising to the right and left of the organ.

Plants may be either monoecious or dioecious. The majority have terminal, compound ovuliferous cones and simpler, lateral pollen-producing cones, although there are notable exceptions, e.g. Taxaceae and Podocarpaceae. The ovuliferous cones have a spiral of fertile ovuliferous scales, each borne in the axil of a bract. The ovuliferous scales have one or more inverted ovules fused proximally on their abaxial surfaces. A pollen-drop mechanism is usually employed to capture pollen grains. All modern conifer pollen germinates to form pollen tubes that grow to the archegonia in the apical region of the embryo sac. As there is no nucellar breakdown, the pollen tube has to grow through this tissue to achieve fertilisation.

Most of the controversy about the evolution of the conifers and their early diversification centres around one main issue, the compound nature of the ovuliferous cone together with the inverted nature of its ovules. Conifers must have had spore-producing ancestors and it is quite possible that they might have had their origins in the progymnosperms (see page 139). If this were the case, there was likely to have been some intermediary group because the enormous difference in reproductive biology between these two precludes any possibility that they are closely related. Much of the recent literature suggests that such a group might have been the Cordaitaleans. This argument is based largely upon the work of the Swedish botanist Florin (1950a,b, 1951, 1958). Although recent re-evaluation of his material has cast doubt on some of the published details, his overall interpretations are still generally accepted.

Cordaitales

Cordaitanthaceae

The Cordaitales first appeared in the Lower Carboniferous of Eurameria, diversifying and persisting through to the Permian. The majority of species were probably arborescent, growing in the drier areas above the water table (Figure 11.1A). *Cordaites* itself grew monopodially, forming trunks supported by an extensive pycnoxylic secondary xylem of tracheids and parenchymatous medullary rays. The primary xylem is a ring of exarch bundles surrounding a relatively large and chambered pith. The leaves were the first organs, called *Cordaites*; although the name is now commonly used for the whole plant. They were large, multiveined and strap-shaped, sometimes reaching about 1 m in length and 15 cm in width. Their veins had parenchymatous bundle sheaths with large patches of fibres above and below them. Stomata were most likely to have been on both surfaces, although some species have been described with them only abaxially. The guard cells are slightly sunken and the epidermal cuticle is thick.

Some plants were of a rather different appearance. Rothwell (1982a) and Rothwell and Warner (1984) have described an Upper Carboniferous species as shrubby, with leaves ranging from the typical strap shape to needle-like forms with single veins (Figure 11.1C). The latter are on those stems with the smallest pith diameters of about 1 mm and characterise buds and branch bases.

Cridland (1964) has described a further form based upon anatomically

Figure 11.1: Reconstructions of Cordaitean Plants. (A, After Scott, 1909; B, after Cridland, 1964; C, after Rothwell and Warner, 1984)

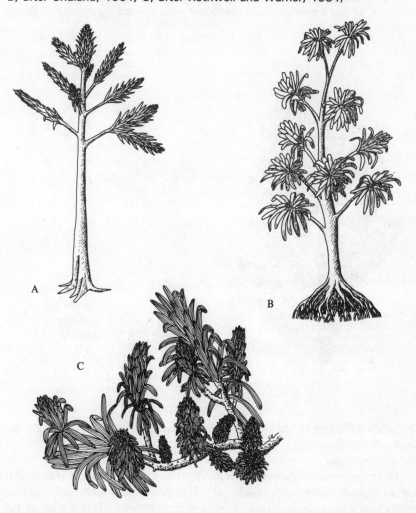

preserved remains. Their roots, called *Ameylon*, had a medullated primary xylem derived from the solid protostele of the young roots, periderm, aerenchyma and lenticels. Lateral roots were formed only on one side of the main roots and were often in clusters. All of these features are found in the roots of living mangrove plants, suggesting that some species of *Cordaites* had such a root system and grew in that kind of environment (Figure 11.1B).

Both pollen-producing and ovuliferous fructifications are so similar that most of them are called *Cordaianthus* (Figure 11.2). They were probably borne between the leaves on the smaller branches, although none has been found attached. Each is a compound structure with lateral shoots in the axils of bracts.

Figure 11.2: *Cordaianthus*. A: Cordaite branch with lateral fertile shoots. B: Portion of fertile axis bearing stalked ovules on its secondary shoots. C: Portion of similar fertile axis with secondary shoots bearing spirally arranged sterile scales and fertile scales with terminal pollen sacs. (A, After Grand'Eury, 1877; B, after Taylor and Millay, 1979; C, after Delevoryas, 1953)

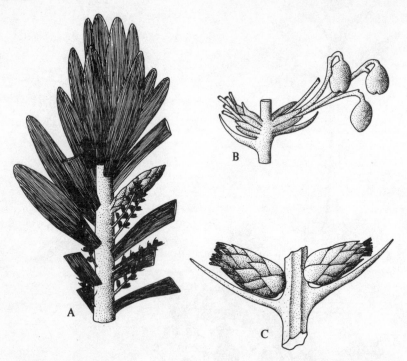

The whole structure is often called a compound strobilus and the laterals are described as cones. The cones have helically arranged scales, of which only a few terminal ones are fertile.

The fertile scales of the pollen-producing cones have a terminal cluster of a few elongated pollen sacs. Each sac is about 1 mm long and dehisces longitudinally towards the centre of the cluster (Rothwell, 1977). Pollen grains are of the monosaccate type that would be called *Florinites* if found dispersed. During pollen development the saccus separates from the central body, the corpus, remaining attached only proximally and distally. The Lower Carboniferous *Gothania* has very similar pollen-producing organs that form larger monosaccate pollen grains. This pollen, called *Felixipollenites*, has its saccus connected only to the proximal surface of the corpus where there is a conspicuous monolete or trilete suture. Ultrastructural studies show that it was the inflation of the outer portion of the sexine layer that resulted in the formation of the saccus. The saccus in these grains, and in the later conifer grains, could have had one or more of several possible functions. Grains with sacci could have been more buoyant and carried further in the air currents, or the saccus, acting as a float, might have orientated the grains in the pollen drop. It could also have controlled water loss or even acted as an insulating layer for the centrally suspended corpus.

Taylor and Daghlian (1980) have suggested through their ultrastructural studies that within the Cordaitales there was a distinct reduction in size that was accompanied by a shift in polarity. Germination in *Felixipollenites* was from the proximal surface, whereas in the later *Florinites* it was most probably distal. Reduction in grain size would presumably have been accompanied by an increase in the numbers produced, thereby resulting in a greater chance of pollination and fertilisation. The reduction in size could be the result of the evolution of haustorial pollen tubes that grew into the nucellar tissue. The facility to absorb metabolites from the nucellus would reduce the need for pollen grains to transport food for the development of their gametes.

The terminal fertile scales of the ovuliferous cones are divided to terminate in two or more pendulous ovules. Taylor (1981) noted that very few structurally preserved seed cones have been discovered, suggesting that the ovules were exposed for only a very short time during pollination before being shed from their stalks. Stratigraphical evidence of the distribution of various species shows that the ovule stalks were larger and more divided in the earliest species, becoming more reduced in the later ones. Isolated seeds are bilaterally symmetrical, with two prominent wing-like extensions of the integuments. *Cardiocarpus* and *Mitrospermum* are common and well-known genera of such petrified seeds, and *Saramopsis* is the name given to similar compression fossils. Their integuments are three-layered, with an inner uniseriate parenchyma, a middle sclerotesta of thickened cells and an outer parenchymatous sarcotesta. A single vascular bundle enters the seed to terminate at the base of the nucellus. It gives off two strands which pass up through the sarcotesta wings. In *Mitrospermum* these two strands further divide to form a fan of bundles within the sarcotesta wing. The nucellus is free from attachment to the integuments and has a simple distal pollen chamber.

Vojnovskyaceae

These Permian plants, often included in a separate order (Vojnovskyales) were shrubby, or possibly arborescent, simple gymnosperms (Figure 11.3). These, and not *Cordaites*, dominated the Angara flora. They were originally thought to have had bisexual fructifications, but Meyen (1984) has re-examined the type material of *Vojnovskya* and recognises there to be only ovules present. He considers the ovuliferous structures to be closely comparable to those of *Cordaianthus*, differing only in having conical axes and more leaf-like bracts. For this reason, and for the fact that *Vojnovskya* is found in association with *Cordiates*-like leaves, he includes them in the same order. The male fructifications were probably of the form referred to *Kuznetskia*. They had much-branched pollen-producing organs and liberated pollen that had a small, single saccus.

Figure 11.3: Fructifications of Permian Angaran Vojnovskyaceae. A: Fertile axis of *Vojnovskya paradoxa* with bracts and multi-ovulate structures covered with minute scales and bearing elongated interseminal scales with interspersed ovules on its flattened apex. B: Associated seed of *V. paradoxa*. C: Microsporangiate structure called *Kutznetskia planiuscula*. D: Group of enlarged sporangia of *K. planiuscula*. (After Meyen, 1984)

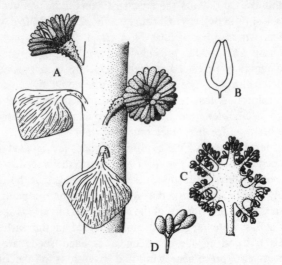

Nevertheless, even with Meyen's reinterpretations, the Vojnovskyaceae remains somewhat problematic. At best it was probably a regional and short-lived group that gave probably no surviving important descendants.

Rufloriaceae

Other Angaran plants with cordaitalean characters showing further variation within the group have been grouped as a separate family, the Rufloriaceae by Meyen (1984) (Figure 11.4). *Rufloria* leaves have been found in association with both ovuliferous and pollen-producing organs. The earliest known ovuliferous forms, included in the middle Carboniferous genus *Krylovia*, have their ovules borne on variable-length axes. The length appears to decrease in successively younger forms until they are very similar to the Upper Carboniferous species of *Gaussia*. At the same time the ovule stalks appear to become increasingly longer. In contrast to this apparent reduction series, the lower Permian *Bardocarpus* has an elongated axis with short stalked ovules borne in a compact spiral along its entire length.

Cones included within *Pechorostrobus* can be either ovuliferous or pollen-producing. Both had their ovules or pollen sacs attached directly to their central axes. In contrast to this arrangement of pollen sacs, *Cladostrobus* had many of them attached in pairs to the abaxial sides of the pedicels of microsporophylls.

The ovuliferous cones were therefore of the same basic design, whereas the pollen-producing ones were of two basic kinds, including variations on the

Figure 11.4: Fructifications of Angaran Upper Palaeozoic Rufloriaceae. A: Reconstruction of *Bardocarpus depressus*. B, C: Reconstruction and vertical section of *Gaussia scutellata*. D, E: Reconstruction and vertical section of *G. cristata*. F, G: Reconstruction of extreme members of a series of *Krylovia sibirica*. (After Meyen, 1984)

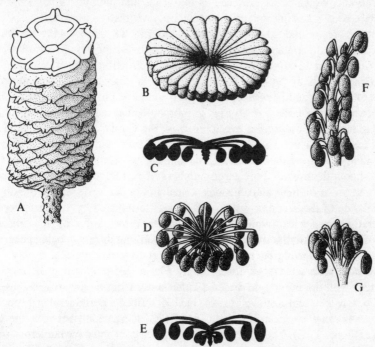

themes. Meyen (1984) therefore suggested that the Cordaitaleans, like the later Pinales, possessed a high variability of male fructification structures and 'mobility' of their pollen sacs.

Voltziales

During the late Carboniferous and the early Permian, when the Cordaitales were most widespread, conifers were making their first appearance. Current ideas group these early conifers into three families, the Walchiaceae, the Voltziaceae and the Cheirolepidiaceae. The order as a whole is thought to be ancestral to all the living conifers. It is the details of their evolutionary pathways that are still very much open to debate.

Walchiaceae

The Palaeozoic Walchiaceae are typified by *Walchia*, *Ernestiodendron* and *Ortiseia* which are known from Upper Carboniferous and Permian localities.

The names *Walchia* and *Lebachia* have been used in different ways, with many authors restricting their use of *Walchia* to sterile shoots. Clement-Westerhof (1984) has summarised their nomenclatural history and concluded that *Lebachia* should be included as a synonym of *Walchia*. This logic is accepted here, so *Walchia* and Walchaiceae are used as the genus and family. *Culmitzschia* can be retained as a suitable genus for exclusively vegetative remains.

Both *Walchia* and *Ernestiodendron* were probably trees with branching patterns and leaf form and arrangement very much like the living Norfolk Island Pine (*Araucaria heterophylla*). Their unisexual fructifications were cone-like and borne terminally on leafy shoots. Florin first described the cone morphologies of *Walchia* and *Ernestiodendron* from compressions. His interpretation of these cones as the earliest in a series of morphological reductions is essential to his classical ideas on conifer evolution from the Cordaitales (Florin, 1951).

Walchia had a compact, elongated ovulate cone with helically arranged, tangentially flattened shoots borne in the axils of bifurcating bracts. These flattened shoots themselves bore large numbers of sterile scales on their abaxial sides and fewer on their adaxial sides. Usually only one scale was fertile on the adaxial side of the secondary shoot. Each was terminated by a solitary, erect, bilaterally symmetrical ovule with an outwardly directed micropyle. *Ernestiodendron* has very similar ovulate cones to *Walchia*, but differs in being generally more lax and in having lateral shoots with three to seven ovules and only very few to no sterile scales. As described by Florin, the ovules may be erect or inverted, with the micropyle directed either away from or towards the base of the shoot. Mapes and Rothwell (1984) have described a permineralised species, *Walchia* (*Lebachia*) *lockardii*, with characters intermediate between the two genera (Figure 11.5). The cones are relatively compact and have numerous scale leaves on their lateral fertile shoots like *Walchia*. In contrast, the fertile shoots may have more than one functional terminal ovule and these ovules are inverted with inwardly directed micropyles. Both of these are *Ernestiodendron* characters.

Ortiseia (Figure 11.6) is similar to *Walchia* and *Ernestiodendron*, and supports Florin's concept of the compound nature of coniferous ovuliferous cones. Clement-Westerhof (1984), however, disagrees with Florin's notion of stalked ovules, believing all three genera to have one inverted ovule directly attached to the abaxial surface of a megasporophyll, the abaxial surface being part of the adaxial side of the dwarf shoot.

The terminally attached pollen-producing cones of the Walchiaceae are very similar to those of modern conifers. They have helically arranged flattened scales, or microsporophylls, each with a stalk, an upturned lamina-like distal portion and two peltate pollen sacs on the abaxial surface. These cones differ from those of the Cordaitales, for those of the latter have terminally erect pollen sacs intermixed with sterile scales.

Pollen sacs referable to *Walchia* have been found to contain monosaccate pollen grains of the *Potonieisporites* type (Rothwell, 1982b). Similar grains have been found in the pollen chambers of *W. lockardii*, suggesting that this cone

Figure 11.5: *Walchia (Lebachia) lockardii*. Reconstruction of a segment of cone showing general features of axis, bract and fertile shoot. The reflexed ovule is drawn in section to show integument, nucellus and megaspore membrane. Integument and vascular tissue of axis in black. (After Mapes and Rothwell, 1984)

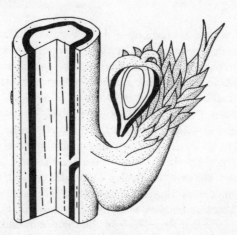

Figure 11.6: Ovuliferous Dwarf Shoot and Ovule of *Ortiseia*. A: Dwarf shoot with ovule, abaxial view. B: Dwarf shoot, ovule detached, adaxial view. C: Ovule, obverse view. D: Longitudinal section through the ovule (i, integument; m, micropyle; mm, megaspore membrane; n, nucellus cuticle; pc, pollen chamber). (After Clement-Westerhof, 1984)

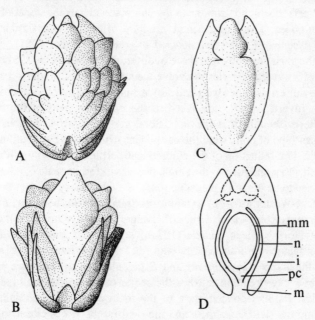

probably employed a pollen-drop mechanism. Mapes and Rothwell (1984) also interpret histological features of the integument, nucellus and megagametophyte as indicative of ovule immaturity, being similar in developmental state to those of many living gymnosperm ovules just after pollination.

Voltziaceae

The Permian Voltziaceae is made up of large numbers of diverse shoots and ovuliferous cones that have been generally accepted to show evolutionary intermediate form between the members of the Walchiaceae and modern conifers. They all have reduced, bilaterally symmetrical short shoots, or single large ovuliferous scales in the axils of bracts.

The late Permian *Pseudovoltzia*, as described by Florin (1951), has bilaterally symmetrical lateral shoots of five fused sterile lobes that were themselves fused to their sterile subtending bracts (Figure 11.7B,C). Three ovules are fused via their short stalks to the central three lobes. Schweitzer (1963) subsequently showed the ovules to be directly attached to what he regarded as the middle portion of a true scale. Clement-Westerhof (1984) reinterpreted the vascular bundles supplying the scales of its dwarf shoots as inverted. This suggests that *Pseudovoltzia*, like *Walchia*, had its ovules on the abaxial surface of its megasporophylls (= part of the adaxial side of the flattened dwarf shoot). *Glyptolepis* from the Permian and Triassic has a recurved ovule on either side of the five or six sterile lobes making up the lateral shoots. The Permian *Ulmannia* has the most reduced structure with large flattened scales in place of the lateral shoots (Figure 11.7D). Each has a single inverted ovule on its upper surface which is similar to the characteristic arrangement of the extant araucarian cones. The integumental structure of both is also similar in having extensive systems of tightly compacted sclereids (Stockey, 1981).

From the variation shown in these ovuliferous cones Florin (1951) postulated his ideas of a reduction series to evolve the conifer cone. Modern conifers have compact ovuliferous cones with helically arranged ovuliferous scales in the axils of bracts. Inverted ovules are fused to the upper surfaces of the ovuliferous scales. Florin thought that the lax lateral fertile branches of the compound cordaitalean type of cone were the equivalent structure to the ovuliferous scales of conifers. The reduction of the cordaitalean fertile branch to the *Walchia* type of flattened shoot and then to the single large ovuliferous scales of the *Ulmannia* form seemed to be the most plausible route of evolutionary change. Schweitzer's and Clement-Westerhof's observations on *Pseudovoltzia*, however, create a gap in the otherwise complete sequence of changes, although they do not completely invalidate Florin's ideas. Harris (1976b) argued that *Pseudovoltzia* shows a close resemblance in both external and internal organisation of its scales with those of the living *Cryptomeria* and *Sciadopitys*. It is therefore easier with Schweitzer's interpretation to trace the sequence to the various kinds of living cone scales. Parallel condensation of the main axes of the cordaitalean cone would bring the flattened shoots and the ovuliferous scales closer together and

Figure 11.7: Reconstructions of Ovule-Scale Complexes. A: *Cordaianthus pseudofluitans*; lateral shoot with spirally arranged scales. B, C: *Pseudovoltzia liebiana*; B, bract and scale with two of its three ovules; C, longitudinal section showing the vascular tissue. D: *Ulmannia bronnii*; single ovule on a large flattened scale. (A, After Florin, 1944; B, C, after Schweitzer, 1963; D, after Florin, 1951)

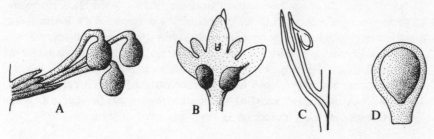

Figure 11.8: Possible Origin of Planated, Unifacial Coniferalean Seed Scale with Adaxial Seed Position from Radial Axillary Complex Consisting of Seed Stalks with Abaxially Attached Seeds (after Meyen, 1984). A: Conifer of the *Sashinia* and *Ernestiodendron* form. B: Hypothetical stage, with the seed stalks basally fused into a flattened organ. There are no seeds on the abaxially facing bracts. C: Further fusion and flattening, with the abaxial sterile stalks only free apically as in *Pseudovoltzia*

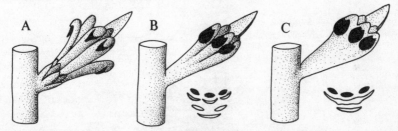

eventually into the compact form of the modern conifer cones.

There is still some controversy over the evolution of modern ovuliferous scales. Meyen (1984) suggests a rather different series of events from Florin. On the basis of his earlier studies on the upper Permian *Sashinia* and his re-examination of much of Florin's material, Meyen proposed that ovuliferous scales resulted from a fusion of ovuliferous stalks. This fusion, spreading upwards from the basal regions, was accompanied by the loss of ovules on those stalks facing the bract. Eventually a *Pseudovoltzia*-like ovuliferous scale would result (Figure 11.8). According to Meyen's view, *Walchia* would then occupy only a peripheral position in conifer evolution.

In this context it is interesting to note two other pieces of work. Miller (1982) expressed his results in a phenogram based on percentage similarity of derived character stages. His analysis based on characters of the bract-scale complex shows that all modern families, except the Cephalotaxaceae, show greater similarity to genera of the Voltziaceae than to those of the Walchiaceae. Harris

(1976b) shows similarities between the ovuliferous shoot of the extant yew (*Taxus*) and that of *Walchia*, suggesting an early divergence of modern conifer families.

Cheirolepidiaceae

Conifers of the Cheirolepidiaceae lived from the Triassic to the late Cretaceous or perhaps early Tertiary. They were especially widespread in the Jurassic and early Cretaceous when many occupied habitats in low palaeolatitudes.

Large logs provide evidence that some were trees, and decorticated branches and twigs indicate a greatly whorled pattern of branching of extension shoots. The evergreen *Pseudofrenelopsis* has been reconstructed as such a tree (Figure 11.9), with cylindrical unbranched and fleshy photosynthetic shoots (Alvin, 1983). Others have been described as low-lying trees or shrubs.

Figure 11.9: Reconstruction of the Lower Cretaceous Conifer *Pseudofrenelopsis parceramosa* with detail of a photosynthetic shoot. (After Alvin, 1983)

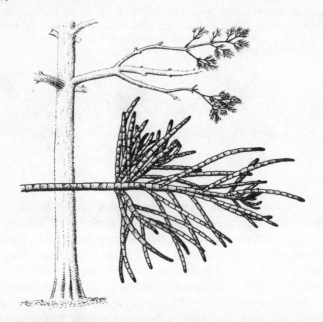

Only a small number of foliage species have been found bearing reproductive organs. It can be impossible to distinguish shoots on the basis of vegetative characters, even with a knowledge of their cuticular features. For this reason Harris (1979) proposed the use of a series of form genera which implied no affinity to families understood on the basis of reproductive characters. For example, *Brachyphyllum crucis* is known to bear cheirolepidiaceous reproductive organs, and *B. mamillare* has been shown to be a member of the extant

Figure 11.10: A,B, *Classopollis*: A. Possible Origin and Evolution. B: Section through a mature pollen grain wall with lamellate nexine and surface ornamentation that is identical to that on the orbicules. C, D: Cone and ovuliferous scale (adaxial view) of *Hirmeriella muensteri*. (A, After Alvin, 1982; B, after Taylor and Alvin, 1984; C, D, after Jung, 1968)

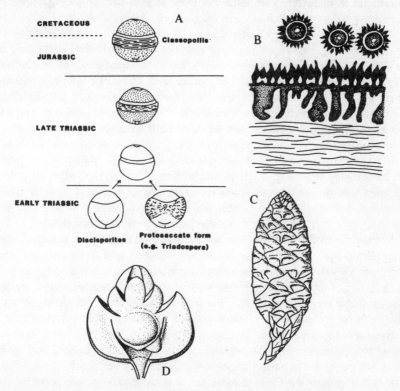

Araucariaceae. Clearly the conifers were diverging in reproductive characters while they were evolving similar vegetative features: not a unique situation, for similarity of leaves and leaf arrangement had been very obvious in the lycophytes, ferns and pteridosperms and would occur again later in the angiosperms.

The shoots known to belong to the Cheirolepidiaceae usually have thick cuticles and sunken stomata (Watson, 1982). The significance of these characters is, however, very debatable. Several species with exceptionally thick cuticles are thought to have grown in maritime environments, but there is no evidence to suggest that others grew in similar habitats. Alvin (1983) and Francis (1984) have suggested that some trees grew in areas of alternating favourable and non-favourable (arid) conditions. Irregular growth rings in the wood and the probable irregular shedding of the ultimate photosynthetic shoots support this idea.

All our information on the female cones is based upon compression/

impression material as no permineralised remains have been found (Figure 11.10C,D). The cones had helically arranged ovuliferous scales subtended in the axils of bracts. The scales had one ovule on their adaxial surface that was covered by a flap-like structure. The pollen-producing cones were small, a few millimetres in diameter. The helically arranged semi-peltate microsporophylls carried a number of pollen sacs on their abaxial surfaces (Alvin, 1982).

The pollen is of the type called *Classopollis* (Figure 11.10A,B). Ultrastructural studies on grains recovered from cones have allowed an understanding of the development of the sporoderm and the functional significance of its structure (Taylor and Alvin, 1984). Exine organisation is unlike that of other gymnosperm grains but is comparable to the complexity of angiosperm pollen. There are clearly defined nexine and sexine layers. The lamellated nexine is a typical feature of gymnosperm pollen, but the outer sculptured layer of the sexine was most probably the result of angiosperm-like deposition of sporopollen rather than from gymnosperm-like gametophytic activity. The pollen grains were almost certainly wind dispersed, although there is evidence to suggest that they germinated on the ovuliferous scales rather than being carried right to the ovule micropyle. It might be that the distinctive sculpture of the outermost sexine was a repository for proteins that were involved in some interaction between the pollen and the ovuliferous scale or ovule. Angiosperm pollen releases proteins from sites in its wall soon after it becomes attached to a flower stigma. This permits recognition of the pollen by the stigma and governs breeding behaviour. By this means incompatibility systems can prevent cross-breeding between species (Heslop-Harrison, 1976). Perhaps the Cheirolepidiaceae might even have been on the threshold of evolving such an incompatibility system.

Even with the development of a seemingly sophisticated breeding system the group was not ensured everlasting success. Its importance declined during the Cretaceous until it was extinct at about the Cretaceous/Tertiary boundary. Like many other groups the Cheirolepidiaceae was to become outcompeted by the angiosperms.

Coniferales

Extant conifers are generally grouped into a number of families that are either all included within the Coniferales or split, with some families grouped as the Taxales. Even though there may be some disagreement over the evolution of the ovuliferous scale, it is most likely that the Voltziaceae gave rise to most of the conifers, while the taxads arose from a *Walchia*-like plant.

Many modern families have long and extensive fossil records. Miller (1982) in his brief review of the major conifer families shows that several modern families appear to have been contemporaneous with their presumed ancestors. The Podocarpaceae are documented from the basal Triassic onwards. The Taxaceae, Araucariaceae, Cupressaceae, Taxodiaceae, Cephalotaxaceae and

Pinaceae have their earliest undisputed records in the Jurassic, although there is good evidence to suggest that some evolved earlier than this. Only one extinct family, the Mesozoic Palissyaceae, is included in the Coniferales.

Some of the earlier fossils of these families share many of their characters with living conifer genera, but they cannot comfortably be included within them. Some modern genera appeared in the early Mesozoic, although many are unknown until the late Mesozoic or the Tertiary.

Many families and genera were much more widespread in the past than their present distribution might suggest. This could be partially due to competition from angiosperms but more likely because of the mass migrations and extinctions that occurred during the climatic oscillations of the Quaternary period. Many conifers now cover vast areas in the Northern Hemisphere beyond the limits of the climatic tolerance of angiosperm trees.

Podocarpaceae

This family of predominantly Southern Hemisphere shrubs and trees is currently classified into seven genera, the largest being *Podocarpus* with about 100 species. Some are monoecious, others dioecious. The ovuliferous strobili are not at all cone-like. Ovules develop, either singly or in pairs, terminally on short axes that have a very few spirally arranged bracts. The pollen-producing strobili are much longer, with many microsporophylls, each bearing two pollen sacs. The bisaccate pollen grains are wind dispersed and trapped on the ovules by pollen drops. After pollination the ovules expand and develop into dry nut-like seeds. The ovuliferous scale also enlarges from its original reduced condition to form a fleshy layer (the epimatium) that partially or completely envelops the seed.

The family extends from the Triassic and is well documented in numerous Jurassic fossils. Most fossils are found in the present-day areas that made up the old Gondwana continent, although a few scattered shoots and pollen grains have been recorded from the Northern Hemisphere.

The available evidence (Figure 11.11) suggests that the Podocarpaceae evolved from the Voltziaceae in the later Palaeozoic and was a separately evolving conifer line by the beginning of the Mesozoic (Miller, 1977). The Triassic *Rissikia* from Africa and Antarctica had leafy shoots and pollen-producing cones that are very similar to modern podocarps. The organisation of the ovuliferous structure is, however, much more complex and on its own could easily be included in the Voltziaceae. Each spike-like cone had 15 to 25 bract-scale complexes arranged in a loose helix. Trifid bracts have three lobed ovuliferous scales in their axes. One or two inverted ovules are attached by their stalks to the basal parts of each lobe of the ovuliferous scales. The Jurassic *Mataia*, with its more simple ovuliferous scales with two inverted ovules, may represent a reduction stage. The tips of these ovuliferous scales are folded over the ovules in *Mataia* as they are in the uni-ovulate scales of *Nipaniostrobus* and *Nipioruha*. If this is a reduction series that continued, it could have led to the arrangement

Figure 11.11: Terminal Ovuliferous Complexes of Fossil and Living Podocarpaceae. A: *Rissikia media* with one inverted ovule cut in median longitudinal section. B: *Mataia podocarpoides*. C: *Nipaniostrobus sahnii* drawn in median longitudinal section. D, E: *Dacrydium taxoides* in adaxial view and in longitudinal section. F: *Podocarpus minor*. b, bract; i, integument; n, nucellus; o, ovule. (A, B, After Townrow, 1967; C, after Vishnu-Mittre, 1958; D–F, after Bierhorst, 1971)

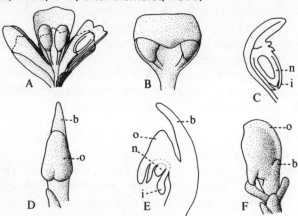

found in the modern *Dacrydium elatum* group (Townrow, 1967). The folding over of the ovuliferous scale might also indicate how the fleshy epimatium evolved.

Podocarpus itself probably evolved at about the Jurassic-Cretaceous boundary. Foliage has been reported from the early Cretaceous of Argentina and the late Cretaceous of Russia, and a petrified cone from the late Cretaceous of Japan is very similar to those of some living podocarps. The genus persisted in the Northern Hemisphere until the Eocene (Dilcher, 1969).

Taxaceae

Five extant genera comprise the Taxaceae: *Taxus, Amentotaxus, Austrotaxus, Pseudotaxus* and *Torreya*. Vegetatively they are very similar to other conifers, forming shrubs and trees with needle-like leaves. They are dioecious and it is in their reproductive organs that they stand apart from other conifers. The small pollen-producing cones are made up of tightly packed peltate sporangiophores, with each bearing up to nine pollen sacs. In the spring the cones expand to liberate great clouds of pollen into any air currents that disturb the tree. Some of these simple, non-saccate grains are carried to ovules where they are trapped by exuded pollen drops. The ovules are solitary and terminate short shoots, bearing only small bracts. These shoots are themselves borne by another short shoot having a hidden aborted terminal bud. The common occurrence of two ovules per shoot in *Taxus* is explained by there being two axillary short shoots in the overall fertile shoot complex. A further distinctive feature is the subsequent activity of a meristematic region at the base of the ovule. It grows up and

Figure 11.12: Ovuliferous Shoots of *Walchia* and some Taxaceae. A: *Walchia piniformis*, adaxial view. B: *Palaeotaxus rediviva*. C: *Marskea thomasiana*. D: *Taxus baccata*; seed surrounded by an aril (partially cut away). E: *T. baccata*; vegetative shoot with an immature ovule on a lateral fertile shoot. (A, After Florin, 1944; B, after Florin, 1951; C, after Harris, 1976b)

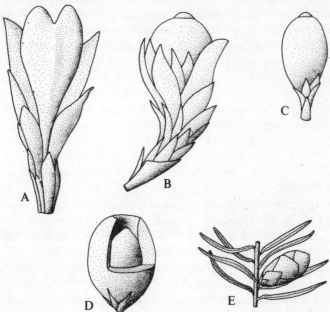

around the fertilised ovule to form a fleshy aril. This aril turns red on seed maturation and, by inducing birds to eat the seeds, furnishes the plant with a means of animal dispersal.

Plants very similar to living Taxaceae existed by the early Jurassic (Figure 11.12). They were clearly evolutionarily separated from the other conifers by that time. The lower Jurassic *Palaeotaxus*, with its arillate ovules on leafy twigs, looks remarkably like *Taxus*. The Middle Jurassic *Marskea*, as revised by Harris (1976b), is another distinct member of the Taxaceae whose important characters are all to be seen in one or other of the three living genera — *Taxus, Amentotaxus* and *Torreya*. As such it usefully unites the three, although it cannot be regarded as a primitive ancestral form.

The crucial point still remains that we do not know how and when the taxads separated from the other conifers. Florin had argued that there was a fundamental difference between lateral sporophyll-borne ovules in the conifers and terminal shoot-borne ovules in the taxads. Acceptance of this widely separates the two groups even to the point of their being put into different classes. Harris, however, believed that there were common features between the ovuliferous shoot of *Taxus* and the fertile short shoot of *Walchia*. Harris imagined that the starting point for the group could have been a conifer, with its fertile shoots scattered over the ordinary leafy shoots instead of being aggregated on special shoots. The fertile shoots would have formed several sterile scales and two or three stalked ovules before the shoot apex ceased activity. A rapid cessation of meristematic division would leave the apex as a minute bulge between the ovules. Even earlier cessation would then have reduced the number of ovules that were formed. The reduction to one ovule per shoot might therefore have been an occasional possibility. A subsequent developmental pattern in which only one ovule was to be permanent might have brought it to an apical position on the shoot. The Indian Permian *Buriaea*, with its scattered ovules (Pant and Nautiyal, 1967), might possibly represent a link in this theoretical evolutionary chain, but as yet it is really not well enough known for us to be certain.

Although the evidence is far from complete, the taxads have enough similarity to other conifers to be retained within the Coniferales. Their separation as a group at such an early time does, however, make them the earliest differentiated family of extant conifers.

Araucariaceae

Both living genera of the Araucariaceae (*Araucaria* and *Agathis*) have southern distributions. In the Mesozoic, however, they spread over both hemispheres, dominating the low latitudinal belt of summer-dry climates and living in mixed conifer communities in the middle latitudes (Krassilov, 1978). The family is known to extend back to the Jurassic, with several earlier but dubious records published from the Triassic (Stockey, 1982). The identification problems result both from *Araucaria* and *Agathis* wood being indistinguishable from each other and from that of the cordaitalean *Dadoxylon* wood, and from araucarian foliage being so similar to that of some other Mesozoic conifers belonging to the Cheirolepidiaceae. The cones are the most diagnostic organs. The ovuliferous cones are large and spherical, with ovuliferous and bract scales partially (*Araucaria*) or completely (*Agathis*) fused (Figure 11.13E,F). The pollen-producing cones have bract-like microsporophylls with many pendulous axially directed pollen sacs (Figure 11.13G). Living plants are dioecious, but we have no idea of the sexual condition of their extinct relatives.

Araucaria is divided into four sections: *Columbea, Eutacta, Intermedia* and *Bunya*. Stockey (1981, 1982), from her studies of fossil araucarians, suggests that *Bunya* is the most primitive. This is based on detailed examination of the cone anatomy, seed and embryo development of *Araucaria mirabilis* from the Cerro Cuadrado (Jurassic) petrified forest of Patagonia (Argentina). This well-known locality has also yielded other cones, wood, twigs and even seedlings showing hypogeal germination. Modern-day South American araucarians comprise the section *Columbea* with *A. araucana*, the Chilean pine or monkey puzzle, being one of the best known species in the whole genus (Figure 11.13C,D). The Tertiary record of *A. nathorstii* from Argentina, belonging to the *Columbea*, suggests that this section evolved from *Bunya* ancestors in this region. Subsequently the *Bunya* section became very limited in distribution,

Figure 11.13: Araucariaceae. A, B: From the Yorkshire Jurassic; A, *Brachyphyllum mamillare* foliage; B, *Araucarites phillipsii* ovuliferous cone scale. C, D: *Araucaria araucana*; C, silhouette of mature tree; D, foliage. E, F: *A. bidwillii* adaxial surface and longitudinal section of ovuliferous/bract-scale complex; o — free end of ovuliferous scale, v — vascular bundles; the ovule is shaded. G: *A. araucana* male sporophyll showing pendulous pollen sacs (p). (A, B, After Harris, 1979)

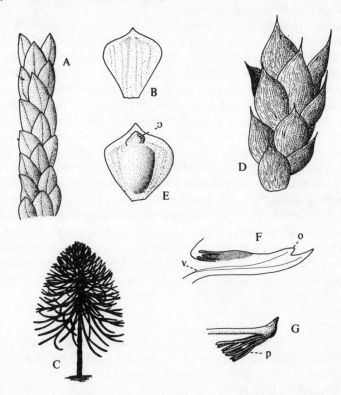

today persisting only with *A. bidwillii* in southern Queensland. The other two sections appear much earlier in the fossil record. *Brachyphyllum mamillare* foliage with its attached pollen cones and *Araucarites phillipsii* cones and cone scales from the Yorkshire Jurassic indicate early members of the *Eutacta* section (Figure 11.13A,B). *Araucaria haastii*, from the Upper Cretaceous of New Zealand, probably belongs to the *Intermedia* section which might have its origin in the *Eutacta* section.

The other extant genus, *Agathis*, appears to have had a much later origin, being documented from Oligocene and Miocene strata in the Southern Hemisphere close to its present-day range in New Zealand, Australia, Malaysia and some Pacific islands. However, its earlier origins are completely unknown.

Cupressaceae

There can be great problems in identifying fossil foliage because of the variability encountered in extant families. For this reason many artificial genera of leafy shoots have been established. Fossil foliage is best referred to the Cupressaceae if the leaves are opposite and decussate, even though a few of its modern genera have alternating whorls of three and some Podocarpaceae and Taxodiaceae also have opposite leaves.

The earliest records are of ovuliferous cones, leafy twigs and wood from the late Triassic of France (Lemoigne, 1967) and the USA (Bock, 1969), although there is great difficulty in deciding if they should really belong to the Taxodiaceae. Perhaps these fossils represent the remains of ancestral conifers before the evolutionary diversity had separated these two families into distinct lineages.

The Cupressaceae is divided into four sub-families (Callitineae, Cupressoideae, Thujoideae and Juniperoideae) with all of them reported from the late Mesozoic. Many of these sub-family records result from extant generic names being given to fossils, and many are based upon very fragmentary evidence. Therefore a large number of these identifications are doubtful not only at the generic but also at the sub-family level. Even so it is clear that there was large diversity in the Mesozoic and that the family was very widespread.

Modern genera of this family, like those of the Araucariaceae, can be thought of as relics of much more widespread ancestral forms. Today most living genera have very few species with limited distributions. In contrast, three genera are quite large. *Callitris* has 14 species, *Cupressus* has 21 and *Juniperus*, with its 60 species, is the second largest family of living conifers. This suggests that at least some of the Cupressaceae are still actively evolving.

Taxodiaceae

This family probably had its origins in early Mesozoic forms that also gave rise to the Cupressaceae. There is no doubt that the family was distinct by the Jurassic when plants bearing *Pararaucaria* and *Romeroites* ovuliferous cones grew in the area of present-day Argentina. One characteristic of such cones is the fusion of their ovuliferous scales with the bract scales into single structures. The petrified Argentinian cones have similar, somewhat peltate, ovuliferous/bract-scale complexes to those of the extant *Sequoia*. *Elatides* from the Middle Jurassic of Yorkshire has much larger bract scales in its cones rather like those of the extant *Cunninghamia*.

Although some Jurassic genera persisted into the Cretaceous, the family had greatly diversified by then. Many of these later forms have been attributed to extant genera giving us the idea that the family existed from the late Cretaceous to the present day with very little evolutionary change. One genus, *Metasequoia*, was even described first of all as a fossil from Cretaceous and Tertiary deposits in middle and high latitudes in the Northern Hemisphere. Only later was a previously unknown conifer growing in China shown to be generically identical, establishing that the genus had survived to the present day. All extant genera, except *Sequoiadendron*, have been reported from the Mesozoic (Miller, 1977).

The geographical ranges of the genera were great throughout the Tertiary, especially in the swampy areas that gave today's Brown Coal. It was only the severe climatic deterioration at the onset of the Quaternary period that eliminated them from so many areas. Today's distribution of the Taxodiaceae shows no genus to be living on more than one continent. Seven of the ten genera are monotypic (*Cryptomeria, Glyptostrobus, Metasequoia, Sciadopitys, Sequoia, Sequoiadendron, Taiwania*). *Cunninghamia* has two species, while *Arthrotaxis* and *Taxodium* have three. Such a disjunct distribution, coupled with a high species-to-genus ratio, is to be expected in a family that evolved early and had a much more widespread past history.

Palissyaceae

This family consists of two, or three, Mesozoic genera that seem to comprise a natural group that is quite distinct from the other families of conifers.

The Lower Jurassic cones *Palissya* and *Stachyotaxus* are long and cylindrical, with distinctive free ovuliferous scales in the axils of bracts (Figure 11.14). *Palissya* has about ten more or less erect ovules on each ovuliferous scale. The ovules are in two rows and each ovule is partially covered by an asymmetrical aril. *Stachyotaxus* is similar but has two ovules on much shorter ovuliferous scales. Pollen cones of *Stachyotaxus* are similar to small catkins. Each bract-like microsporophyll has two abaxial pollen sacs that produce round, bladderless pollen.

Figure 11.14: Ovuliferous Scales of Mesozoic Palissyaceae. A: *Palissya*. B: *Stachyotaxus*. (After Schweitzer, 1963)

Several authors have noted the resemblance of *Palissya* and *Stachyotaxus* to both *Cephalotaxus* and *Dacrydium* (Podocarpaceae). Florin (1951, 1958) and Schweitzer (1963) suggested that these ovuliferous scales could have evolved from a Permian *Ernestiodendron* type of cone by reduction and planation of its secondary fertile shoots. Further reduction could give the type of highly reduced ovuliferous scale with two ovules characteristic of *Cephalotaxus*. The Upper Triassic *Metridiostrobus* described by Delevoryas and Hope (1981) might represent an intermediate form. The evidence for this line of evolution is however far from overwhelming, and perhaps Meyen (1984) has taken the better line in describing this family as absolutely incomprehensible.

Figure 11.15: *Cephalotaxus drupocea*. A: Female cones near the apex of a vegetative branch. B: Enlarged female cone. C: Mature ovule on a female cone. D: Male cones borne on short lateral shoots of a vegetative shoot. E: Enlarged short lateral shoot. F: Male cone in the axil of a bract. G: Single microsporophyll. (After Singh, 1961)

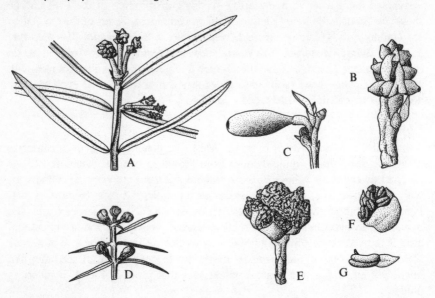

Cephalotaxaceae

The family consists of six species of *Cephalotaxus*, all living in eastern Asia (Figure 11.15). They are dioecious, with very characteristic ovuliferous cones. Each has a few pairs of opposite and decussate scales subtending a pair of erect ovules on a secondary shoot. However, only one ovule develops, forming a large olive-like seed with an outer fleshy layer enclosing a stony layer.

The fossil record is very limited, with only a few known reproductive organs of the *Cephalotaxus* form. The problem here is to distinguish these ovules, called *Cephalotaxites*, from those of *Podocarpus* and *Torreya* (Miller, 1977). So as yet the fossil record does not tell us much about the family's immediate ancestry. As mentioned above, members of the Palissyaceae might illustrate the type of evolutionary intermediate that could have led to the Cephalotaxaceae, but there is no firm evidence to substantiate this at present.

Pinaceae

This is the largest family of modern conifers and was probably the last to evolve. There are ten living genera with about 200 species. *Pinus* is the largest, with nearly 100 species. The family is restricted to the Northern Hemisphere except for one species, *Pinus merkusii*, which grows in Sumatra.

Nearly all Pinaceae are arborescent, with needle-like leaves on short lateral shoots. Their ovuliferous cones have bracts subtending free ovuliferous scales that become woody after pollination. The pollen-producing cones have two abaxial pollen sacs on each scale and produce bisaccate pollen.

A substantial number of Mesozoic cones have been referred to this family. Their reference to modern genera on the basis of their gross morphologies has been questioned by Miller (1977) who has re-evaluated the evidence.

The family most probably evolved in the early Mesozoic as ovuliferous cones, called *Comsostrobus*, from the late Triassic of the USA have the basic features of the Pinaceae (Delevoryas and Hope, 1973). Winged seeds, the structure of the pollen-producing cones and the bisaccate nature of the pollen confirm the relationships with this family. Other early fossils of the Pinaceae-type are known from the Upper Triassic and the Middle Jurassic, showing that the family was most probably well established by then. *Pinus* itself had evolved by the Lower Cretaceous, being first known as the lignitic ovuliferous cones called *Pinus belgica* (Alvin, 1960). *Pinus* was likely to have been the centre of an ancestral complex containing other forms that are better referred to other genera such as *Pityostrobus* and *Pseudoaraucaria*. Many of the *Pityostrobus* cones are very like cones of *Pinus*, while those of *Pseudoaraucaria* appear to represent a natural but extinct genus.

By the late Cretaceous two-, three- and five-needle forms of *Pinus* had evolved, and there is also a little evidence that *Cedrus* and *Larix* had appeared by then. The other genera of modern Pinaceae did not appear in the fossil record until the Tertiary.

There is much evidence of evolutionary diversity in the Pinaceae, probably reflecting geographical isolation and adaptation to different environmental pressures. *Pinus, Cedrus, Larix, Ketleeria, Pseudolarix* and *Cathaya* have long and short shoot development, whereas *Abies, Picea, Tsuga* and *Pseudotsuga* have no such differentiation. *Larix* species are deciduous.

Pollen germination is also variable in different Pinaceae. In conifers pollination generally occurs after the pollen has been drawn through the integument micropyle to the nucellus tip by the reabsorbed pollen drop. In some Pinaceae this is not necessary as germination can occur at some distance from the nucellus. *Pseudotsuga* and *Cedrus* pollen can germinate on the integuments, while *Tsuga* pollen is capable of germinating on the adaxial surface of the ovuliferous scale. The only other genus in which this can occur is *Araucaria*.

Extant conifers are clearly a heterogeneous group of plants. They must surely represent the products of divergent lines of evolutionary change, but there is no one clear picture of how this occurred. Ideas of a relatively few years ago laid out an evolutionary sequence from the Cordaites through the Walchiaceae and the Voltziaceae to modern conifers. The discovery of new material and the reworking of original specimens have shown this idea to be too simplistic.

The Cordaitales remains as the prime candidate for the ancestral group, but the evolution of ovuliferous cones from such cordaitalean fertile shoots may have happened more than once and in several different ways.

Clement-Westerhof's (1984) interpretation of the seemingly adaxial position of ovules as representing original abaxial positions on megasporophylls could lead to much rethinking of our ideas. If both ovules and pollen sacs are abaxial, then the microsporophyll and the megasporophyll could be thought of as homologous; an idea that could be extended to the pollen-producing cone and a single ovuliferous dwarf shoot.

Much remains to be discovered about the generic relationships and the evolutionary histories of extant families. Some genera of Mesozoic plant fossils are included in extant families: *Rissikia* and *Mataia* (Podocarpaceae), *Palaeotaxus* and *Marskea* (Taxaceae), *Pararaucaria, Romeroites* and *Elatides* (Taxodiaceae) and *Comsostrobus* (Pinaceae). Others are so like living conifers that they are included in the same genera (*Pinus* and *Araucaria*). In contrast, there are some plant fossils that are so different from extant genera that they are used as the bases for families. By the very nature of doing this we are saying that the family must have become extinct. This is not inherently bad, providing we do not exaggerate differences between families that could lead to taxonomic inflation, thereby concealing any taxonomic and evolutionary relationships. For example, the evolution of the Taxaceae from *Walchia*-like ancestors suggests a very early divergence of the famiy. If this was used as supporting evidence for separating the plants once more into the Taxales, then it could influence further thoughts on the evolutionary relationships of the earlier members.

12 THE LIMITED SUCCESS OF OTHER GYMNOSPERMS

Although conifers represent the most significant group of gymnosperms, they were not alone in becoming dominant members of parts of the world's vegetation. Other groups developed to very prominent positions, some still successfully surviving today. If they had all become extinct, it would have been very easy to think of them as unsuccessful or primitive plants that failed to survive climatic change or competitive onslaughts from other more vigorous species. Those that do still live demonstrate the ability of plant groups to persist for very long periods of time, to compete successfully in their own ecological niches and in some instances to continue active evolutionary change.

Certain comparisons may be made between the groups discussed in this chapter. Some evolutionary parallels can be drawn but there are no overall close relationships between the groups. They have been included together for convenience.

The Cycads (Cycadales)

Cycads live today in scattered tropical and warm-temperate regions of the world. There are nine genera, with about 100 species, but only *Cycas* has widespread distribution, occurring in East Africa, India, China, Japan and Australia.

Cycads have vertical stems that range from small tubers to stout, usually unbranched, palm-like trunks up to about 18 m tall. The South African *Stangeria* alone has a subterranean stem which probably resulted from an adaptation to unfavourable conditions that were threatening the survival of its ancestors. The stems have relatively little conducting tissue, with small, endarch, primary xylem continuing into the narrow, radial files of secondary xylem tracheids. To the outside of this manoxylic xylem is the bifacial cambium that forms only a little secondary phloem. Surrounding the vascular cylinder is a broad zone of parenchymatous cortex. Secretory canals are present in both the pith and the cortex, being interconnected by others running through the broad medullary rays in the vascular cylinder. Leaf traces are visible in the cortex following a rather unusual and characteristic path. They depart from the vascular cylinder and girdle the stem to the other side before turning outwards to a leaf petiole. Cycads have a very slow growth rate and very few species produce any recognisable annual rings in their xylem. However, they can live for a long time and some plants have been estimated to be about 1000 years old (Chamberlain, 1935).

Leaves are formed in spirals from the single apical meristem and have their leaflets characteristically rolled up in their early stages of development. The leaves exhibit many xeromorphic characters. Their epidermal cells, and sometimes those of their hypodermis, are thick walled and strongly cutinised, stomata are sunken and mostly on the abaxial surface, the bundle sheaths have thick-walled cells, and many fibres run through the densely packed spongy mesophyll. The petioles are very fibrous and the larger plants retain their bases on the trunks after the laminal portions have been lost. These bases give an armoured, protective covering to the axes.

The earliest plants recognisable as cycads were probably slender plants with small, widely spaced leaves that did not give rise to armoured trunks. The leaf shape of these earliest forms is still, however, a matter for debate. Henricks (1982) has suggested that the cycads might have been polyphyletic in origin and that the different types of leaves were probably derived from separate lines. On other evidence this does not seem very likely.

Mamay (1976) has argued for the derivation of the cycads from Upper Carboniferous pteridosperms. A progressive incision of the lamina margin evolved the modern cycad pinnate leaf from the entire leaves of the *Taeniopteris* type. The various forms of modern cycad leaves have arisen via different numbers of such incisions. The incorporation of only one vein per segment would have given the *Cycas*-like leaf, and the incorporation of several parallel veins would have resulted in the *Zamia* kind. A compound taeniopterid leaf could have given an *Stangeria*-like leaf with little modification, and marginal incision would have resulted in the *Bowenia* type of leaf.

In contrast to the idea of leaf incision, Delevoryas (1982) has suggested that a dissected leaf could have been the primitive kind for cycads. The occurrence of many Triassic pinnately compound cycad and cycad-like leaves can be argued to be supporting evidence for it being the ancestral condition.

The difference in opinions about the ancestral shape of leaves is reflected in the reconstructions of early cycads (Figure 12.1). The Upper Triassic *Bjuvia* and *Leptocycas* have entire and divided leaves respectively. *Leptocycas* was at least based on the evidence of leaves and a pollen cone being attached to a cycad stem. *Bjuvia's* trunk is, however, entirely imaginary. Florin (1933) based it only on the association of *Taeniopteris*-like leaves and megasporophylls called *Palaeocycas*. Harris (1961a) similarly used an imaginary upright stem for his reconstruction of the Jurassic cycad *Beania*, which bore leaves referable to *Nilssonia*. Kimura and Sekido (1975) have subsequently published an alternative reconstruction to that of Harris. Their *Nilsonniocladus* from the Lower Cretaceous of Japan had slender stems that gave off spirally arranged shoots of limited growth bearing terminal clusters of three to seven or more *Nilssonia* leaves. There are also indications that *Nilsonniocladus* was deciduous.

Living cycads are dioecious and there is no evidence to suggest that any ancestral cycads were not. The female plants of *Cycas* form crowns of sporophylls arising in a spiral in place of the leaves. The upper part of the sporophylls are leaf-like, and ovules are attached to the rachis-like basal area. This condition may well be primitive as all other genera produce compact megasporangiate cones.

Figure 12.1: Reconstructions of Cycads. A: The Upper Triassic *Bjuvia*; about 1.5 m tall. B: The Jurassic cycad bearing 10 cm long *Beania* ovulate cones (C). D: The Upper Triassic *Leptocycas* about 1.5 m tall. E: An ultimate shoot of the Lower Cretaceous *Nilssoniocladus* with leaves about 12 cm long. (A, After Florin, 1933; B, C, after Harris, 1961a; D, after Delevoryas and Hope, 1971; E, after Kimura and Sekido, 1975)

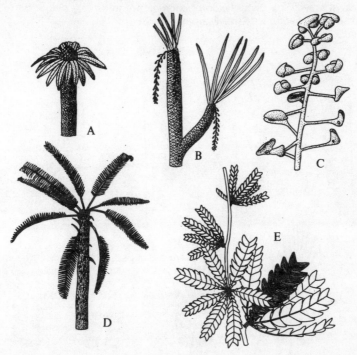

There is however a further difference that suggests that the alternation of leaves and megasporophylls in *Cycas* is not primitive. In every other type of stem, including the microsporangiate *Cycas*, the main axes cease growth when forming a cone. Further growth is through the initiation of a new meristem that forms near the base of the cone's pedicel. This pushes the cone to one side and the stem appears to grow in an uninterrupted manner. It is in fact a sympodial pattern of growth. The ancestral *Cycas* may also have formed terminal cones of looser sporophylls that also caused the main axis to stop growing. But if the main axis recommenced to grow without the initiation of a new meristem, then new photosynthetic leaves would be formed inside the sporophylls. A regular alternation in this way gives the pattern of growth in *Cycas* (Harris, 1976a).

There are several late Palaeozoic putative cycads. Mamay (1976) has suggested that the cycad megasporophyll could have arisen from Upper Carboniferous pteridosperms. A likely ancestral condition is shown by *Spermopteris* which has two abaxial rows of superficial ovules on simple *Taeniopteris*-like leaves. Reduction of the lamina together with a restriction of the ovules to

Figure 12.2: Hypothetical Evolutionary Development of *Cycas*-like Megasporophylls. A–E: From *Spermopteris*-like pteridosperm by reduction of lamina and numbers of seeds. The ultimate *Cycas*-like derivative (= *Phasmatocycas*) has the basal part of its lamina completely reduced and all its seeds attached to the megasporophyll stalk. F–I: From *Archaeocycas* by progressive reduction of the basal lamina leaving all seeds attached to the sporophyll stalk. The distal lamina becomes progressively pinnatifid to like that of *Cycas revoluta*. (After Mamay, 1976)

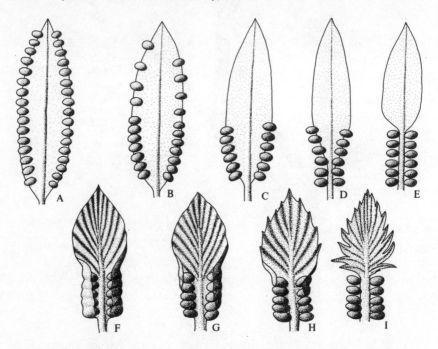

its lower part could have resulted in the condition shown by the Permian *Phasmatocycas* from Kansas and Texas (Figure 12.2A–E). However, there is also the contemporaneous *Archaeocycas* from Texas. This has its ovules partially enclosed by the inrolled margins of the lamina. This might well represent an intermediate condition between the *Spermopteris* and *Phasmatocycas* kinds that itself gave rise to the cycad megasporophyll (Figure 12.2F–I).

The evolution of the cycad megasporophyll via a *Phasmatocycas* type of ancestor seems highly likely. But it is as well to remember that Mamay bases his reconstruction of the laminal tip of *Phasmatocycas* on the evidence of associated foliage rather than on any attached laminal structure. Kerp (1983) stressed this very point when he interpreted a different possible intermediate between *Spermopteris* and *Phasmatocycas*. His *Sobernheimia* is reconstructed as a laminated organ bearing two rows of ovoid seed-like bodies. The laminal margins are lobed, the lobes alternating with the 'seeds'. Kerp's hypothesis rests partially on the belief that it is more logical to suggest an overall reduction

Figure 12.3: Cycad Reproductive Organs. A: Terminal crown of megasporophylls of *Cycas*. B: Single leaf-like megasporophyll of *Cycas* with six ovules. C: *Zamia* cone of megasporophylls. D: *Zamia* megasporophyll. E: Longitudinal section through a cycad ovule. F: Longitudinal section through the tip of an ovule just before fertilisation. One of the four pollen tubes has shed its sperm into the pollen drop. G: *Ceratozamia* microsporophyll with numerous pollen sacs. H: Lateral view of a sperm with its apical spiral of cilia (a, archegonia; e, endosperm; l, abscission layer; m, micropyle; n, nucellus; p, pollen tube; s, sperm). (A–G, After Chamberlain, 1935; H, after Brough and Taylor, 1940)

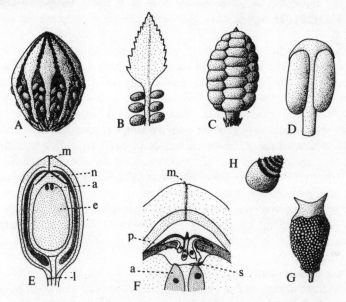

of the *Spermopteris* lamina than a partial reduction of only the basal part.

The ovulate cones of the extant genera other than *Cycas* are usually thought to have evolved by reduction of the sporophylls and a condensation of the axis bearing them. The various genera show differences in sporophyll shape that might indicate the direction of evolutionary reduction. *Cycas* itself shows variation, *C. revoluta* having a pinnate upper sporophyll and several pairs of ovules and *C. circinalis* having only a serrated sporophyll and a single pair of ovules (Figure 12.3A,B). *Macrozamia* has the sporophyll rachis remaining as a tapering spine. *Ceratozamia* and *Encephalartos* have a suppressed rachis but the serrations that are often present probably represent the pinnae. The broad *Dioon* sporophylls do not even have the serrations, and are only loosely compacted into a cone. *Zamia* appears to be the most reduced with its small sporophylls forming a very compact cone (Figure 12.3C,D). The Jurassic *Beania* probably represents an intermediate form in being a lax pendulous cone with a loose spiral of megasporophylls similar to those of *Zamia*.

Some of these living cycads produce enormous cones that must be as large as any that have ever existed. *Macrozamia* cones can be nearly a metre in length and can weigh over 40 kg. The ovules are correspondingly large. In *Cycas circinalis* and *Macrozamia densonii* they are about 6 cm long, comparable in size to the larger pteridosperm seeds such as *Trigonocarpus*. Each ovule is in fact very similar to those of the pteridosperms. They are radially symmetrical with a three-layered integument that is largely fused to the nucellus. Development of the gametophyte tissue is centripetal, wall development commencing at the periphery. Archegonia are formed at the distal (micropylar) end and become exposed by the autolysis of the nucellus tip as it forms the pollen chamber (Figure 12.3E,F). However, their final exposure only occurs if the ovule is pollinated.

All cycads, including *Cycas*, have pollen-producing cones or loosely organised strobili. The abaxial pollen sacs are in clusters of three to six and are joined together at a papilla rather like the sporangia of *Angiopteris*. The number of pollen sacs on each microsporophyll varies from about 25 in *Zamia* to over 1000 in *Cycas* (Figure 12.3G). Stewart (1983) argues that the cycad microsporophyll could have evolved from the simpler type of medullosan fertile frond. Reduction of the *Aulacotheca* and *Halletheca* form of synangia could have given rings of cycad-like microsporangia on the abaxial surface of the frond. Unfortunately we have very little direct information on which to base any ideas of the evolutionary development of these cones. The best understood example of a Mesozoic cone is the Jurassic *Androstrobus* which is very much like those of living cycads. It had helically arranged, imbricated microsporophylls, each bearing numerous rings of pollen sacs on its abaxial surface (Harris, 1964).

Pollen is usually transported to the ovules of another plant by wind. There is some evidence to suggest that beetle pollination can occur, although no species relies upon it. Pollen drops exuding from the micropyles catch the pollen grains and as the drops are withdrawn the grains are carried through the micropyle into the shallow pollen chamber at the tip of the nucellus. There, a haustorial tube grows out from the pollen grain into the nucellus. This stimulates further autolysis of the nucellar tip to expose the archegonial necks. Then two spermatozoids, that have developed in the body of each pollen grain, are shed into the fluid remaining in the pollen chamber (Figure 12.3F,H). These spermatozoids do not differ greatly in their development from that of cryptogam spermatozoids. They do, however, differ in containing only small mitochondria and in lacking a large anterior mitochondrion, plastids and starch. This probably reflects their shorter life and a much smaller need for energy expenditure in order to reach the archegonia (Norstog, 1982). The spermatozoids swim to an archegonium by means of their spiral band of flagellae, and one enters to fuse with the egg cell. More than one egg cell may become fertilised in this way, but only one embryo will develop to maturity.

Cycads seem to have changed very little since the Mesozoic when they were much more widespread and more numerous than the living forms. This present-day distribution has been explained many times as representing that of isolated genera which have been left from more widely distributed ancestors. Cycads have no seed dispersal mechanism of their own to facilitate geographical migration or expansion, although they are often spread by birds. Fertilisation can sometimes also be a problem for these dioecious plants as pollination is rare at over 100 m from a male plant (Chamberlain, 1935).

Cycads should not be thought of as evolutionary remnants. They are specialised rather than primitive. The only real primitive character that cycads possess is their free-swimming spermatozoids. Cycads and *Ginkgo* are the only living gymnosperms that have retained them. There is also plenty of evidence to suggest that the group is still actively evolving.

The Quaternary Ice Ages may have restricted the cycads' distribution, especially in Africa and the Americas, but the great cycad regions of the world are still Mexico and the West Indies, Australia and southern Africa. Many species persist in vast numbers. There are about 14 species of *Encephalartos* in southern Africa, and many are locally very abundant. Many species of *Macrozamia* are similarly widely distributed throughout the eastern coastal areas of Australia. *Macrozamia spiralis* is especially common, sometimes forming almost impenetrable thickets extending down to the high-tide line (Chamberlain, 1935).

The Quaternary climatic oscillations also seem to have stimulated evolutionary change through migration and geographical isolation. This is especially true in the West Indies and Florida where land/sea boundary changes were especially significant. Subtly different populations of *Zamia pumila* live throughout the area, but they are not sufficiently different to be regarded as true species. The rare Cuban monotypic species *Microcycas calocoma* is probably the result of similar geographical isolation, but in this case rapid evolutionary change has given rise to a very distinctive plant (Eckenwalder, 1980).

The Cycadeoids (Cycadeoidales)

The cycadeoids consist of a few morphologically different forms of gymnosperms that possess similar cone-like reproductive organs. The fact that these reproductive organs have a superficial resemblance to angiosperm flowers led to a great deal of interest in the group. This in turn provoked authors to voice highly speculative ideas on angiosperm evolution from the cycadeiods; ideas that were later to be firmly discredited.

Cycadeoids presumably evolved from Palaeozoic pteridosperms but, as we shall see later, their exact origin is still a mystery. They flourished in the middle Triassic, throughout the Jurassic, and into the early or middle Cretaceous. For about 100 million years they were important elements in much of the world's vegetation before abruptly disappearing. Their extinction was rather rapid and probably the result of a combination of inbreeding and competition from the early angiosperms.

The vegetative structures of the cycadeoids are very reminiscent of those of

Figure 12.4: Cycad and Cycadeoid Leaves. A–D: *Ptilophyllum caytonense* (Cycadeoid); A, leafy shoot; B, leaf; C, lower epidermis of the leaf; D, single stoma with the underlying subsidiary cells outlined with a dashed line. E–H: *Nilsonnia compta* (cycad); E, leafy shoot; F, leaf; G, lower epidermis of the leaf; H, single stoma with the outlined guard cells. (After Harris, 1942)

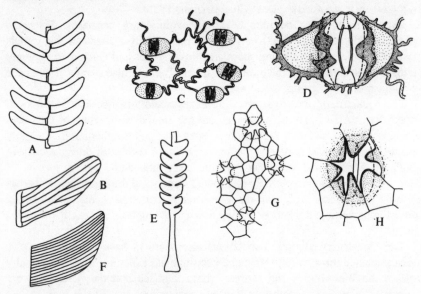

the contemporaneous cycads. Some had cycad-like stems, although the secondary xylem is more compact and there are no girdling leaf traces. Their leaves were originally thought to belong to cycads and it was not until differences in epidermal cellular arrangements were discovered that there was a ready means of separating them. The stomata of cycads have no subsidiary cells (they are haplocheilic) whereas those of cycadeoids have them (syndetocheilic) (Figure 12.4). We should remember, however, that the latter term describes the origin of guard cells and subsidiary cells from the same mother cell. We have no knowledge of the development of cycadeoid stomata and use the term only to describe the mature stomata. No inference of their development should be made, for the similar syndetocheilic-like stomata of the extant gymnosperm *Gnetum* acquire their subsidiary cells from divisions of the surrounding epidermal cells (Maheshwari and Vasil, 1961). Some rather different leaves have also been shown to be cycadeoid. *Eoginkgoites* was originally thought to belong to the Ginkgoales on the basis of it being a genus of fan-shaped, deeply divided leaves with short rachises and long petioles. Their anastomosing venation and syndetocheilic stomata are however more indicative of its being a cycadeoid (Ash, 1976).

The earliest and most primitive cycadeoids had elongated, branched or unbranched, slender stems with numerous loosely arranged leaves. They were

Figure 12.5: Mesozoic Williamsoniaceae. A: *Williamsonia sewardiana* reconstruction, about 2 m tall. B, C: *Williamsoniella coronata*; B, cone in longitudinal section with bracts (b), microsporophylls (m) and ovule-bearing receptacle (o); C, microsporophyll with embedded microsporangia. D, E: *Weltrichia whitbiensis*; D, reconstruction of the fructification; E, microsporophyll with synangia. (A, After Sahni, 1932; B, C, after Harris, 1964; D, E, after Nathorst, 1911)

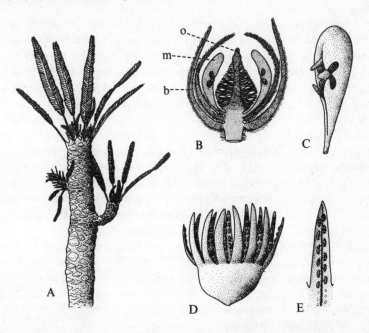

periodically deciduous, and when the leaves were shed no persistent bases were left on the stems. There is fossil evidence of these slender stems, although many plant reconstructions are largely imaginary. Some are based on association of plant organs. For example, Harris (1969) reconstructed a much branched cycadeoid with slender distal axes on the evidence of the common association of *Bucklandia* stems, *Ptilophyllum* leaves and *Williamsonia* ovuliferous cones (Figure 12.5). Harris (1973a) subsequently suggested that the plants might even have been trees making forests on the flat but moderately dry land beyond the river-banks where they were competing on equal terms with conifers and ginkgos. This would explain how the leaves can often be locally very abundant and are the commonest organs in the Yorkshire Jurassic deltaic sediments.

These early slender-stemmed cycadeoids are grouped together to form a family, the Williamsoniaceae. The later forms constitute the second family, the Cycadeoidaceae. The Williamsoniaceae is based on the ovulate reproductive organ *Williamsonia* together with some other morphologically similar forms. *Williamsonia* cones have their ovules borne on fleshy receptacles that may be elongated and conical, or short and almost globular. Interseminal scales are interspersed between the ovules packing them into a tight mass over the whole surface of the receptacle. Mature ovules became elevated on stalks in some species whereas in others they remained sessile. The whole receptacle is surrounded by spirally arranged bracts. The corresponding pollen-producing organs are called *Weltrichia*. These are cup-shaped with up to 20 flattened distal extensions that bear synangia on their inner surfaces (Figure 12.5D,E). The pollen is monocolpate.

Some hermaphrodite cones can be referred to the Williamsoniaceae. *Williamsoniella* is the name given to the forms known from Yorkshire Jurassic rocks (Figure 12.5B,C). The cones were borne on slender stalks that arose in the axils of leaves. Their central receptacles were covered with up to 300 ovules and innumerable interseminal scales. About 12 wedge-shaped microsporophylls are attached in a whorl below the receptacle. Their pollen sacs are on the inner surface in the form of paired synangia that are partially covered by finger-like projections. Harris (1974) has described the pollen from the type specimen of *W. lignieri* as resembling the dispersed spore *Exesipollenites scabratus*. All the reproductive organs are enclosed by hairy bracts that probably functioned as bud scales. Similar hermaphrodite cones have been described from the earliest member of the family: *Wielandiella* from the Upper Triassic rocks of Sweden. The parent plant has been thought to have had forked axes that give a false dichotomy (dichasium) whenever a cone terminated the axes.

Members of the Cycadeoidaceae exist in Upper Jurassic and Cretaceous strata. Their growth habit was rather different from that of the Williamsoniaceae. Instead of having slender trunks or branching axes they had massive trunks that seldom branched. The leaves left persistent bases that densely covered the surface of the trunks. These plants were therefore very much like the cycads in overall appearance.

The vegetation of the Upper Jurassic and Lower Cretaceous suggests more aridity than does that of the Lower and Middle Jurassic. This led Harris (1973a) to suggest that the cycadeoids looked rather like the living cycad *Encephalartos* and grew in stunted savannah together with a low scrub of conifers and herbaceous pteridophytes (Figure 12.6A).

Complex flower-like cones were borne on the trunk surface, and it is these reproductive organs that were particularly thought to represent an ancestral form of the angiosperm flower (Figure 12.6B–D). Each had a central fleshy receptacle densely covered in ovules and interseminal scales. Surrounding the receptacle were the microsporangiate structures. Weiland (1906, 1916) believed there to be a whorl of pinnate frond-like organs that bore two rows of sporangia on their pinnately divided lateral branches. These microsporophylls and their subtending bracts were thought to open out prior to pollination rather like angiosperm flower buds. Subsequent work by Delevoryas (1968) and Crepet (1974) reveals instead that the cycadeoid organ could not have opened out. The microsporangiate region was fused basally around the receptacle, and although there was sporadic separation distally into free units, this did not occur all at

Figure 12.6: *Cycadeoidea*. A: Reconstruction of plant with a stem about 0.5 m tall. B: Reconstruction of mature cone partly cut away to show the fused segments of the sporophylls with their pollen sacs; in the centre is the fleshy receptacle covered with ovules and interseminal scales. C: Longitudinal section of part of a younger cone than is shown in B; the sporophylls have free pinnae. D: Longitudinal section through an ovule and two interseminal scales. (A, After Delevoryas, 1971; B, after Delevoryas, 1968; C, D, after Crepet, 1974)

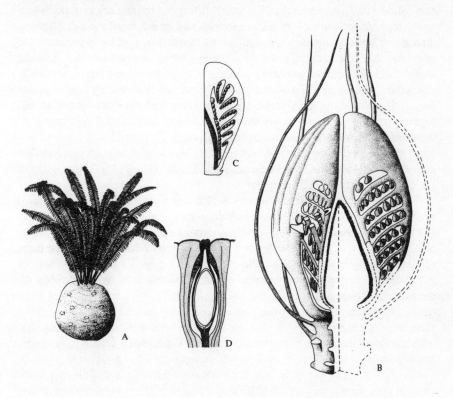

one level. During the early stages of development the microsporophylls became revolute and then through maturation their recurved rachises became fused together. Their pinnae were folded inwards to become radially aligned, becoming closely adpressed or even fused at their distal ends. This gave rise to the 'trabeculae' of mature cones that Delevoryas envisaged as linking the inner and outer portions of the pollen-producing apparatus.

Such a microsporangiate zone was probably the phyletic result of a partial fusion of a whorl of microsporophylls similar to those found in members of the Williamsoniaceae. This partial fusion of the microsporophylls, together with the fact that their rachises developed parenchymatous thickening at the point of maximum curvature, strongly suggests that they never altered their position after maturing. Such structural modifications would have prevented the cycadeoid cone from opening out in the same manner as those of the Williamsoniaceae. The whole structure must therefore have remained closed within its protective covering of bracts.

Crepet (1974) suggested that insect predation was a factor that could have selected for enclosure of the pollen sacs and the ovules. If this was the case, a transition series would have evolved, presumably reinforced by natural selection. Stidd (1980) has postulated a rather different evolutionary process which could account for the lack of such intermediates in the fossil record. Neotony through macroevolutionary change is the basis for Stidd's argument. He suggests that the sexually mature pollen organ represents the juvenile morphological condition persisting into the adult stage of the plant. This condition could have come about abruptly without any intervening intermediary forms. The closed cycadeoid cone might then have had the evolutionary advantage of protection from insect predation but it was an accidental change that gave the plant its superiority over the others. Self-pollination would have ensured that some of the next generation would have the same new advantage of closed cones. Natural selection might then have acted relatively quickly, bringing about the complete change in the population.

There are three possible ways in which the cycadeoid cones may have evolved from ancestral medullosan pteridosperms (Delevoryas, 1968). The fact that the cones' vascular strands originate from leaf traces in the cortex rather than from the main stem stele suggests that cones may be considered to be parts of leaves (Figure 12.7A,B). This is also consistent with the idea of pteridosperms being the ancestors of cycads for these bear ovules directly on their leaves. The second manner in which the cones could have arisen was from the condensation of a fertile branch in a leaf axis (Figure 12.7C). Both of these ideas seem plausible provided that we can accept the notion of some phyletic telescoping of the stem. The vascular traces of the cones depart from traces leading to other leaves than those in whose axils they are positioned.

The third possible origin is from a branch arising from a leaf. This may seem unusual but, although it has never been shown to occur in pteridosperms, it has been described in ferns. Delevoryas (1968) quotes the examples of the coenopterid ferns *Anachoropteris* and *Botryopteris* in which leaves can bear small branches that duplicate the parent plant in miniature. Telescoping of the main axis would have to be invoked again to account for the separation of the two vascular traces.

Delevoryas (1968) and Crepet (1974) both suggest that the pinnate microsporophylls are more primitive and similar to the fertile organs of medullosan pteridosperms. Some, like *Williamsoniella* and *Weltrichia*, are reduced so that their pinnate origin is not obvious. Fusion of tubular synangia, as in *Aulacotheca* and *Halletheca*, could have given the synangial structures of the cycadeoids.

The pollen also suggests a medullosan ancestry for the cycadeoids as both

Figure 12.7: The Origin of Cycadeoid Cones. A: Portion of idealised ancestral Palaeozoic pteridosperm with the base of a leaf bearing fertile structures. B: Possible hypothetical intermediate stage between pteridosperms and cycadeoids with the basal pinna of the leaf partially 'embedded' within the fleshy stem. C: Possible axillary branch origin of a cycadeoid cone. D: Diagrammatic longitudinal section of a portion of a cycadeoid stem showing its vascularisation. (After Delevoryas, 1968)

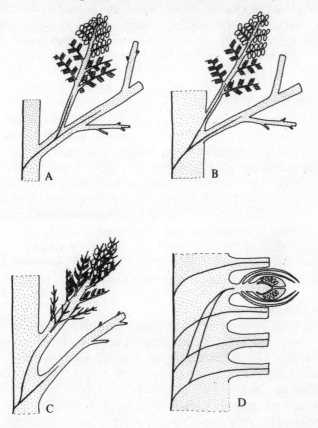

produce monosulcate grains. Fine structure of the pollen walls is also similar (Taylor, 1982a).

But just as the origin of the cycadeoids is still a little obscure, so is their disappearance. Self-pollination must have occurred in the vast majority of the later cycadeoids. Gradual disintegration of the internal organs would have permitted pollen to reach the ovules in the same reproductive bud but would not have facilitated cross-pollination. Borings in many specimens suggest that beetles may also have been acting as pollinators and could possibly have brought about cross-pollination (Delevoryas, 1968). No animal remains have been found in any of the damaged cones, although the nature of the borings allows the possibility that they were caused by beetles. This would parallel the situation in living *Encephalartos* where *Phlaeophagus* beetles are attracted to the microsporangiate cones by the odour of the pollen. Both sexes utilise the pollen as their main source of food and the females visit ovuliferous cones to oviposit. Fertilisation can be effected in this way.

What is more certain is that self-pollination would ultimately have resulted in increased homozygosity within the population. This would have led to a loss of potential genetic adaptability to climatic or other changes. Alternatively the loss of any beetle or other insect pollinators to more attractive gymnosperm plants or to angiosperm flowers might have hastened their demise during the Cretaceous. Whatever the reasons, the cycadeoids became extinct leaving no recognisable descendants.

Ginkgoales (the Maidenhair Trees)

Ginkgo biloba, the maidenhair tree, is a graceful deciduous tree that will grow almost anywhere in a temperate climate. Mesozoic ginkgos had a virtual world-wide distribution although they seem to have been more abundant in north-eastern Siberia than anywhere else (Harris, 1976a). Today the monotypic *G. biloba* is naturally restricted to a small part of southeastern China. Since Darwin's time the plant has been described as a living fossil which can invoke ideas and thoughts that can make it seem more important than it really is. It is naturally tempting to take the survival of one species as marking the end of an evolutionary line, but there is sufficient morphological variation to suggest that the species is not genetically stagnant. It is of course the last surviving species of a group that we can recognise from the late Palaeozoic, but there are many other gymnosperms and pteridophytes that are similar remnants of groups that once were more important.

Shoot dimorphism is characteristic of the living plants but we do not know when this feature evolved as no comparable leafy shoot fossils have ever been found. This type of branching is rather unusual, giving the trees a strange growth pattern. Spur shoots do not elongate, resulting in crowding of leaves and tightly packed leaf scars (Figure 12.8A). Some of these short spur shoots can in subsequent years grow to form long shoots with elongated internodes and lateral spur shoots of their own. Conversely, long shoots may change to a spur-shoot type of growth for a year or more. Lateral branches may also even grow from the trunk and older branches.

Leaves are fan-shaped and often bilobed with two veins extending from the petiole to give the characteristic, finely dichotomous venation. There is much variation in size and in the amount of leaf dissection, which has some relation-ship to the position of the leaves on the shoots. Larger and more divided leaves are on the long shoots, and the smaller entire or less divided leaves are on the spur shoots. *Ginkgo biloba* is deciduous. If its ancestors were also deciduous, this could account for the fact that their leaves are usually in great quantities when they are found. *Ginkgo*-like leaves are known from the early Mesozoic onwards and many are so similar to those of *Ginkgo* that they have been included

Figure 12.8: *Ginkgo biloba*. A: Vegetative shoot; long shoot with two short shoots; one short shoot also bears paired ovules on stalks. B: Longitudinal section through a young ovule. C: Longitudinal section through a seed after the stony layer has become hard and the inner fleshy layer dry and papery. D: Single male cone with each sporophyll bearing two pollen sacs. E: Single sporophyll. e, Embryo; m, megaspore mother cell; n, nucellus; p, pollen chamber; s, stony layer (black) surrounding the inner fleshy layer (white)

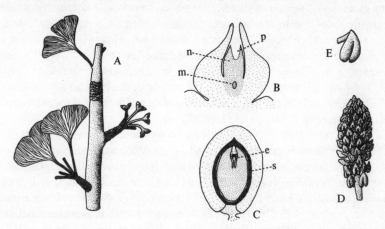

within the extant genus. However, in using a name in this way there is a great danger of misleading the reader to think that the plants were similar in every way. There is every reason to believe that all plant organs do not evolve at the same rates and there is little evidence to prove that the Mesozoic forms had reproductive organs comparable to those of *Ginkgo biloba*. Therefore it is preferable to use the less emotive genus *Ginkgoites* for these fossil leaves.

The additional problem of separating the many similar Mesozoic and Tertiary leaves can be overcome by using epidermal characters. By this means a series of different leaf species of *Ginkgoites* can be distinguished which exhibit a steadily changing morphology from the highly dissected Jurassic leaves to the entire and bilobed forms known from the Tertiary. This evolutionary series can be taken back to presumed Palaeozoic ancestors. *Sphenobaiera*, which extended from the Permian to the Cretaceous, had deeply dissected leaves that were borne on both long and spur shoots. The Lower Permian *Trichopitys* had shoots with spirally arranged dichotomous leaves. All of them may have had their origins in the Upper Carboniferous *Dichophyllum* that Remy and Remy (1977) have referred to the pteridosperm *Callipteris*.

The evolutionary history of the group is, however, better discussed on the basis of what we know about reproductive organs. *Ginkgo biloba* is dioecious, with the ovules and pollen organs borne in the axils of the foliage leaves or bud scales of spur shoots. Erect ovules are generally in pairs at the ends of stalks. Each has a nucellus surrounded by an integument that grows up and around it to form a micropyle. The cells at the nucellar tip break down to form a pollen chamber while liberating the fluid for a pollen drop (Figure 12.8B). The pollen-producing organs are catkin-like with stalked sporangiophores bearing two, or rarely more, pollen sacs (Figure 12.8D,E).

Pollen grains are wind dispersed when they are shed in spring. Any that are trapped by pollen drops are drawn inside the ovule and into the pollen chamber with the receding drop. Once there, a short haustorial tube grows out from the pollen grains into the side of the nucellus pollen chamber. After several months, two multiciliated sperm are liberated into the pollen drop and swim to one of the two or three archegonia that have developed. Meanwhile the ovule has enlarged and the integument has become three-layered with a watery inner sarcotesta, a middle stony sclerotesta and an outer fleshy sarcotesta (Figure 12.8C). *Ginkgo* is therefore very like the cycads in ovule structure, pollination and fertilisation as both have retained motile sperm. These are, however, merely similarities and not evidence of close affinity.

Fossilised reproductive organs of the ginkgoaleans are rare, but we must be careful not to assume that they were all alike. Some were similar to those of *Ginkgo*. For example, paired ovules have been found associated with *Ginkgoites* leaves in the Yorkshire Jurassic (Harris, 1976c). A small catkin-like organ has also been found in the same bed and it yielded *Ginkgo*-like monosulcate pollen (Van Konijnenburg-van Cittert, 1971). It is quite possible, however, that some extinct forms had different fructifications that have not yet been found or correctly interpreted (Hughes, 1976). Many dispersed ovules can be found in association with other *Gingkoites* leaves but they need not have been borne on the plants in similar ways. Archangelsky (1965) has described *Karkenia* as having over a hundred small ovules attached singly on short stalks to a central axis. This fructification, from the Cretaceous of Argentina, was found in association with *G. tigrensis* so they probably belonged to the same plant. Further specimens of *Karkenia* have been found in Siberia in association with leaves. Krassilov (1970) called these leaves *Sphenobaiera* although they are very similar to the Argentinian ones. It therefore seems likely that any number of ovules may have been attached to fertile axes and that there are many ways in which *Ginkgo*-like leaves could have evolved from early forms.

The origin of the Ginkgoales probably lies in the Palaeozoic pteridosperms although it is difficult to decide which of these groups contained the most likely ancestors. Meyen (1984) has suggested that the Callistophytales are the initial ancestral forms which gave rise to the peltasperms. This possibly heterogeneous group might have given rise to the Gingkoales together with the Leptospermales and even the Caytoniales. This is feasible based upon the evidence of similarities of both vegetative and reproductive organs, but the evidence is not yet suffi-ciently overwhelming to be accepted without question.

The virtual extinction of the group is perhaps easier to understand. As temperate deciduous trees they must have been occupying the same habitats that the early angiosperms were able to colonise successfully. The ginkgos gradually became restricted to the more northerly temperate forests during the late

Cretaceous and early Tertiary. A subsequent cooling of the climate then led to a concertina-ing of plant zones southwards. The heightened competition that resulted brought about their extinction in much of the Northern Hemisphere. They disappeared from North America during the Miocene and from Europe at the end of the Pliocene. The only survivors were in eastern Asia where they still grow today.

Gnetophytes

Ephedra, Gnetum and *Welwitschia* are highly specialised extant gymnosperms that are often included together as the Gnetales. They do, however, show certain striking differences in habit and geographical distribution that suggest they are better separated into families of their own. Some even claim that there is no close relationship between the genera and segregate them into three distinct orders (Eames, 1952). The only fossil record we have is of pollen grains from the Triassic onwards that resemble those of *Ephedra* and *Welwitschia* (Scott, 1960), and a few debatable Tertiary stem fragments that resemble *Ephedra*.

There are about 35 species of *Ephedra*. All are xerophytic shrubs with photosynthetic stems and reduced scale-like leaves. *Gnetum* also consists of about 35 species, mainly growing in tropical rainforests. Most of them are vines although a few are trees. All have opposite leaves which are very similar to those of dicotyledonous angiosperms in having broad laminae and reticulate venations. The monotypic *Welwitschia* grows in the deserts of south-west Africa. It is a strange plant with most of its short axis underground and tapering down to its main taproot. Only two leaves ever grow, extending throughout the life of the plant by a basal meristem and autolysing at their tips. Some plants live for over 1000 years, but long before then the leaves have split down to their meristems resulting in a tangled mass of segments that appear to be separate leaves. *Welwitschia* has been referred to as a persistent seedling because of this peculiar growth form. If this is true we have no way of knowing what was the appearance of the original ancestral form.

The three genera have anatomical features that show them to be the most evolutionarily advanced of all gymnosperms. They have vessels with foraminate or simple perforation plates. The many types and different numbers of pores in these end-plates suggest that these vessels evolved from pitted tracheids. Angiosperm vessels in contrast are generally accepted to have evolved from tracheids with scalariform thickenings. Some species of *Gnetum* also have sieve tubes and companion cells that appear to be rather like those of angiosperms. However, they arise from two independent cambial initials rather than from the subdivision of one sieve-tube mother cell. The stem apices of *Ephedra* and *Gnetum* are also unusual in having a double-layered cellular arrangement. They have the outer tunica layer that is generally thought of as an angiosperm character, but this is not as important a character as it may first appear for it is also found in the Araucariaceae.

The three plants all have their ovules and pollen organs borne in cones that

Figure 12.9: A: Ovuliferous Strobilus of *Ephedra*. B: Longitudinal Section of a Mature Ovule of *Ephedra*. C: Compound Male Inflorescence of *Ephedra*. Microsporangiophores, bearing groups of septate pollen sacs, are in the axils of bracts; one bract has been cut away to reveal the two-lipped perianth. D: Longitudinal Section of an Ovule of *Gnetum* Prior to Fertilisation. E: Longitudinal Section of an Ovule of *Welwitschia* (the outer two integuments have been omitted). (a, Archegonia; c, pollen chamber; f, fertilisation; g, gametophyte tissue; i, integument; n, nucellus; p, pollen tube.) (After Bierhorst, 1971)

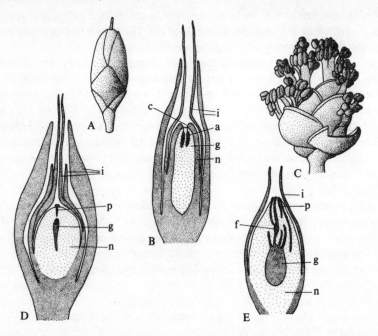

are somewhat flower-like (Figure 12.9). This, together with the angiosperm-like conducting cells, has been taken to suggest that these plants might have been the ancestors of the flowering plants and led some authors to use angiosperm terminology. The reproductive organs are, however, really gymnospermous so gymnosperm terminology is more appropriate.

All three genera are dioecious although the male plants of *Gnetum* also produce abortive ovules. The arrangement of the ovules and pollen sacs varies between the genera although there are certain similarities in the way in which they are all partially protected by bracts or sheaths. *Ephedra* has the most flower-like reproductive organs. Its cones arise in the axils of scale leaves on its stem and have opposite and decussate bracts. In the male cones each bract has in its axil a pair of smaller bracts enclosing a microsporangiophore with its several terminal pollen sacs. Ovules are similarly partially enclosed in a pair of small bracts although they only develop in the axils of the terminal bracts of the cones.

All the ovules have an integument extended into a tube-like micropylar tip. This is surrounded by another integument in *Ephedra* and *Welwitschia* and by two integuments in *Gnetum*. Pollination is by wind and possibly insects and involves a pollen-drop entrapment mechanism. Fertilisation is achieved by the growth of pollen tubes, but the actual sexual fusion of nuclei occurs in rather different ways. *Ephedra* is the only one that has archegonia. It is therefore only in *Ephedra* that sexual fusion is between one of the two sperm nuclei of the pollen tube and an archegonial egg-cell nucleus. In *Gnetum*, fusion is between a sperm nucleus and one of the several free larger nuclei in the apical region of the embryo sac. *Welwitschia* has the most unusual fertilisation process of all the gymnosperms. As the pollen tubes grow through the nucellus some of the multinucleate cells of the embryo sac grow towards them. Contact and nuclear fusion therefore take place outside the embryo sac.

The widespread distribution of *Ephedra* and *Gnetum* suggests that they are highly specialised survivors from different ancient gymnosperm groups; *Ephedra* from Euramerian ancestors and *Gnetum* from Gondwanan ones. Eames (1952) proposed that the cordaites were the probable ancestors of *Ephedra*, basing his ideas on what he believed to be comparable ovuliferous organisations. The discovery of smaller herbaceous cordaites adds a little support to Eames's idea, but as yet there are no known fossil intermediates.

Welwitschia is highly specialised to live in the coastal deserts of southwestern Africa. It obtains its water supply from the coastal fog that regularly covers the area. The condensing water vapour is absorbed through its leaves, the shallow roots acting mainly as anchoring organs. To aid this absorption its enormous leaves have very large numbers of stomata on both surfaces. When the fog lifts, transpiration is reduced by the sunken stomata closing very tightly and through the reflectance of the sun's radiation by crystals of calcium oxalate grown in the epidermis (Bornman, 1972). These adaptations suggest that the plants have lived in the unusual conditions of the coastal deserts for some time.

PART IV: THE THIRD PHASE OF DIVERSIFICATION

13 THE FIRST FLOWERING PLANTS

What is an Angiosperm?

In many respects the angiosperms, or flowering plants, may be considered to be the dominant terrestrial plant life on Earth today. They grow in a greater range of environments, exhibit a wider range of growth habits, display more morphological variation, and are represented by more species than any other living vascular plant group. In many environments they are also ecologically dominant although there are large regions where they play a subordinate role in the vegetation. All the world's major food plants are angiosperms and their economic importance far exceeds that of any other plant group. In spite of this, there exists no satisfactory single definition of what an angiosperm is.

This might seem a surprising statement to make, in view of the fact that we all have a mental image of a flower, so we need to examine the various features that characterise angiosperms. These features are:

(1) ovules enclosed in a conduplicate carpel;
(2) presence of vessels;
(3) reticulate leaf venation;
(4) a double integument;
(5) tectate pollen;
(6) the presence of a flower;
(7) double fertilisation;
(8) phloem companion cells derived from sieve-tube mother cells.

To begin, let us consider the degree to which the ovules are enclosed during fertilisation. In contrast to gymnosperms, or 'naked-seed' plants, the ovules of angiosperms are enclosed within carpels and pollen germinates on a stigmatic surface instead of having direct access to the micropyle. Although this is true for the vast majority of angiosperms there are exceptions. In *Drimys* (Magnoliaceae), for example, the carpels are not completely closed when pollination takes place (Figure 13.1).

Another characteristic of most angiosperms is the presence of vessels rather than just xylem tracheids. Unfortunately this character is even less restricted to the angiosperms than the previous one. The gymnosperms *Gnetum, Ephedra* and *Welwitschia* possess vessels and there are instances of angiospermous plants, particularly in the Magnoliales (Winteraceae) and Hamamelideae (*Tetracentron*) where vessels are absent.

A third feature of many angiosperms is the net-like or reticulate venation of their leaves. By no means all flowering plants possess this character because the

215

Figure 13.1: The Basic Form of a Conduplicate Carpel. A: Paired stigmatic crests or lips are shown running the length of the left side. B: Carpel opened out so that the stigmatic lips are displayed peripherally; the ovules are shown with pollen tubes growing towards them. (After Bailey and Swamy, 1951)

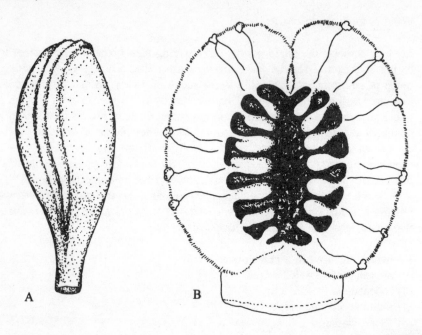

A   B

majority of monocotyledonous angiosperms are parallel veined. Reticulate venation is also a feature of some pteridosperms (e.g. *Sagenopteris*), is an obvious character of the gymnosperm *Gnetum*, and is even found in ferns (*Hausmania*).

A double integument is found in most angiosperm seeds but here again it is not universal. Single integuments occur in the dicotyledonous Sympetalae (those with fused petals). There are numerous instances among the gymnosperms where a wholly or partly enclosing structure around the seed has been interpreted as homologous to the outer integument of most angiosperms. The cupule of some pteridosperms and the 'bractioles' of *Welwitschia* fall into this category.

The pollen-grain wall (exine) of most angiosperms consists of pillar-like structures called columellae supporting an outer covering or tectum. Pollen grains of this type are described as tectate (Figure 13.2). However, non-tectate pollen is found in some angiosperms (Walker, 1976) and some conifers, notably the Cheirolepidiaceae, are known to have produced tectate pollen (page 216) (Pettitt and Chaloner, 1964; Taylor and Alvin, 1984).

We cannot even use the most obvious angiosperm feature, the flower, to define angiosperms because its organisation varies so much. Whereas many flowers have whorls of well-differentiated sepals and petals, some do not.

Figure 13.2: Section through an Angiosperm Pollen-grain Wall Showing its Multilayered Structure. The outermost layer or tectum is often supported by pillars or columellae resting on a footlayer. Some authors (e.g. Erdtman, 1963) refer to the tectum as the sexine. The footlayer rests on the endexine, which together, in Erdtman's (1963) terminology, form the nexine. The inner pollen wall is called the intine. The tectum/footlayer/endexine terminology is that of Faegri and Iversen (1950)

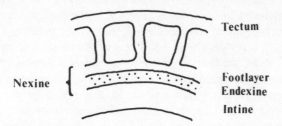

Among the predominantly wind-pollinated flowers (the 'Amentiferae' and grasses, for example) such structures are generally lacking. Most angiosperm stamens are organised into a filament surmounted by a two-lobed anther. Each anther is usually divided into two so that there are four locules or microsporangia. There are, however, numerous variations on this 'normal' set of four sporangia. For example, in some members of the Metastomataceae and Onagraceae, two sporangia may abort, and within the Loranthaceae and Rhizophoraceae some species possess sterile plates of tissue dividing up the sporangia and producing an apparently multilocular condition.

The most consistent angiosperm character so far found is that of double fertilisation. In angiosperms, two sperm nuclei from the microgametophyte are involved in fertilisation. One unites with the egg nucleus while the other fuses with two so-called polar nuclei to form a triploid endosperm which acts as a food store. At the time of writing this process appears to be unique to angiosperms. However, it has only been observed in a very small proportion of all flowering plants. It could be considered premature, therefore, to base a definition of angiosperms on this feature. It is also worth remembering that any definition not based on morphological criteria is impossible to apply to the fossil record. This does not invalidate the definition, of course, but it does make it inappropriate for all but hypothetical evolutionary studies.

Sporne (1974) considers that companion cells in the phloem, which are formed from the same mother cells as the sieve-tube elements, are also apparently unique to angiosperms. Again, however, this is difficult to demonstrate in fossil material, particularly as phloem tissue is rarely preserved.

The concept of modern angiosperms envelops a plexus of characters only some of which may be present in a given species. The recognition of angiosperms in the fossil record is consequently extremely difficult and the problem becomes more acute when we are searching for possible angiosperm origins.

Angiosperm Ancestors

Various candidates for the ancestral angiosperm group have been proposed and have included the Gnetales, cycadeoids, cycads, and pteridosperms.

The existence of vessels among members of the Gnetales has been cited as evidence that the Gnetales are close to the pre-angiosperm stock. Muhammad and Sattler (1982) review the range of features found in the wood of *Gnetum* and conclude that the variation in vessel structure is so great that it encompasses the typically circular pits of the conifers as well as the patterns of pits and perforation plates found in angiosperms. These authors are of the opinion that because of the large number of similarities between *Gnetum* wood and angiosperm wood the two wood types are unlikely to have arisen by chance, and that *Gnetum* may represent a group close to that which gave rise to the angiosperms. If this is true, the vessel-less angiosperms must have undergone evolutionary retrogression or have evolved from a separate group in which vessels were absent.

Ephedra has been compared to the woody angiosperm *Casuarina* but the similarities are only superficial. Although a kind of double fertilisation is said to occur in *Ephedra*, no triploid endosperm is produced and the reproductive process is no more advanced than that seen in *Pinus*. In *Welwitschia* an endosperm is produced but not by a double fertilisation process.

The flower-like *Cycadeoidea* reproductive structure was the basis of the idea that the Cycadeoidales had some direct part to play in angiosperm evolution. The suggestion of Weiland (1906) that the whorled microsporophylls opened at maturity encompassing the ovule-bearing receptacle has subsequently been shown to have been unlikely (Delevoryas, 1968; Crepet, 1974). The supposed similarity to an angiosperm flower is apparently therefore without foundation.

The carpel of the angiosperm *Drimys piperata* has been interpreted as a leaf-like structure with the lamina folded along the midrib so as to bring the marginal stigmatic surfaces into close proximity. The ovules are deemed to be super-ficially attached to this lamina in a non-marginal position. The Permian cycadophyte *Archaeocycas* exhibited a similar arrangement of two superficial rows of ovules on a partially inrolled lamina. Mamay (1969, 1976) has suggested that the hypothetical complete closure of the lamina so as to enfold the ovules would have produced a structure with some similarity to an angiosperm carpel. It must be stressed, however, that Mamay's reduction sequence was hypothetical and any derived phylogenetic significance based upon it is therefore highly speculative. This point is emphasised by Mamay (1969) but apparently is often overlooked by other authors (see also p. 196).

Among the pteridosperms, the Caytoniales and Glossopteridales have provided the most fertile areas for speculation concerning angiosperm origins. When *Caytonia* ovules were originally discovered, it was thought that they were enclosed (Thomas, 1925). However, reinvestigation subsequently showed that pollen had direct access to the micropyle (Harris, 1940) implying a gymnospermous relationship, and speculation regarding possible angiosperm affinities was curtailed. Since then Krassilov (1977) has reinterpreted both the female (*Caytonia*) and male (*Caytonanthus*) reproductive organs of the *Caytonia* plant. He is of the opinion that *Caytonia* may be regarded as a more-or-less closed many-seeded capsule which lacked a true stigma but had a protostigmatic mouth. Instead of a style there was a channelled pollen transfer tube. *Caytonanthus* paralleled the pollen-producing organs of angiosperms in that it was a four-loculed synangium. In Thomas's (1925) interpretation the *Caytonia* cupule was seen as homologous with the angiosperm carpel whereas Krassilov (1977) views the *Caytonia* cupule as a modified leaf. There can be little doubt that the organisation seen in *Caytonia* is one of the most sophisticated in the pteridosperms and one that closely approaches that of the angiosperms. Even the leaves of the *Caytonia* plant (*Sagenopteris*) had reticulate venation. Perhaps *Caytonia* is best seen as representing a degree of evolutionary advancement within a plexus of pteridosperms some of which may have given rise to the angiosperms. Perhaps the relationship of *Caytonia* to the angiosperms is analogous to that of *Lepidocarpon* to the true seed: that is, they have evolved a similar structure in response to the same kinds of evolutionary pressures but are quite distinct. If, as seems likely, *Caytonia* was not directly on the phylogenetic line that produced the angiosperms, then the evolutionary pressures favouring angiosperm organisation and reproduction must have been strong and largely ubiquitous during the latter part of the Mesozoic.

Another group of pteridosperms that has attracted a lot of attention regarding possible angiosperm origins is the Glossopteridales. Melville (1960, 1962) proposed that the angiosperm conduplicate carpel arose from a planated webbed telome truss that became a leaf. The sporangia phyletically slid to the adaxial leaf surface and then became enclosed by inrolling of the leaf lamina. In support of this 'gonophyll' theory, Melville pointed to *Glossopteris* with its adaxial ovuliferous scutum subtended by a spatulate leaf with reticulate venation.

That the angiosperms may have had a glossopterid ancestry was an idea first proposed by Plumstead (1952). This was soon rejected, however, as the initial interpretations of supposed bisexual fructifications were not substantiated by later work. Subsequent discovery of structurally preserved glossopterid fructifications (Gould and Delevoryas, 1977) and increased knowledge of early angiosperm reproductive structures have led Retallack and Dilcher (1981a) to believe that glossopterids display some morphological characters that are homologies for structures seen in early angiosperm flowers. They also suggest that the pollen, wood, leaves, habit and geological age of the glossopterids do not preclude them from being angiosperm ancestors.

The classical interpretation of a flower is that it is fundamentally a strobiloid structure, analogous to a gymnosperm cone, derived from a condensed shoot or apical bud (see Chapter 15). This has been challenged by Melville (1983) who envisages the flower as consisting of a tuft or fascicle of organs each of which consists of a leaf-like structure bearing two fertile branches, one male and one female. In support of this idea Melville cites numerous examples of this kind of organisation in a variety of putatively primitive and advanced living angiosperms. The similarity between this type of structure and the fertile units of glossopterids such as *Eretmonia* or *Ottokaria* forms the basis of Melville's ideas of a glossopterid ancestry for angiosperms, although all glossopterid reproductive structures so far discovered are unisexual. To overcome this difficulty Melville suggests that either bisexual glossopterids existed but have not been found because they represented only a very small proportion of the group as a whole, or that the bisexual flower could have been derived by condensation of a bisexual inflorescence. Relative evolutionary development of male and female structures could determine subsequent floral sexuality and positioning of the sex organs, the angiospermous enclosure of the ovules could have been brought about by enveloping growth of the subtending capitulum or scutellar margins, and the petals and sepals could have been derived from the leaf-like structures supporting the fertile branches. Melville points out that the venation of many angiosperm petals is very similar to that seen in glossopterid leaves. This implies that petals have in general retained their primitive organisation rather than being secondarily derived from vegetative leaves as is normally accepted. Melville also suggests that the parallel-veined leaves of monocotyledons and some dicotyledons (e.g. in the Proteaceae) could have been derived from *Gangamopteris*. If the angiosperms were derived from the glossopterids, it would require a major re-evaluation of the time of origin of the flowering plants and of their relationships one to another. Their origin would have to be in the Permian, and a long gestation period in upland regions would have to be invoked to explain their absence from the fossil record; an idea that seems to have little credibility in the light of currently available evidence.

Of all the gymnosperms the most likely contenders for angiosperm ancestors would appear to come from among the pteridosperms. This group displayed a wide range of morphologies, habits and reproductive strategies and some closely approached angiosperm organisation (e.g. *Caytonia*). Failure to pinpoint 'an ancestor' or ancestral group is not surprising when one considers the nature of angiosperms. The character combinations which we use to describe the angiospermous condition would not have arisen simultaneously. Moreover the features that proved advantageous to angiosperms were apparently also evolving in other groups probably in response to the same environmental pressures.

The angiosperms as we know them today may well be descended from a Mesozoic group of plants that were the first to reach a kind of 'critical mass' or character combination threshold in the accumulation of angiospermous characteristics. Thereafter the particular type and number of angiospermous characters enabled this pro-angiosperm stock to proliferate and outcompete its less angiosperm-like contemporaries. A consequence of this scenario is that we might expect to find a number of diverse Mesozoic plants, each with some angiospermous features, that may or may not belong to the pro-angiosperm group. This also means that we are never likely to find a single archetypal angiosperm ancestor and that we will not be able to define a point in time when angiosperms arose.

Although not all angiosperms exhibit all the features characteristic of angiosperms, every angiosperm possesses most of them. This is generally taken to suggest that all flowering plants are derived from the same ancestral stock: that they are monophyletic. A monophyletic origin has been implied above by referring to 'a Mesozoic group' but the possibility exists under the character threshold hypothesis that angiosperms could be polyphyletic in that the angiosperm character combination threshold could have been passed more or less simultaneously by more than one group of plants. Furthermore, if the origin of the angiosperms was the result of convergent evolution, such an origin would be obscured both in the fossil record and in today's living descendants.

At the time of writing most palaeobotanists favour a monophyletic origin for the angiosperms, but notable among the champions of a possible polyphyletic origin are Krassilov (1984) and Hughes (1976).

The Cladistic Approach

Recently there have been several attempts to determine more rigorously the relationships between plant groups, and in particular the position of the angiosperms, using cladistic analysis (Hill and Crane, 1982; Doyle and Donaghue, 1986). Cladistics is a classificatory technique that is largely derived from and based on the work of Hennig (1966). It is believed to have considerable relevance to our attempts to reconstruct phylogeny. The advantage of cladistics over more traditional approaches is that it attempts to provide a logical framework for establishing relationships between individuals or groups of individuals. The rigour of the analysis and the application of the principle of parsimony (the simplest solution is the best solution) try to ensure that the method by which a particular classification is achieved is widely understood and that the likelihood of ambiguity is reduced. The result of any particular analysis is usually a graphical representation of relationships known as a cladogram which provides a working model that can be tested in the light of additional information.

In the past, phylogenetic 'trees' have been constructed using concepts of similarity between taxa and their times of first occurrence in the fossil record. The fossil record, as we have already seen, is far from complete and the time of first appearance in the record and the true biological first occurrence may be very different. Similarity is also difficult to assess in that it is often a highly subjective concept. There are two quite distinct approaches to assessing relationships based on similarity: we can either look at overall similarity (known as a phenetic approach) or we can adopt a cladistic approach by seeking the relationship and accumulation of individual homologous characters.

It is possible to express relationships between taxa by linking them all in a complex network, but relationships are often more easily expressed and visualised if they are arranged in a hierarchical tree-like structure. In a hierarchical scheme, classes or groups containing fewer individuals are nested within larger classes or groups that are at 'higher levels of universality'. In other words, individuals within these larger classes are united by more generalised characters. In cladistics a hierarchical scheme is constructed of nested classes and subclasses based on discriminating the similarities shared by individuals (or groups of individuals) from differences that separate them at progressively higher ranks of 'levels of universality'. The branch points of the cladogram are positioned by seeking similarities and differences. Some characters are more generalised (widely distributed) than others. In order to test at which level a character should be used to discriminate between classes, the level of universality is altered for the purpose of comparison. A given character will discriminate and be unique to a subclass only at one level of universality. At this level it may be used for defining subclasses. This is known as 'outgroup comparison'.

Parsimony and Homology

One of the fundamental principles of cladistics is that a classification diagram (cladogram) should be constructed in the simplest, least complicated and most efficient way. This is the principle of parsimony. In general the optimum classification involves the fewest branch points necessary to accommodate the available data.

In any classification or phylogenetic scheme, it is essential to distinguish between homology and analogy. Homology indicates some kind of perceived essential similarity, whereas analogy refers to a superficial resemblance. Unfortunately distinguishing between homology and analogy requires that we have prior knowledge of the true (phylogenetic) relationship between characters, and this is what we are trying to determine. To circumvent this problem, the recognition of homology is formalised in terms of methodology. In cladistics, homology is defined as similarity at a given level of universality which is consistent with parsimonious arrangement of similarities of other characters. Analogy on the other hand is inconsistent with parsimony (Hill and Crane, 1982).

In cladistics, the less generalised (or derived) components of homology (synapomorphies) are used for defining taxa. In phenetics, synapomorphies are also used but only in addition to more generalised homologies (symplesiomorphies).

So far this discussion has centred around classification. If the classification is a natural one, it will reflect phylogeny and under these conditions relative degrees of similarity may reflect ancestry. Generalised and relatively less generalised characters can correspondingly be regarded as 'primitive' and 'advanced' (derived). Homologous similarity indicates common ancestry whereas analogous similarity indicates evolutionary convergence or parallelism. Groups defined on the basis of homologies are therefore interpreted as monophyletic, having apparently descended from a single common ancestor. Those united by analogy are interpreted as polyphyletic. Polyphylesis is not permitted in cladistic methodology. Cladistic analysis can be carried out by hand or, in the case of numerical cladistics, by computer. This brief and simplistic account of cladistics is only meant to serve as an introduction. A more comprehensive explanation in relation to palaeobotany is given in Hill and Crane (1982).

Doyle and Donaghue's (1986) numerical cladistic analyses of seed plants is given in Figure 13.3. At the base of the cladogram and most primitive in all the characters considered is *Aneurophyton*. *Archaeopteris* is linked with seed plants largely on the basis of heterospory. Lyginopterids are the sister group of all other extant and extinct seed plants except the early Carboniferous *Pitys*. These taxa may be divided into two major clades. The late Carboniferous seed-fern *Callistophyton* is a basal taxon in a clade which included the corystosperms, *Peltaspermum*, the glossopterids and *Caytonia*, as well as the cordaites, ginkgos and conifers. This *Callistophyton* clade is distinct from the other clade which has *Medullosa* as its basal taxon. The medullosan clade includes the cycads as a sister group to a clade comprising the angiosperms, cycadeoids, *Pentoxylon*, *Ephedra*, *Welwitschia* and *Gnetum*.

According to this cladogram the closest living relatives to the angiosperms are the Gnetales but they should not be thought of as an ancestral group. Doyle and Donaghue's (1986) cladistic analyses support the ideas of Arber and Parkin (1907) that the Gnetales are relatives of angiosperms and the Bennetitales (Cycadeoidales) which responded to wind pollination by severe floral reduction and aggregation. The computer-based cladistic analysis that Doyle and Donaghue employed allowed them to experiment with modified data sets and even 'dummy' data. Significantly, when these analyses were run, the relationship of the angiosperms to the Cycadeoidales and Gnetales proved remarkably robust.

The Recognition of Early Angiosperms

Surprising as it may seem the lack of a suitable simple definition of angiosperms does not preclude the use of the fossil record in elucidating the evolutionary history of the flowering plants. Palaeobotanists have long accepted the incomplete nature of the material they have to work with and arrive at conclusions concerning the affinity of remains often on a probabilistic rather than an absolute basis. Furthermore, if a single definition of angiosperms did exist, the chances are it would rest upon such a criterion as double fertilisation which could not be tested in fossil material. We have to learn to live with the limitations of the fossil record in this respect in much the same way as we have come to accept the rather nebulous concept of angiosperms as applied to extant plants. Accepting the fact that angiosperms cannot be simply defined we do not abandon the concept of flowering plants. Instead we evaluate the characters possessed

Figure 13.3: A Cladogram Summarising the Relationships of the Major
Groups of Seed Plants.

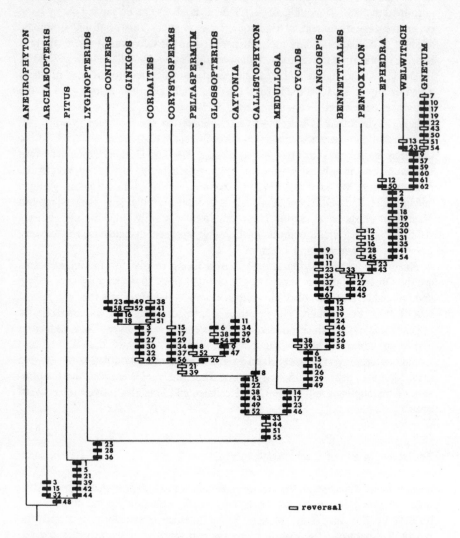

☐ reversal

Source: Doyle and Donaghue (1986) reproduced with permission of the Systematics
Association
Notes: The cladogram was produced using numerical methods and represents one of
four 130-step most parsimonious cladograms derived from the data matrix given in
Table 13.1. The characters important in defining the classification are shown as
numbers on the cladogram and are explained as follows: 1, axillary branching; 2,
accessory axillary buds; 3, leaves on homologues of progymnosperm penultimate-
order axes; 4, opposite phyllotaxy; 5–7, 000 = all leaves dichotomous, 100 =
simple or dissected leaves with pinnate venation and cataphylls, 110 = simple or
dissected leaves with pinnate venation and cataphylls, XX1 = linear or dichotomous

leaves and cataphylls, or scale leaves only; 8, rachis bifurcate — simple; 9, finest vein order reticulate; 10, several vein orders reticulate; 11, guard-cell poles raised — level with aperture; 12, syndetocheilic stomata; 13, tunica corpus; 14-15, 00 = protostele, 10 = eustele with regular internal secondary xylem, X1 = eustele with (mostly) external secondary xylem; 16, all stem bundles endarch; 17, each leaf supplied by > 2 stem bundles; 18, metaxylem without scalariform pitting; 19, scalariform pitting in secondary xylem; 20, vessels; 21, multiseriate rays; 22-23, 00 = no secretory cavities or canals, 10 = secretory cavities, X1 = secretory canals; 24, Maule reaction; 25-27, 000 = megasporangia/ovules on dichotomous structures on radial axis, 100 = on pinnately compound sporophyll, 110 = on once-pinnate sporophyll (or simple but with lateral ovules), XX1 = on one-veined stalk or sessile; 28-30, microsporangiate structures, same coding; 31, ovule appendicular — terminal; 32, (homologues of progymnosperm) fertile appendages on (homologues of) last-order axes; 33-35, 000 = ovule(s) in radial cupule, 100 = directly on more or less planate sporophyll, X10 = in circinate cupule, or anatropous bitegmic ovule, 101 = in second integument derived from two appendages; 36, several–one ovule per radial cupule; 37, several–one ovule per circinate cupule/ovule; 38, microsporangia abaxial; 39, microsporangia more or less fused; 40, microsporophylls whorled; 41, compound male and female strobili; 42-43, 00 = no seeds, 10 = radiospermic, X1 = platyspermic; 44, lagenostome; 45, micropylar tube; 46, vascularised nucellus; 47, nucellar cuticle thin–thick; 48, heterospory; 49-50, 00 = tetrad scar, no sulcus, 10 = sulcus, pollen tube, 11 = pollen inaperturate; 51, pollen radial/mixed–bilateral; 52, pollen saccate; 53, exine alveolar–granular; 54, striate pollen sculpture; 55, megaspore tetrad tetrahedral–linear; 56, megaspore wall thick–thin or lacking sporopollenin; 57, microgametophyte with prothallial but no sterile cell; 58, siphogamy, non-motile sperm; 59, tetrasporic megagametophyte; 60, egg a free nucleus; 61, cellular early embryogenesis; 62, embryo with feeder.

by a plant against our experience and knowledge of other plants. The same is true for fossil material but unfortunately these comparisons cannot be taken very far.

The earliest undisputed angiosperm remains are known from the Cretaceous, and those which have been studied for the longest time and upon which early concepts of angiosperm evolution were based are leaves. Attempts at identifying these leaves have traditionally centred around comparisons with living forms in an effort to find the closest match. It was assumed that similarity of form indicated some kind of taxonomic and phylogenetic relationship. Comparison with leaves of living plants was often based on gross morphological similarities and vein courses, which inevitably led to the identification of extinct leaf forms as representatives of extant genera or even species. The implications of 'picture matching' have been discussed by Dilcher (1971), Wolfe (1972, 1973) and Cronquist (1968) who observed that, at best, matchings cannot 'provide new or independent information on the evolutionary diversification of a group, or on the transition between groups; they merely document the existence (at best) of a particular group at some time in the past' (Cronquist, 1968).

Erroneous identifications based upon comparisons with living forms have grossly exaggerated the antiquity of modern taxa and have given support to the theory that angiosperms must have evolved prior to the Cretaceous in some remote 'homeland' away from any environment in which they could have become fossilised. The angiosperms are then supposed to have burst forth on the world already diversified into extant taxa approximately 100 million years ago

Table 13.1: Data Matrix Used to Generate the Cladogram Shown in Figure 13.3. 0: primitive state; 1: derived state; –: information not available or questionable; X: not applicable (character not present), or precursor state unknown (in multistate characters and autapomorphies).

```
                 1          2          3          4          5          6

Aneur   0X00000X00---0000000000-000000X0xxxxxX0X0000Xxx0000000xx-0xx--
Archa   0X10000X00---x1000000---x0000x0x1xxxxx000000xx1000000000-00x--
Pitus   --0010000000-0000000100-00000000000x010x10100-10000X000-000--
Lygin   10001000000-0000000100-10010000001x010x10100-10000X000-000--
Medul   10001000100-0-x100001x1-10010000100xx010x10001-1001000-0-0-0--
Calli   10x0110x10-0-x1-00000xx-xx0xx0001x0x0000xx10001101101-0--0
Gloss   -0-100100-0------1101000100x0100x10x-011010-0-0----
Pelta   --0010000000-100100010-10011000x10x1100xx10x-011011-0-1------
Corys   --00100x1010--------11010000x10x0110xx10001110x1001100-1----
Cayto   -xx0110x00000x111000x101101100x100x100x10001011010001000010000
Cycad   10x0110x0001-x1100101x1-xx11100x0001x011x100100110101011--0--0
Benne   10x0110x0000-1000010100--xx10100x100xx00x1xx100--1100-0---
Pento   1010xx1x0000-x100000010-x01x0101100xx0001x1001-1000100-000-0--
Corda   1010xx1x00000x1101000100x01x0101100x10x1000x10000110100010000000
Ginkg   1010xx1x000000x1101000x10x01x0101100xx1000x1000010010010010000
Conif   11x1xx1x00001x110101001xx1xx1x101xx0111x101001110111011110100000
Ephed   11x1x1x10010x11-1011x1xx1xx11x101xx0111x101001101011x1111111
Welwi   11x1110x11011x11111111101xx11x101xx11x101xx0x11100100111000X1111111
Gnetu   10x0110x11111x11101x10011101100xx10x1x10x1000011101011011x10010
```

(Axelrod, 1959; Takhtajan, 1969; see also Melville, 1983). In spite of the variety of new environments to which they were then exposed, the early angiosperms were supposed to have undergone little or no evolution since that time. It is easy to see that there are some fundamental flaws in this scenario.

In Chapter 5 we saw how isolated populations tend to undergo genetic modification but generally speaking heterogeneity within a population only arises when there are barriers to breeding and gene flow. Most speciation events require some form of break-up of a breeding population into separate isolated communities. Fragmentation of this type, sufficient to bring about diversification to the level of modern genera and families, would have been impossible in the hypothetical homeland, unless it occupied a significant proportion of the Earth's surface. If this were the case, we would expect to find some fossil evidence for the existence of the homeland even if it were only in the pollen record. This we do not find. Perhaps even more unlikely is the implication that after the angiosperms spread from their homeland they underwent very little evolution even though they migrated into new environments and would have experienced an inevitable increase in the frequency of geographic isolation and diverse environmental selection. The opportunity for evolutionary innovation in the angiosperms must have been enormous while they competed with the pre-existing conifer, cycadophyte and fern vegetation, and yet according to the scenario, evolutionary change was minimal. Supporters of the 'homeland' hypothesis required the angiosperms to have become preadapted to all the environments they were later to invade and dominate.

Suggestions as to the location of the original angiosperm homeland have ranged from the Arctic (Heer, 1868; Saporta, 1877), the Indo-Australian region (Takhtajan, 1969) or the Permo-Triassic tropical uplands (Axelrod, 1952, 1960).

Work by Doyle and Hickey (1976) and Hickey and Doyle (1977) showed, however, that it is not necessary to hypothesise about such homelands. Almost certainly they did not exist. Pollen evidence indicates that it is extremely unlikely that angiosperms existed before the early Lower Cretaceous or that they were evolving far from sites of deposition. Instead both pollen and leaves show evidence for diversification in early Cretaceous times, and additional work has shown that such patterns of diversification are seen in widely separated areas (Doyle, 1978).

Pollen Evidence

The earliest generally accepted angiosperm pollen grains are monosulcate (that is, they have a single furrow) and are included in the genera *Clavatipollenites*, *Retimonocolpites* and *Stellatopollis*. These grains exhibit either an exine structure composed of radial rods, or columellae, which connect the inner exine layer (or nexine) with an outer perforated tectum (Figure 13.2) or, in grains with a coarsely reticulate tectum (*Retimonocolpites*), the columellae may be absent. Whether or not the columellae are present, the walls of angiosperm grains are distinct from the honeycombed or alveolar exine of gymnosperms such as the cycads and saccate conifers, or the granular exine of the Gnetales and most non-saccate conifers (Van Campo, 1971; Hickey and Doyle, 1977). Another feature that appears to site angiosperm pollen apart from that of gymnosperm groups is the lack of a laminated endexine.

Using these criteria the earliest undisputed angiosperm pollen records are in the Barremian (Couper, 1958; Kemp, 1968; Doyle, 1969; Doyle and Robbins, 1977; Doyle *et al.*, 1977; Hughes *et al.*, 1979) and are found in Europe, Africa and South America. However, it is unlikely that these records represent the earliest angiosperms because there is already evidence of floral provincialism. Pollen from Africa and South America has a predominantly tectate, possibly granular, exine structure and that from Laurasia has a reticulate columellar organisation such as is seen in *Clavatipollenites*. At higher stratigraphic horizons the divergence and frequency of angiospermous grains increases (Doyle *et al.*, 1982). This pattern of occurrences strongly suggests that the pollen record is witness to a main evolutionary radiation of the angiosperms within the Cretaceous (Müller, 1970; Doyle and Hickey, 1976; Hughes, 1976; Hickey and Doyle, 1977; Doyle *et al.*, 1982). Continuous early Cretaceous angiosperm pollen successions are best known in England (Kemp, 1968; Laing, 1975; Hughes, 1977), the Atlantic coastal plain of the USA (Brenner, 1963; Doyle, 1969; Wolfe and Pakiser, 1971; Doyle and Hickey, 1976; Doyle and Robbins, 1977), western Canada (Norris, Janzen and Awai-Thorne, 1975; Singh, 1975), Brazil (H. Müller, 1966; Herngreen, 1973; Brenner, 1976), equatorial Africa (Jardiné and Magloire, 1965; Jardiné, Kieser and Reyre, 1974; Doyle, Biens, Doerenkamp and Jardiné, 1977), and Australia (Dettmann, 1973). However, there is more diversity, and several important pollen types appear earlier, in Africa and South America, which suggests that this was a centre for the early angiosperm radiation and could even be their centre of origin.

Although first to appear, small monocolpate grains are not typical of the majority of angiosperms. Most angiosperm pollen is of the tricolpate (having three furrows) or triporate (three pores) type. Grains with a tricolpate kind of organisation first occur somewhat later than the monosulcates in pre-Upper Aptian beds of Brazil (H. Müller, 1966; Brenner, 1976), equatorial Africa (Jardiné *et al.*, 1974; Doyle *et al.*, 1977), and Israel (Brenner, 1976). But it is not until the Early Albian that tricolpates are found in Europe or North America. At high latitudes they first occur even later. Details of this apparent poleward spread of the tricolpate producers will be discussed in the next chapter, but it is significant that the centre of origin was at low latitudes and apparently in the Africa/South America region. Similarly the first occurrences of grains with pores are found in the middle and late Albian of Brazil and equatorial Africa (Jardiné and Magloire, 1965; Boltenhagen; 1967; Brenner, 1968; Herngreen, 1973). Currently available data suggest that the most likely area for the origin and diversification of the angiosperms was in northwestern Gondwana, probably in the rift valley separating South America and Africa (Retallack and Dilcher, 1981a).

Leaf Evidence

The leaf record of very early angiosperms is less clear than that of pollen. This is because leaf remains are less abundant than pollen and represent communities growing close to the depositional environment. Also, and perhaps most importantly, the true affinity of detached leaves is often difficult to determine with any certainty. Hickey and Wolfe (1975) list several characters, the presence of any of which they consider to be strong evidence of an angiosperm affinity. Briefly these characters are:

(1) blade expansion predominantly the result of intercalary growth;
(2) the presence of stipules;
(3) venation consisting of several discrete orders;
(4) the presence of freely ending veinlets;
(5) the anastomosing of veins between two or more orders of veins.

On this basis, a small (25 mm), simple, pinnately veined leaf from the Neocomian of Siberia (Vahkrameev, 1973) may be a candidate for the earliest angiosperm leaf record. Of more certain angiosperm affinity are pinnate leaves with toothed margins that have been reported by Pons (1982) from the Aptian of Colombia. By Aptian times leaves of the Potomac Group on the Atlantic Coastal Plain of the USA exhibit a variety of forms ranging in shape from oblanceolate to reniform and lobate. Apart from having a basically pinnate pattern, the venation of all these leaves is poorly organised with irregular course, spacing, branching and looping of the secondary veins and poor differentiation of the vein orders (Doyle and Hickey, 1976; Hickey and Doyle, 1977). Similar types of leaves are evident in deposits of equivalent age from Portugal (Saporta, 1894). The level of organisation exhibited by these leaf forms is fully in accordance with the pollen evidence, suggesting an initial radiation of flowering plants in the late early Cretaceous.

The poorly organised venation of these leaves would seem to indicate that there was a major reorganisation of developmental processes just prior to the first entry of angiosperms into the fossil record. This of course means that the angiosperm ancestor (or ancestors) need not have had a leaf form even remotely resembling that of the descendants, and that the possession of an angiosperm-like leaf, with for instance reticulate venation (e.g. *Sagenopteris, Gnetum*), is poor evidence for angiosperm ancestry.

Stebbins (1965) suggested that primitive angiosperms might have been semi-xerophytic shrubs, and to date the early angiosperm leaf record seems to lend some support to this concept. Small, entire-margined spatulate forms with poor differentiation of petiole from lamina are typical of the earliest leaves and are precisely the kinds of leaves that are suited to physiologically arid environments.

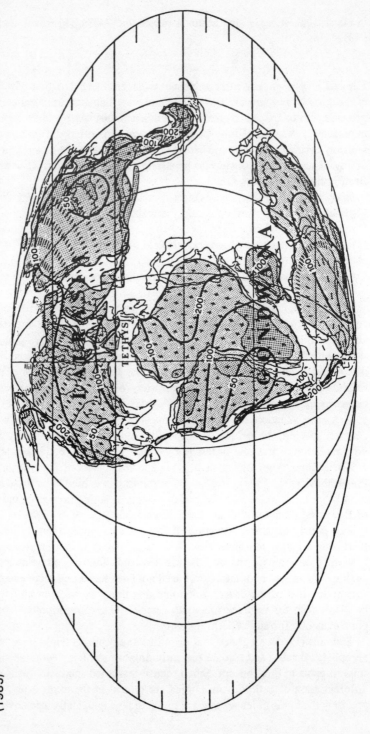

Figure 13.4: Barremian Rainfall Map. This map shows the positions of the continents during Barremian times (late early Cretaceous) and predicted relative rainfall patterns are shown by the contour lines. No specific units are implied. Note the relative wetness of northern Africa and the range of rainfall along the South American/African rift system. Map from Parrish (1985)

Doyle and Hickey (1976) and Hickey and Doyle (1977) have discussed the implications of this at some length and envisage a xeromorphic 'bottleneck' in angiosperm evolution. According to this hypothesis the angiosperms arose from a mesic gymnosperm ancestor under selective pressures for efficient reproduction and growth in an unstable semi-arid environment. Reduced leaf size and architecture would have been well suited to such conditions and expansion of the lamina and elaboration of form and venation would have been associated with the development of intercalary growth during leaf development and subsequent invasion of a variety of more mesic environments.

Environment, and in particular environmental change, plays a crucial role in plant evolution. Any serious consideration of the origin and early radiation of a major group like the angiosperms therefore has to examine environmental conditions, in so far as they can be determined, that prevailed at that time.

Figure 13.4 is a palaeocontinental reconstruction for Barremian times overlain by contour lines representing relative annual rainfall units. (For an explanation of how maps of this type are produced, see Parrish, 1982; Parrish, Ziegler and Scotese, 1982). In Northern Africa and bordering the Tethyan seaway the annual rainfall was quite high and apparently remained so through into the Cenomanian (Parrish *et al.*, 1982). This is in accordance with the suggestion by Doyle *et al.* (1982) that there was a wet equatorial belt in the mid-Cretaceous. Further south along the South America/Africa rift-valley system the climate became increasingly drier. If this rift-valley system was a cradle for early angiosperms, they were clearly exposed to a range of climatic conditions very early in their radiation: a range of climatic conditions which may have stimulated their subsequent dynamic evolutionary development.

Until fairly recently it was generally agreed that the fossil record had little to offer in the way of detailed information relevant to the initial radiation of the angiosperms: comparative morphological studies of living plants seemed the only way forward. Since the early 1970s, however, there has been a dramatic change in the level of information supplied by palaeobotanical studies. This has come about not as the result of any new fossil finds (although new material has resulted from intensified activity) but from a change in philosophy and the way the fossil record is interpreted. Responsibility for these new insights must rest in large part with J.A. Doyle and L.J. Hickey who demonstrated for the first time that we do have a record of some early evolutionary trends. They did this by combining palynological and megafossil studies, by making detailed comparative morphological studies of fossil material in stratigraphic sequence, and by paying careful attention to the depositional environments in which their assemblages had accumulated (Doyle and Hickey, 1976; Hickey and Doyle, 1977). We will return to this topic later, but first it is pertinent to examine the global environment immediately prior to the rise of the angiosperms.

The climate appears to have been more uniform than it is today, with a less steep latitudinal temperature gradient. The poles were comparatively warm and may well have been free of ice (Barron, Thompson and Schneider, 1981; Barron and Washington, 1982). Any glaciation there may have been was limited to small areas of high elevation at high latitudes.

The vegetation, however, was not uniform and the early Cretaceous was a time of significant floral provincialism. Brenner (1976) identifies four palynofloral provinces: Northern and Southern Laurasia and Northern and Southern Gondwana. Assemblages of the Laurasian provinces are dominated by gymnospermous bisaccate pollen (Pinaceae and probably Podocarpaceae), pollen of taxodiaceous-cupressaceous types, Araucariaceae, Cheirolepidiaceae, Caytoniaceae, Cycadales and Cycadeoidales, some Gnetales, and spores of schizaeaceous, gleicheniaceous and cyatheaceous ferns, some lycophytes and bryophytes. The Northern Laurasian province can be distinguished from the Southern by its lack of *Classopollis* (Cheirolepidiaceae), ephedroids, and many fern spore types. Overall it has a lower diversity than the Southern Laurasian province.

In the Northern Gondwana province bisaccate grains and the taxodiaceous-cupressaceous forms are rare and there is lower diversity and abundance of spores. Instead there is evidence that the Cheirolepidiaceae, Araucariaceae, Gnetales and Cycadeoidales were abundant. In the Southern Gondwana province the Podocarpaceae and ferns were more prevalent.

On geochemical and other evidence Doyle *et al.* (1982) conclude that there was some aridisation, although perhaps local, in the Northern Gonwanan rift-valley system during Barremian to early Aptian times. This coincides with the first appearance of angiospermous pollen in the region in the form of coarsely reticulate grains that we call *Afropollis*. Initially one might think that this seems to confirm Stebbins' (1965) proposal that the angiosperms arose in a semi-arid environment. However, angiospermous pollen appears elsewhere at the same time, or even earlier, in regions which appear to have been uniformly wetter throughout the Cretaceous. Moreover species of *Afropollis* occur in greatest numbers in the equatorial wet region during the Upper Aptian. In fact, *Afropollis* appears to be associated with a wide range of environments, including moist lowlands, intercontinental rift systems and coastal habitats. Whatever kind of environment the angiosperms first arose in, they very rapidly became adapted to a variety of others.

Retallack and Dilcher (1981c) have proposed that many early angiosperms were coastal, even mangrove-like plants, and that the geographic spread of the group was enhanced by widespread marine transgressions in the mid-Cretaceous. Such a hypothesis, however, seems somewhat simplistic and perhaps too restrictive in view of the early angiosperm occurrence in both coastal and inland sites.

An important key to understanding the angiosperm radiation is to be found in the work of Doyle and Hickey on the Potomac group (Doyle and Hickey, 1976). Using well-defined pollen zones, and fossils from associated marine rocks, they were able to correlate their non-marine rock units with those that were better dated in Europe and provide a stratigraphic framework for the interpretation of the leaf fossils. Another significant feature of their work was the attention paid to the association of particular fossil forms with particular rock types and their original environments of deposition. This facies analysis has proved to be of great importance when trying to reconstruct early angiosperm communities.

One of the reasons why early angiosperm pollen grains were so stratigraphically useful was that the angiosperm pollen appeared to reflect the evolution of the group as a whole, and changes in pollen morphology occur frequently throughout the section. Added to this is the fact that pollen and spore assemblages change as the pre-existing fern and gymnosperm vegetation was infiltrated and in many areas eventually replaced by an angiosperm-rich flora. These community changes appear to be independent of climate or other physical environmental factors and largely reflect inter-plant competition.

At the base of the stratigraphic interval studied by Doyle and Hickey the angiosperm pollen is rare, small and monosulcate, and falls within the *Clavatipollenites-Retimonocolpites-Liliacidites* complex (Figure 14.1). Most of the pollen and spores of the Barremian to lower Albian zone I assemblages were produced by ferns, cycadophytes, ginkgophytes and conifers.

At the top of zone I (lower Albian) the first tricolpate pollen appears. This is significant because tricolpates are highly distinctive of dicotyledonous

Figure 14.1: Generalised Pollen Types from the Potomac Group Showing Evolutionary Trends and Stratigraphic Ranges. The dashed line indicates evolutionary transformations for which the fossil record provides indirect evidence only. a, Angiospermous tectate-columellate monosulcates of the *Clavatipollenites-Retimonocolpites* type. b, Monocotyledonoid reticulate monosulcate grains of the *Liliacidites* type. c, Tectate-reticulate tricolpates broadly of the *Tricolpites* type. d, Generally tectate-reticulate tricolporoidates of the *Tricolpites* and *Tricolporoidites* type. e, Small, generally smooth-walled tricolporoidates broadly of the *Tricolporoidites* type. f, Small, oblate-triangular tricolporoidate grains with smooth walls (*Tricolporoidites*). g, Larger tricolp(oroid)ates, often with more sculptured surfaces (*Tricolporoidites, Tricolporopollenites* spp.). h, Larger oblate-triangular, often highly sculptured, tricolp(oroid)ates (*Tricolporoidites, Tricolporopollenites* spp.). i, Primitive triangular triporate members of the Normapolles complex. (After Doyle and Hickey, 1976)

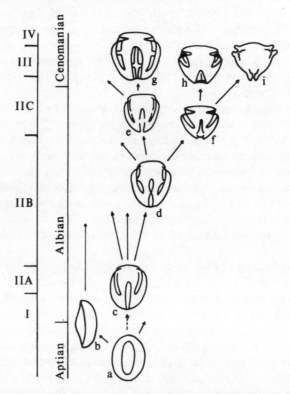

angiosperms. Both tricolpates and monosulcates diversify in subzones IIA and IIB, and many tricolpates develop thin areas in the centre of the colpi. These 'tricolporoidates', as they have been called, diversify further in size, shape and sculpture in subzone IIC and zone III (upper Albian to Cenomanian). These forms presumably gave rise to the truly tricolporates of zone III. In zone IV (Cenomanian), grains with round pores rather than furrows appear. These are the first triangular triporates of the so-called Normapolles complex which becomes a significant element in assemblages in eastern North America and Europe during the late Cretaceous and early Tertiary (see Chapters 15 and 16).

Much of the Potomac pollen zonation was based on subsurface borehole data as these provide good vertical stratigraphic control. The surface outcrops yielding the leaf fossils were positioned stratigraphically using the pollen/spore assemblages they contained. With due regard to the taphonomic influences governing the formation of pollen and leaf assemblages, and the resulting facies biases, Doyle and Hickey (1976) were able to derive a reasonably complete picture of the range of leaf morphologies and ecological heterogeneity of the angiosperm flora at several points in their stratigraphic sequence. Furthermore, instead of comparing the fossil leaves with living forms, they examined the fossils in their own right relative to one another and their stratigraphic position. Leaf architectural features (venation pattern, margin characteristics and overall leaf organisation) were used to document evolutionary trends in stratigraphic sequence. Conventional taxonomic partitioning was largely ignored because previously it had been misleading and it proved to be largely irrelevant to the study of evolutionary change.

In the lower Albian zone I assemblages of Virginia and Maryland, entire-margined pinnately veined leaves and irregularly lobed forms occur with poorly organised venation. Anastomosing veins enclose elongate rhombic areas of lamina (Figure 14.2a). In middle subzone IIB (middle to upper Albian) palmately veined forms occur for the first time and exhibit more regularity in their vein courses than the leaves of zone I. In such forms as *Menispermites virginiensis* we find even the tertiary veins have a tendency towards a degree of regularity. By the middle of subzone IIB we also see leaf margins with double convex glandular serrations and both pinnately and palmately lobed forms (Figure 14.2h,i,j).

Upper subzone IIB (upper Albian) sees a possible elaboration of the cordate-reniform complex as typified by the genera *Menispermites* and *Populophyllum* (Figure 14.2k,l).

Truly pinnately compound leaves first appear in upper subzone IIB and the leaflets have secondary and tertiary vein orders that are readily distinguishable.

Palmately lobed leaves are also present in upper subzone IIB and later become locally dominant in subzone II (uppermost Albian to lower Cenomanian). Confined to the coarser fluvial facies, they are typified as *'Sassafras' potomacensis* Berry (Figure 14.2m).

Mid-Cretaceous angiosperm leaves (Figure 14.3) may be broadly grouped into a number of rather loosely defined categories only some of which we will be discussing here. Architecturally the most simple are the pinnately veined forms with entire margins. Lateral branches of the midvein (secondary veins) usually loop near the margin (Figure 14.3a) and are described as being brochidodromous. In the earliest forms the venation is poorly organised. Later more regular venation does occur, but there are extant leaves of this general

Figure 14.2: Examples of Mid-Cretaceous Angiosperm Leaves from the Potomac Group Illustrating Increasing Levels of Vein Organisation and Overall Leaf Morphology from Successively Higher Zones. a, *Rogersia augustifolia* Fontaine; b, leaf of the *Celastrophyllum latifolium* Fontaine complex; c, *Proteaephyllum reniforme* Fontaine; d, *Vitiphyllum multifidum* Fontaine; e, *Acaciaephyllum spatulatum* Fontaine; f, *Proteaephyllum dentatum* Fontaine; g, *Menispermites virginiensis* Fontaine; h, *'Populus' potomacensis* Ward; i, *Sapindopsis magnifolia* Fontaine; j, member of the *Araliaephyllum obtusilolum* Fontaine complex; k, leaf similar to some members of the *Populophyllum reniforme* Fontaine complex; l, *Menispermites potomacensis* Berry; m, *'Sassafras' potomacensis* Berry. (After Doyle and Hickey, 1976)

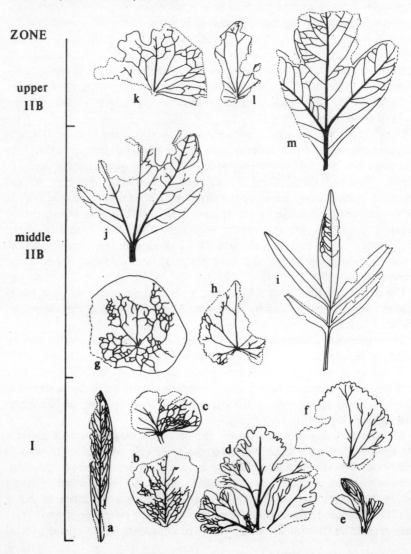

Figure 14.3: Examples of Some Common Late Cretaceous Angiosperm Leaf Forms Based on Actual Specimens. a, Pinnately veined, entire-margined leaf with secondary veins that loop near the margin (brochidodromous). This is typical of a range of leaf forms broadly categorised as being magnoliid-like. In this specimen the secondary veins are not well organised in that they are somewhat erratic in course and strength. b, Palmately veined leaf with a toothed margin. In addition to a primary midvein two lateral primary or pectinal veins (Spicer, 1986a) depart from near the base of the leaf. The venation runs into the teeth (craspedodromous), which have a hollow terminal gland and are of the platanoid type (Hickey and Wolfe, 1975). c, Similar to 'b' but with a lobate base and stronger pectinal veins. Both 'b' and 'c' are typical of late Cretaceous platanoid leaves. d, Detail of the platanoid-type tooth. e, Detail of a tooth type known as chloranthoid (Hickey and Wolfe, 1975) with rounded apices and acute sinuses. The apical gland has a rounded apex and the medial vein is served by a series of looping veinlets. f, An acrodromously veined leaf of a type architecturally similar to those of the extant *Trochodendron*. In acrodromous leaves the pectinal veins curve inwards towards the leaf apex. The secondary veins and their subsidiaries are brochidodromous but give off veinlets that run into the chloranthoid teeth. This type of marginal venation is known as semicraspedodromous (Hickey, 1973). g, A palmately veined leaf with several sets of pectinal veins. The teeth are of the chloranthoid type and the leaf is of the *Menispermites* type. For a more comprehensive account of leaf architectural features see Hickey (1973) or Dilcher (1974)

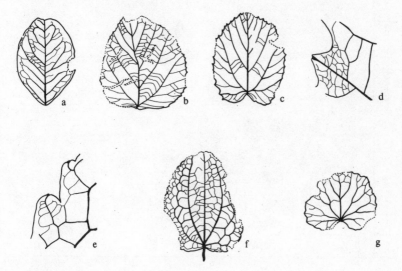

form with also a low level of vein organisation. These vein patterns are seen in modern members of the Magnoliidae and for this reason the fossil forms may be termed magnoliid-like — not because they are necessarily related to the magnoliids but because they display similar vein organisation.

The second major type of leaf is typified by the so-called *Menispermites* forms (Figure 14.3g). These are more or less circular or reniform, with a series of major veins that radiate from a single point. Some may represent aquatic plants, but the majority appear to be climbers.

The third major form, and one in which there is considerable morphological diversity, is that of the platanoid-like leaves (Figure 14.3b,c,d). These are fundamentally palmately veined, they often have toothed margins (although some of the early forms had entire margins), and are typified by leaves of the *Araliopsoides, 'Sassafras',* and *Platanophyllum* types.

Intergradation of Leaf Form

A marked characteristic of mid- and early late Cretaceous leaves (particularly the platanoids) is a high degree of morphological intergradation. Many of the leaf forms exhibit considerable variation in size, shape and venation. So numerous and subtle are the variations that they often appear as part of a morphological continuum (Figure 14.4).

Figure 14.4: An Example of the Range of Morphological Variation Exhibited by Some Platanoid-like Leaves from the Cretaceous of Alaska (Spicer, 1986b)

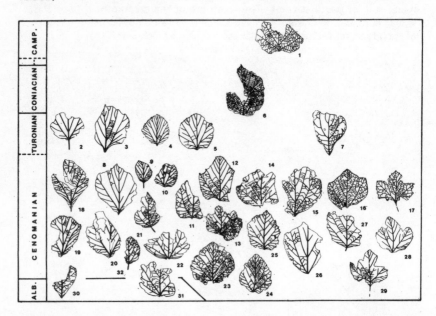

This intergradation has posed serious problems for those attempting to seek the closest matches with extant leaves, and often very similar fossils have been assigned to diverse taxa. Furthermore, adherence to formal taxonomic partitioning and nomenclature can obscure subtle architectural modifications which may prove phylogenetically significant. Because of these problems terminologies have been developed to facilitate concise analyses of leaf architecture (for example Hickey, 1973, 1979; Spicer, 1986b) that allow detailed comparative studies to be made of leaf form (Doyle and Hickey, 1976; Spicer, 1986a).

Heterophylly

In many angiosperms young plants produce 'juvenile' leaves morphologically different from those of mature plants. Also environmental factors, particularly the availability of moisture and the amount of light, can strongly influence the size and morphology of leaves in many taxa (e.g. *Ipomea*).

Plants that can modify their morphological development, especially their leaves, in response to environmental conditions are able to maximise their capacity for photosynthesis. This ability is clearly advantageous to the plants' survival and success. In some cases variation in leaf morphology (heterophylly) is determined by the reproductive or vegetative nature of the shoot. For example, in the ivy, *Hedera helix*, the leaves on the flowering shoots are simple and pinnate whereas those on the vegetative shoots are palmately lobed (Figure 14.5).

Figure 14.5: Heterophylly as Seen in *Hedera helix*. A: Simple leaf from a flowering shoot. B: Palmately lobed leaf from a vegetative shoot. Scale bar represents 1 cm

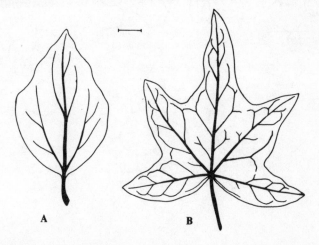

Heterophylly and phenotypic plasticity are generally more common in angiosperms than in any other plant group. Unfortunately it is very difficult to determine when this capability might have evolved due to the problems of separating heterophylly from morphological leaf variation in fossil assemblages. However, as the extent of phenotypic plasticity differs considerably between taxa, and different organs exhibit varying degrees of plasticity, it seems likely that this ability arose independently a great number of times. Phenotypic plasticity could well have evolved in some mid-Cretaceous groups and even contributed to the success of the angiosperms.

Cuticle Studies

Angiosperm leaf cuticles, like those of gymnosperms and pteridophytes, are resistant to decay (and thus have a high fossilisation potential) and display a large number of characters that are useful in taxonomic and evolutionary studies (Dilcher, 1974; Kovach and Dilcher, 1984). Angiosperm leaves often possess small hairs or trichomes that may be simple or highly branched, single celled or multicellular. The pattern of epidermal cells also provides useful character suites but perhaps the most informative characters regarding early angiosperm evolution are those related to stomatal organisation.

Figure 14.6: Patterns of Variation Seen in the Stomatal Organisation of Early Angiosperm Leaves of the Potomac Group. A: zone I; B: pattern seen in *Menispermites potomacensis*; C: stomatal organisation in *Sapindopsis*/platanoid leaves; D: stomatal organisation seen in an unnamed leaf from subzone IIB. (After Upchurch, 1984)

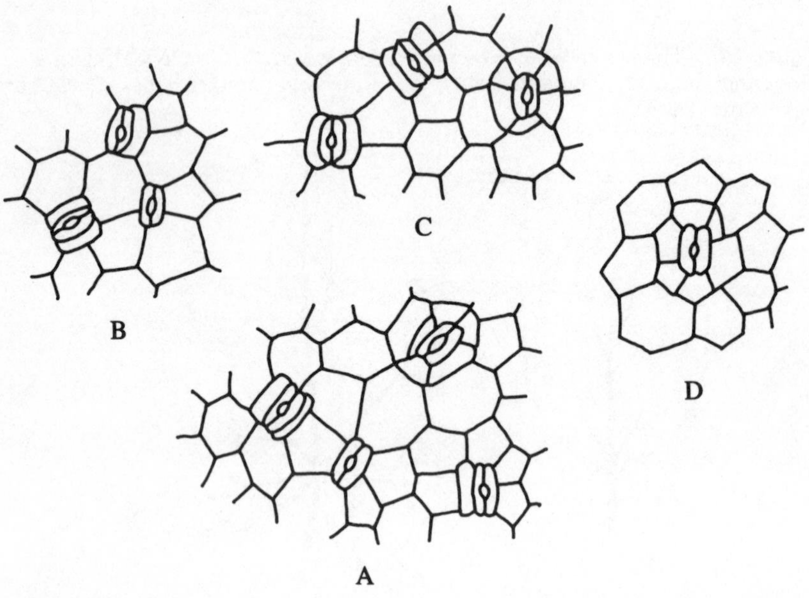

Upchurch (1984) has examined the cuticles of leaves from the Potomac group and has found that zone I cuticles exhibit little interspecific diversity but have a highly plastic pattern of variation in stomatal organisation. From this melange of plasticity emerged more structured stomatal diversity in the cuticles of zone IIB, and the stomatal organisation became more complex (Figure 14.6). The earliest cuticles are reminiscent of those found in extant Magnoliidae whereas the younger forms show affinities with the Platanaceae and Rosideae: a finding consistent with the supposed affinities based on leaf architecture.

The Causes of Early Morphological Intergradation

The high level of morphological variability exhibited by angiosperm leaves during the major radiation and geographic spread of the group is probably attributable to several factors. Morphological similarity between species may well be high early in the history of a group, but it may also be due to depositional mixing of genetically separate populations of plants (species) that produced leaves with overlapping characteristics.

Morphological variability could also be largely the result of high levels of interbreeding. The low level of specialisation in floral morphology (Dilcher, 1979) and the widespread and indiscriminate dispersal of copious pollen could have facilitated the potential for genetic exchange between members of the group. The probability that two individuals would be genetically compatible is perhaps higher in the initial stages of an actively radiating group because there will be a great likelihood of overall similarity of genetic complement between individuals in the population. These factors will tend to increase the frequency of interbreeding, with the result that biological species limits, and therefore taxonomic partitions, will be blurred.

Widespread frequent hybridisation might be expected to depress evolution rates in that characters will tend to become blended. However, during early phases of major radiations the opposite may be true. Hybridisation may produce novel combinations of characters that are better suited to their environment than those of the parental stocks. The hybrid population may then outcompete parental populations and eventually displace them. Nearer the origin of a novel group there is 'more room for improvement' and successful hybrids are likely to occur more frequently than later when the group is more highly evolved. The improved adaptation of the hybrid populations to their specific environments eventually may lead to population segregation, isolation and more well-defined species boundaries.

Facies Associations and Plant Communities

The Potomac group is composed of sediments that were laid down in a mesic floodplain environment. Coarse fluvial deposits representing stream levee, point-bar and channel-fills are present, as are finer grained deposits of the flood basins and back-swamps. Doyle and Hickey (1976) observed that the zone I leaves are largely restricted to the coarser fluvial sediments, suggesting that the plants that produced them grew along stream margins and were pioneering plants able to cope with fluctuating water tables and unstable immature soils of river-banks and bars. This fits well with the contention that early angiosperms arose from mesic gymnosperms via unstable, semi-arid, transitional environments where the comparatively rapid reproductive cycle and morphologically more flexible vegetative features gave the angiosperms comparative advantages (Stebbins, 1965, 1974; Doyle and Hickey, 1976).

From their first appearance the pinnately lobed *Sapindopsis* and palmately organised platanoid forms of Potomac subzones IIB and IIC are also strongly associated with channel sand and stream levee deposits. *Sapindopsis* is the first angiosperm apparently to dominate the flora at any one locality. This suggests that the platanoid leaf producers were riparian trees, and like many colonisers of disturbed sites formed more or less pure stands. An association between platanoid leaves and fluvial facies has also been reported from the early Cenomanian of the Dakota group (Retallack and Dilcher, 1981c), Albian-Cenomanian deposits of north-eastern USSR (Lebedev, 1976) and Cenomanian sediments of the Yukon-Koyukuk Basin and North Slope, Alaska (Spicer and Parrish, 1986).

This consistent facies association suggests two things. The first is that to a large extent the leaf architectural group we informally refer to as 'platanoids' appears to have some real biological coherence in that the plants which produced these leaves had a certain limited range of ecological tolerances. Secondly, the platanoid ecological unit remained stable during the initial geographical spread of the angiosperms. Doyle and Hickey suggest that, if the first primitive angiosperms were relatively small, weedy plants, they might have begun their invasion of forest environments not as canopy trees but as understorey shrubs. Some of the large, simple low-rank (poorly organised) leaves of the Potomac zone I may possibly represent the first example of adaptations to an understorey existence, not only on the basis of leaf morphology but also on the grounds of their comparative rarity even when they are at their most abundant in predominantly gymnosperm/fern assemblages.

In general it was the magnoliid-like producers that grew in the forest environments. Evidence from this comes from the association of magnoliid-like leaves with fern and gymnosperm remains, all of which are restricted to fine-grained facies of the interfluvial swamps and lakes. The angiosperms remain a subordinate element in these forests until well into the late Cretaceous and in some areas remain so to the present day.

These facies associations are consistently evident in other mid- and late Cretaceous assemblages. Retallack and Dilcher (1981c), in their description of the plant fossils of the Dakota Formation in Kansas, report angiosperm remains in a more diverse range of environments. They identified a scrubby angiospermous mangrove-like community, or mangal, characterised by *Ascerites multiformis*, growing on distributory margins of tide-dominated deltas, while freshwater-influenced swamp woodlands were apparently dominated by plants bearing *Magnoliaephyllum, Liriophyllum* and *Sapindopsis* leaves. The swamp woodland understorey was mainly composed of ferns. Platanoid leaf producers were more abundant on channel levees and around lakes, whereas the drier floodplain forests were conifer dominated.

Figure 14.7: Diagram Showing the Latitudinal Relationship of First Occurrences of Angiosperm Pollen. Open circles, monosulcate grains; closed circles, tricolpate grains; split circles, monosulcate and tricolpate grains. Geographic areas: 1 — New Zealand, 2 — SE Australia, 3 — Patagonia, 4 — Congo, 5 — eastern Brazil, 6 — Peru, 7 — Ivory Coast, 8 — Israel, 9 — Gabon, 10 — Portugal, 11 — Oklahoma, 12 — Potomac group, 13 — Maryland, Virginia, USA and southern England, 14 — SW Siberia, 15 — locality not given, 16 — Denver Basin, 17 — Lower Greensand, 18 — Swan River Group, 19 — Youngstown area, 20 — Fall River Formation, Wyoming, 21 — western Central Siberia, 22 — SE Alberta and SW Manitoba, 23 — SE Alberta, 24 — Loon River, 25 — Central Alberta, 26 — Disco Island, 27 — Ellesmere Island and NE Siberia, 28 — Yukon, 29 — Kuk River, Alaska, 30 — Umiat Region, Alaska. (After Hickey and Doyle, 1977)

PALAEOLATITUDE

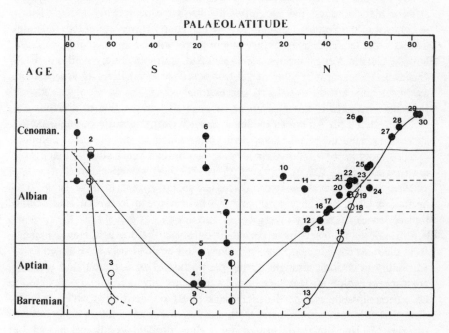

The Poleward Spread

The poleward spread of angiosperms, first postulated by Axelrod (1959), is documented by both pollen and leaf records. Figure 14.7 shows the first occurrence of angiospermous pollen at different latitudes, and the poleward trend is clearly demonstrated. The concurrent rise in platanoid leaf abundances and tricolpate and tricolporoidate pollen zones IIB and IIC of the Potomac sequence (Doyle and Hickey, 1976) strongly suggests that a significant proportion of the platanoid group were producers of tricolpate and tricolporoidate grains. In western Canada tricolpate grains first appear in sediments of late Albian age (Janzen and Norris, 1975; Singh, 1975), but in the Canadian Arctic and northern

Alaska they do not occur until the very latest Albian or early Cenomanian (Hopkins, 1974; Brenner, 1976; Scott and Smiley, 1979). Collections are still somewhat inadequate at these high latitudes, but it seems that the angiosperm leaves that first appear in any abundance are allied to the platanoid complex. It is only later that the magnoliid forms appear. Facies associations for the early phases of angiosperm influx on the Arctic Slope of Alaska, and just south of the Brooks Range in the Yukon-Koyukuk basin, show precisely the same pattern as in the Potomac. The platanoids are found in fluvial sands, and the magnoliid-like leaves and *Menispermites* forms are restricted to fine paludal sediments together with a strong gymnosperm presence (Spicer and Parrish, 1986).

Most palaeogeographic reconstructions for the late Cretaceous show the Arctic Slope of Alaska at about 75°N to 85°N. When angiosperms do appear at these high latitudes, the leaf floras display some interesting characteristics. In contrast to the Potomac sequence, in which zone I leaves are of low rank and thereafer the leaves display progressively more ordered architecture, the high-latitude platanoid angiosperms are first found with moderately well-organised venation. The earliest developmental trends seen at lower latitudes are missing, as presumably are the weed and successional ecological development stages. Instead forms akin to those of upper subzone IIB are the first to arrive.

From the North American evidence a northward migration of angiosperms appears to have taken place and leaf forms allied to the platanoid complex reached latitudes of 75°N within at most two million years after similar forms are first found at palaeolatitudes of 35°N (Potomac sequence).

Many of the leaves we have discussed so far appear to have been from deciduous trees. We have to distinguish here between trees that abscise and replace leaves continually (evergreens), and those trees that abscise their leaves synchronously followed by a period of dormancy before new leaves expand from the buds (deciduous). The deciduous habit could clearly have arisen as an adaptation to periodic drought during the postulated xerophytic phase of early angiosperm evolution. Having evolved such a mechanism it would undoubtedly have been advantageous to the pioneering platanoid weed trees which grew in disturbed sites and were subject to fluctuating water-table levels along stream margins. During times of low water, when drought conditions would be experienced on the well-drained sand bars and river gravels, leaf loss and dormancy would have ensured survival. This ability to survive temporarily adverse conditions must also have been a distinct advantage during the phase of extensive geographic radiation in the early late Cretaceous because it allowed the colonisation of a wide range of environments and facilitated geographic spread to higher latitudes where light and temperature may otherwise have been limiting at certain times in the year.

In presenting his poleward migration hypothesis, Axelrod (1959) suggested that flowering plants only became adapted to swamp sites secondarily, and he envisaged their initially occupying well-drained upland regions. This theme was adopted by Smiley (1967), who was of the opinion that deciduous angiosperms migrated north along upland routes and only later, in response to a deteriorating climate, migrated down slope.

The word 'upland' can be misleading. Often it is taken to indicate high hills or mountainous terrain. This is clearly implied by Smiley's (1967) suggestion that there was sufficient relief for there to be a significant decrease in mean temperature and possibly increased precipitation. Taphonomic studies (Spicer, 1980, 1981) have shown, however, that plant material from high-elevation floras is extremely unlikely to be represented in coastal floodplain depositional environments. If stream-channel distances are short, the gradients are high and the energy of the system destroys the plant remains. However, if the gradient is low, the channel lengths are great and elements from the distant highlands are massively diluted out by plant material from vegetation local to the depositional site. Upland in this context more properly refers to relatively well-drained soils, perhaps a metre or so above the local water table.

The platanoids preserved in the fluvial sands of the floodplain are not derived from highland sources but were probably levee trees. The riparian habit of the platanoids nevertheless would enable them to migrate up and down slope comparatively easily, in contrast to the magnoliid-like leaf producers that were associated with more stable environments. Furthermore, the capacity to tolerate fluctuating water-table levels, which is often necessary for a riparian existence, would have facilitated the crossing of inter-basin highs. These highs need not have been of any great elevation: just sufficient to separate the drainages on a coastal plain. Angiosperm migration routes north were therefore probably not confined to the mountain ranges but followed riparian routes which included disturbed environments and moderately well-drained soils. No doubt the migration of some angiosperms adapted for coastal dwelling was enhanced by fluctuating sea levels during the late Cretaceous, but this would only have applied to a fairly restricted and somewhat specialised group. Far more effective migration was probably over lowland floodplains.

15 THE EVOLUTION OF ANGIOSPERM FLORAL MORPHOLOGY

A flower is an axis upon which spore-bearing assemblages and specialised vegetative structures are borne. There is therefore no fundamental difference between a flower and a strobilus. In general terms the angiosperm flower consists of an axis or pedicel, the distal end of which is enlarged into a receptacle supporting four kinds of floral structure (Figure 15.1). Basal or external to the other organs are the sepals which together form the calyx. The corolla, internal or distal to the calyx, is composed of petals. The calyx and the corolla together form the perianth which surrounds the sexual parts of the flower. In some flowers (for example *Magnolia*) the perianth is composed of structures which are intermediate between sepals and petals. These are known as tepals. The androecium, or set of male organs, is usually composed of a whorl of microsporophylls or stamens, consisting of stalk-like filaments supporting microsporangia or anthers. Fundamentally these are quadriloculate synangia. Centrally located and distal to the other floral parts is the megasporangial apparatus or gynoecium. The pistil or carpel is essentially a megasporophyll enclosing one or more ovules. The megasporophyll or ovary is connected to the pollen — receiving stigma by a stalk-like style.

Figure 15.1: Diagram of a Generalised Angiosperm Flower and Detail of the Ovule. a, Stigma; b, style; c, anther surmounting a filament; d, ovary; e, petal — part of the corolla; f, sepal — part of the calyx; g, receptacle; h, chalaza; i, antipodal cell; j, polar nuclei; k, integuments; l, ovary wall; m, synergid cell; n, funicle

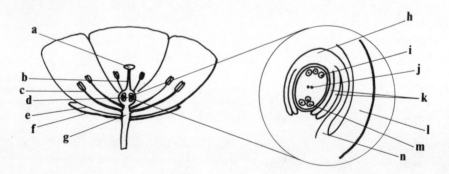

This general plan of the bisporangiate flower is by no means representative of all angiosperms. Whereas many flowers are radially symmetrical or actinomorphic (for example rose and magnolia), others are bilaterally organised or zygomorphic (for example members of the Leguminosae — the pea family). Incomplete flowers are those that lack one or more of the floral parts (for example, grass flowers lack a perianth). The success of the angiosperms is largely due to the considerable variety of floral forms allied in many instances to specific pollination vectors. These in turn can result in pollen targeting and directed gene flow, even in complex communities. It is not only the flower that is important, however, because the angiosperm life cycle embodies numerous competitive advantages over gymnosperm and pteridophyte reproduction.

The Angiosperm Life Cycle

Pollen and Male Gametophyte Formation

Within each developing anther one or more cells give rise to sporogenous cells which become differentiated as microsporocytes suspended in fluid. Meiosis takes place and tetrads of microspores are formed. These tetrahedral, quadrilateral or linear tetrads usually (but not always) break down into individual microspores, each with a single nucleus. At this point pollen wall formation begins. As in other seed plants the male gametophyte begins to form from the uninucleate microspore before it is shed from the miscrosporangium or anther. The single nucleus divides to form two daughter nuclei of different sizes. The larger is known as the tube nucleus and the smaller as the generative nucleus. The cytoplasm then divides. A prothallial cell is not formed. The generative nucleus may divide to form two sperm nuclei before the pollen is shed, or this division may take place in the pollen tube.

Development of the Megaspore and Female Gametophyte

Soon after the ovule begins to develop, a cell near the micropyle differentiates into a primary archesporial cell which may give rise to several sporogenous cells or it may function directly as a megasporocyte. Although some flowering plants differ, the megasporocyte usually undergoes two successive nuclear and cell divisions, during which meiosis takes place, and a linear tetrad is produced. As in most gymnosperms the three megaspores near the micropyle degenerate and the remaining megaspore develops into the female gametophyte.

The female gametophyte is produced as a result of three consecutive divisions yielding eight free nuclei. Four nuclei then migrate to one end of the cell and four to the opposite end. At the micropylar end an egg cell and two synergid cells differentiate as the egg apparatus, while at the opposite, chalazal end three of the four nuclei develop into antipodal cells. The remaining two nuclei from each end, known as polar nuclei, migrate towards each other to form a centrally located binucleate cell (Figure 15.1).

Most angiosperm female gametophytes therefore consist of seven cells, one of which is binucleate, and no archegonia are formed.

Pollination and Fertilisation

In angiosperms, pollen grains usually germinate on a stigmatic surface. The means by which grains arrive at this surface are varied, but in living angiosperms insects and wind are the usual vectors. Complex protein recognition processes and stimuli usually ensure that only pollen of the appropriate type germinates on a given stigma. The criteria for determining which pollen should germinate and which should not are often complex themselves, but the result is that high levels of outcrossing may be achieved. Such sophistication is not a general feature of the gymnospermous pollen-drop mechanism.

Pollen germination is usually rapid and takes place through a thin area of the grain wall known as the germ pore. As the pollen tube grows it absorbs nutrients from the stigma and style. More than one tube may be produced by a single grain.

The pollen tube usually enters the ovule through the micropyle and penetrates the wall of the female gametophyte. The synergids usually disintegrate as this occurs. The two sperm cells and sometimes the tube nucleus are discharged into the female gametophyte.

Both sperms are involved in the fertilisation process. One sperm unites with the egg and so forms the zygote. The other sperm unites with the binucleate cell formed from the polar nuclei and a triploid cell results. Subsequent division of this cell produces the triploid endosperm which functions as the food store in the seed for the developing embryo. This double fertilisation process is, as far as we know, unique to the angiosperms.

In the angiosperm life cycle we find a number of evolutionary advancements over that observed in living gymnosperms. There is a total lack of a moisture requirement for fertilisation, but more importantly the abandonment of the pollen-drop mechanism, and the evolution of a stigmatic surface, also allow complex outcrossing mechanisms to be developed.

The immediate germination of the pollen and the rapid growth of the pollen tube result in fertilisation soon after pollination. This is in contrast to the much slower process in some gymnosperms (for example *Pinus*). The angiosperm life cycle can therefore be completed in a comparatively short space of time, which is important in pioneer plants and those growing in temperate or polar environments.

A flower is an ephemeral structure with delicate parts which may be discarded with a minimum of wastage if fertilisation does not occur. The enclosure of the ovules within carpels protects them from predation and desiccation and allows pollination and fertilisation to occur while the ovules are still comparatively immature with highly reduced megagametophyes. The development of the seed food supply, the endosperm, only takes place after fertilisation has been achieved. Energy wastage in terms of investment of resources in infertile ovules is therefore minimised. In gymnosperms the resources in the cellular megagametophyte are formed prior to fertilisation. They are then wasted if fertilisation is not achieved.

Neoteny, in general terms, means the retention of juvenile characters in the adult state. Takhtajan (1976) argues that the angiosperm flower, and the condensed reproductive cycle, including the double fertilisation, are the result of neotenous simplification from the ancestral gymnosperm condition. The flower represents, for example, not a modified mature shoot but a modified apical bud. The most extreme reduction to a juvenile 'state' is seen not in the sporophyte part of the plant but in the gametophytes. The female gametophyte is extremely simple and is the result of only three mitotic divisions preceded by meiosis. No archegonia are formed. The male gametophyte is an even simpler structure.

The net effect of all these features is to make the angiosperm reproductive cycle not only more efficient but also more rapid. Such reproductive efficiency is at a premium in unstable disturbed environments (Stebbins, 1965); precisely the kind of situation which favours neotenous forms (Takhtajan, 1976) and in which early angiosperms lived.

Pollination Ecology

Pollination in angiosperms is very complex, and any discussion must, of necessity, be in simple terms. In many instances the sophisticated adaptations displayed in extant flowers are the product of co-evolution between the flowering plants and the biotic pollen vectors. Not all co-evolution is manifest in morphological adaptations; important factors might be the timing of flowering and the behaviour of the pollen vector. Under these circumstances the fossil record can have only a limited contribution to the study of pollination biology but, as we shall see, important advances are being made.

Before embarking on a discussion of pollination ecology relative to the palaeobiology of angiosperms it is necessary to define certain terms. In this respect we shall follow closely the terminology of Faegri and van der Pijl (1979). The term 'blossom' refers to the pollination unit in an ecological context, whereas the word 'flower' is a morphological term. A blossom may therefore be an inflorescence, a flower, or part of a flower.

Although cross-pollination or outbreeding (allogamy) is evolutionarily advantageous in that it promotes variation in the gene pool and distributes advantageous mutations (Stebbins, 1957), inbreeding is common in some angiosperms. Angiosperms, in addition, reproduce by vegetative means, can set seed without fertilisation (apomixis), or by self-pollination (autogamy). An autogamous plant may set seed without exposing its anthers or stigmas to a pollinating agent (the term 'anthesis' here is preferable to the term 'flowering'), in which case it is said to exhibit cleistogamy. Seed setting without anthesis is known as chasmogamy. Allogamous plants are by definition chasmogamous, but in addition they can be geitonogamous, where cross-pollination occurs between two flowers on the same plant, or xenogamous where cross-pollination is between two flowers on different plants. In a genetic sense geitonogamy is equivalent to autogamy, and even xenogamy may be equivalent to autogamy if the plants involved belong to the same clone. These distinctions are critical when considering gene flow (Levin, 1978).

The extent to which autogamy, geitonogamy and xenogamy take place depends on reproductive barriers which are themselves a product of the evolutionary process. It is therefore inevitable that the role of the various forms of autogamy and allogamy will have changed with time. From a morphological point of view autogamy is impossible to detect unless the structural organisation of the flower prevents anthesis (cf. *Cycadeoidea*, p. 203). The fossil record can at best therefore document comprehensively only the evolution of reproductive features likely to promote allogamy. Allogamous devices may be categorised in relation to maturation and positioning of sexual organs or blossoms. In cases of geitonogamy all the blossoms are identical, but the timings of pollen production and maturation of the stigma are staggered. All flowers have the potential for producing seeds, but at different times. This includes the condition of protandry where pollen is produced before the stigma is receptive, and protogyny where the stigma is receptive before pollen production.

Faegri and van der Pijl (1979) point out that because the flower is a modified stem apex, centripetal development where bracts would mature before perianth, perianth before corolla, and so on to the apical gynoecium, protandry is the 'normal' condition. However, for protandry to prevent self-pollination, all self-produced pollen must be removed from a flower before the stigma becomes receptive. This does not always occur, however, and protogyny, involving as it does a reversal of the usual maturation sequence, is more effective at promoting allogamy. This selective premium on protogyny may explain why it is found even among otherwise primitive families such as the Ranunculaceae, but it does not explain the dominance of protandry in more advanced families such as the Papillionaceae.

Herkogamy is the spatial separation of anthers and stigma. This of course is the usual condition, but in some flowers (for example orchids) it is more pronounced than in others.

Heteromorphy is a term used to describe the situation where, although both anthers and stigmas are present at the same time for each blossom, there is a different, functionally complementary and interdependent blossom. An example of this is the long-styled and short-styled (pin and thrum) *Primula* (Figure 15.2).

The most extreme morphological condition favouring allogamy is dicliny where anthers and stigmas are separated on different blossoms, either on the same plant (monoecy) or on different plants (dioecy). Dioecy will favour outcrossing to a high degree, but at a price. Half the population do not bear seeds and there is a serious risk that pollination may not occur at all. However, with

Figure 15.2: Heteromorphy as Seen in *Primula veris*. A. Long-styled flower (pin) with anthers below in the corolla tube. B. Short-styled flower (thrum) with anthers at the top of the corolla tube above the stigma

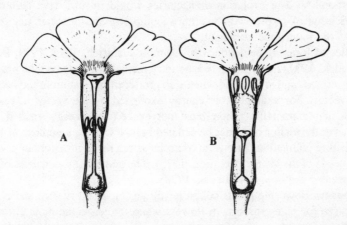

certain types of blossom, notably those that are wind-pollinated, bisexual flowers would inevitably lead to a mutual interference of anthers and stigmas, and self-pollination would be inevitable. Monoecy has the advantage that potentially all individual plants are seed producers.

These devices are not mutually exclusive, and intermediates exist. For example, functional dioecy may occur as a result of staggered maturation of anthers and stigmas in monoecious blossoms. Such 'heterodichogamy' has the effect that it not only prevents autogamy but, unlike normal monoecy, the genetic equivalent of geitonogamy is also prevented.

Pollination may be effected by either biotic (animals) or abiotic (wind and water) vectors, and while by no means absolute, a particular floral morphology is often characteristic of a particular vector. The greatest range of specialisation is seen among the biotically pollinated blossoms, but the morphological adaptations favouring pollination by wind or water can also be extreme.

Pollination by Animals

Although insect pollination (entomophily) is thought to be, or to have been, significant in some gymnosperm groups, there can be no doubt that it is most highly developed in the angiosperms. The often highly complex flower organisation seen in angiosperms is thought to be the result of co-evolution between biotic pollen vectors and the flowers they visit. Basically the plant-pollinator relationship is one in which the pollinator receives a reward, often nutritional, in return for visiting the blossom and incidentally effecting pollination. Early in the development of plant/animal relationships the reward or attractant was the spores or pollen themselves, and relatively unspecialised insects with mandibulate mouthparts could have effected pollen transfer. The evolution of less destructive rewards, such as nectar, was clearly advantageous and, coupled with protective coverings such as cupules, integuments and carpels (which may originally have developed in response to desiccation), improved the efficiency of entomophily. The evolution of nectaries would in turn have favoured the diversification of insects with sucking mouthparts, and in extant angiosperms nectar is the main food attractant.

Studies of foraging behaviour (for example Heinrich, 1976, 1979; Best and Bizychudek, 1981) have illustrated the intricate balance between the amount of nectar secreted, and where and when it is produced in relation to the behaviour of the vector. For example, the energy expended by the vector to reach the blossom and extract the nectar must not exceed the energy yielded by the reward, yet the visitor must not be satiated before visiting a number of flowers and effecting pollination. Provision of nectar is regulated in accordance with the energy needs of the visitor (Heinrich, 1975). The genetic consequences of foraging behaviour are complex (Levin, 1978).

In modern flowering plants pollen is still an important reward and is a vital food source for many insects. In flowers where pollen is the main attractant it is usually produced in large quantities, and some plants (for example *Papaver*) produce a sterile form for nutrition in addition to the fertile grains.

Primary attractants or rewards, in modern angiosperms, include nectar, pollen, fats, oils, sexual stimuli and shelter, and secondary attractants, such as flower shape, colour, motion, scent and temperature, advertise the rewards (Faegri and van der Pijl, 1979).

Five main classes of anthophilous (flower-loving) insects have been identified: Hymenoptera (bees, wasps, ants and sawflies), Lepidoptera (butterflies and moths), Coleoptera (beetles), Diptera (flies, midges and hoverflies) and Thysanoptera (thrips). The degree to which both primary and secondary attractants result in pollination by specific insect vectors varies a great deal. Some flowers (for example *Calluna*) are visited by representatives of the five main classes of anthophilous insects, whereas other flowers are only pollinated by single species. Thus Faegri and van der Pijl (1979) distinguish between generally adaptive (or polyphilic) and specifically adaptive (monophilic) floral organisation. The evolutionary consequences of these two strategies are markedly different. The generalists have their pollen transferred to an appropriate stigma less efficiently, but they are not dependent on the welfare of a single vector. The specialists, however, have their pollen 'targeted' extremely efficiently and there is a high probability of reproductive isolation between populations. This in turn is likely to lead to speciation. However, population extinction rates are also likely to be high because of acute vector dependency. The demise of the vector can prove disastrous for the plant and *vice versa*.

Several authors have attempted to classify blossoms according to the primary pollination vectors involved. The most satisfactory way of doing this is based on functional structure which relates to presentation and reception of pollen and, in the case of biotic vectors, the attractants involved (Leppik, 1957; Faegri and van der Pijl, 1979). The diverse blossom adaptations resulting from co-evolution with biotic vectors necessarily mean that typological classifications of functional structure can hardly be expected to reflect phylogeny. Moreover, such classifications are generalisations and should not be applied too strictly. However, they can be useful in determining the apparent first occurrences of the various pollination syndromes.

Following Faegri and van der Pijl (1979) blossoms that are open during anthesis and are conspicuous (visually attractive) include the following forms: dish- or bowl-shaped; bell- or funnel-shaped; head- or brush-shaped; gullet-shaped; flag-shaped; and tube-shaped. Dish- or bowl-shaped flowers (for example members of the Magnoliaceae and Rosaceae) are almost always actinomorphic and the nutritional rewards (pollen and nectar) are easily access-ible. Sexual organs are centrally or diffusely located and are openly exposed. Relatively unspecialised insects can effect pollination merely by walking among the anthers and stigmas. Large dish- or bowl-shaped flowers are therefore often pollinated by beetles, but the simple open structure does not prevent pollination being brought about by other insects or even bats.

The unsophisticated structure of the dish/bowl blossom makes it a prime candidate for the type of flower likely to have been possessed by primitive angiosperms, and the hypothetical primitive angiosperm of some authors (for example Arber and Parkin, 1907; Takhtajan, 1969) is indeed of this type.

Dilcher and Crane (1984) have described a floral structure (Figures 15.3A and B) from the Cenomanian of Kansas which was probably of the dish/bowl type. The fruiting structure, *Archaeanthus*, which developed from the flower, consisted of an erect reproductive axis bearing a cluster of 100 to 130 follicles on an elongated receptacle. The follicle was in the form of a conduplicate carpel, enclosing 10 to 18 seeds, with a groove running the length of its adaxial surface. This groove was bounded by a rounded lip, which is interpreted as being a stigmatic surface. An unusual characteristic of the carpel is that it contained a number of resin bodies.

At the base of the receptacle were small scars representing the position of attachment of the stamens. Below these were two additional sets of scars where the inner and outer perianth parts were attached. Dilcher and Crane (1984) consider *Archaeanthus* to be closely related to the Magnoliidae with a flower that was open, actinomorphic, visually conspicuous and pollinated by insects.

Associated with the floral remains were leaves (Figure 15.3C) somewhat reminiscent of those of extant *Liriodendron*. These *Liriophyllum kansense* leaves also contained the same resin bodies as the carpels, which suggested to Dilcher and Crane that both leaves and reproductive structures were from the same type of plant.

Other dish/bowl type flowers consisting of a five-parted (pentamerous) calyx and corolla surrounding five stamens, which in turn surrounded a ring of five carpels, have been recovered from the early late Cretaceous Dakota Formation (Basinger and Dilcher, 1980) (Figure 15.4), but this type of organisation is not

Figure 15.3: A. Reconstruction of *Archaeanthus*. The scars where the stamens and perianth parts were attached is indicated by 'a'. B illustrates a cut through a follicle and an enclosed seed showing the conduplicate nature of the carpel and the adaxial stigmatic lip. C shows a leaf of *Liriophyllum kansense* which is thought to have been borne by the same plant that produced *Archaeanthus*. (A, after Dilcher, 1979; B and C, after Dilcher and Crane, 1984)

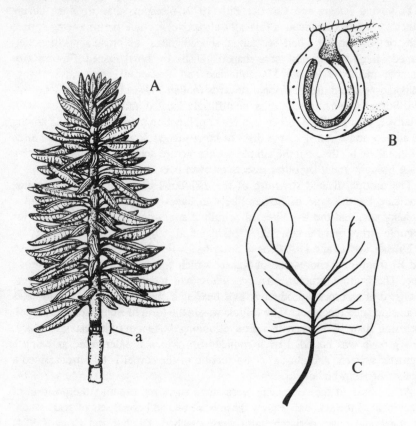

Figure 15.4: A Generalised Actinomorphic Mid-Cretaceous Flower. (Redrawn from Stewart, 1983 after illustrations by Basinger and Dilcher, 1980)

Figure 15.5: Examples of Floral Organisation. A: A spadix of *Arum nigrum*. At the base of the stem within the basal chamber are the reduced female flowers; distal to these are sterile bristle-like flowers; above these are reduced male flowers; above which are sterile bristle-like flowers. The topmost part of the stem is sterile and produces an odour attractive to flies. B: The bell-shaped blossom of *Campanula* at early anthesis. C: Zygomorphic flag blossom of *Cytisus* prior to pollination. (After Faegri and van der Pijl, 1979)

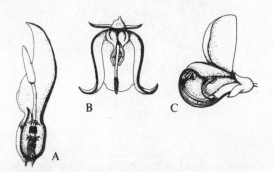

typical of all floral remains recovered from Cenomanian rocks.

In the Middle Eocene Claiborne Formation of the southeastern United States, large dish/bowl blossoms (6 cm in diameter), characteristically pollinated by beetles, occur together with much smaller, similarly organised flowers typically pollinated by small flies (but again not exclusively). Another Middle Eocene flower, this time from the Allenby Formation of British Columbia, is *Paleorosa* (Basinger, 1976). Some 2 mm in diameter, it is pentamerous with 13 to 19 stamens encircling five free carpels, each containing two ovules.

Bell- or funnel-shaped flowers (Figure 15.5B) differ morphologically from each other only in that bell-shaped blossoms have a convex outer surface of the corolla, whereas funnel-shaped forms are flat or concave. In both types the pollinator is forced into direct contact with the floral sexual apparatus by the confines of the blossom. Typical pollinators are bees, wasps, flies and even bats. No good bell-shaped blossoms are known from the Cretaceous, but they do occur in the Middle Eocene Claiborne Formation (Crepet, 1979).

Brush-type blossoms are typically lax and have a divided perianth interspersed between exserted reproductive parts. *Eomimosoidea* (Crepet and Dilcher, 1977) is a bisexual brush-type blossom of Eocene age consisting of an axis bearing four-parted perfect flowers with stamens projecting well beyond a rounded stigma supported by a hairy style (Figure 15.6). Pollen remains as a tetrahedral tetrad and the inference is that the blossom was pollinated by animals. Modern brush-type blossoms may be pollinated by bees, moths, butterflies, birds and bats.

Gullet-type blossoms are zygomorphic, with the sexual organs restricted to the upper (adaxial) half of the pollination unit so that pollen is deposited on the

Figure 15.6: An Eocene Mimosoid Inflorescence, *Eomimosoidea*. (Redrawn from Crepet and Dilcher, 1977)

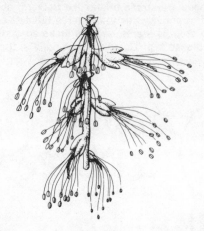

back or upper side of the head of the pollinator. The flag type (Figure 15.5C) is morphologically the reverse of the gullet type in that the sexual organs are found in the lower part of the pollination unit and pollen is deposited on the abdomen of the pollinator. Again, the flower is strongly zygomorphic. Zygomorphic flowers had evolved by Middle Eocene times (Crepet, 1979), although the level of organisation appears to be primitive compared with that seen today. Members of the Hymenoptera (bees and wasps) are typical pollinators of extant zygomorphic flowers, as also are moths, butterflies and birds.

Tube blossoms, according to Faegri and van der Pijl (1979), are a continuation of the morphological sequence dish-bowl-bell-funnel, with the floral unit becoming progressively narrower. However, the main difference with the tube blossom is that the pollinator cannot enter the tube bodily. It must be specially adapted (by having a suitably long tongue or proboscis) to extract nectar from the tube, and if a landing platform is not provided, by, for example, special adaptations of the corolla, feeding (and pollination) is restricted to animals with the ability to hover. Extant blossoms exhibit a range of variation that may exclude all but the appropriately specialised visitors. Tube blossoms are visited by birds and moths, but in particular bees and butterflies. Flowers with the corolla fused into a tube are known from the Middle Eocene Claiborne Formation (Crepet, 1979).

Faegri and van der Pijl (1979) recognise two groups of blossom in addition to those open during anthesis. From a palaeobotanical point of view cleistopetalous blossoms (those closed during anthesis) would be extremely difficult, if not impossible, to recognise in the fossil state. An apparent cleistopetalous blossom may just be immature. Trap blossoms, as their name suggests, entice the pollinators into the blossom where they become imprisoned temporarily (although occasionally permanently) while pollination takes place. The length of imprisonment varies and release usually occurs when the detaining structures cease to function. Modern examples include the spadix of *Arum* which is fly-pollinated (Figure 15.5A). Aroid spadices have also been described from the Claiborne (Crepet, 1978), showing that the spadix was well developed by the Middle Eocene.

As Crepet (1979) points out, by the Middle Eocene, blossoms exhibiting features likely to favour pollination by members of the Coleoptera, Diptera, Hymenoptera and Lepidoptera had evolved. Further studies will probably establish considerably earlier first occurrences of the various blossom types.

Pollination by Wind

The dispersal of pollen and spores by wind (anemophily) is of course commonplace in the pteridophytes and gymnosperms. In the angiosperms abiotic pollination, in particular anemophily, is usually considered to be secondarily derived and an example of retrograde evolution. The implication here, of course, is that all angiosperms are fundamentally biotically pollinated, the most primitive blossom/vector relationship being that between dish/bowl units and beetles.

Anemophilous blossoms are typically diclinous and characterised by a reduced calyx and corolla, exserted or prominent sexual organs, and a lax organisation. Pollen is usually produced in copious amounts and tends to be small and smooth-walled.

Among the earliest well-documented floral structures suited to wind pollination are catkin-like inflorescences from the Cenomanian of Kansas (Figure 15.7) (Dilcher, 1979). Also from the Cenomanian of Kansas are seed balls or heads described by Lesquereux (1892) as *Platanus primaeva*. Their similarity to the seed heads of *Platanus kerrii* is striking (Dilcher, 1979) and there is every reason to believe that the Cretaceous forms, like modern *Platanus*, were wind pollinated.

Trees bearing anemophilous blossoms are generally referred to as belonging to an artificial agglomeration of taxa called the Amentiferae. Extant members of the Amentiferae mostly belong to the Hamamelideae and include the Betulaceae, Fagaceae, Ulmaceae and Juglandaceae. Their modern distribution is essentially temperate and northern. The fossil history of the Amentiferae is quite well documented but somewhat ambiguous.

Inflorescences of betulaceous affinity are known from the Palaeocene Golden Valley Formation (Hickey, 1977), and staminate and pistillate inflorescences of casuarinaceous affinity are known from the Middle Eocene of Australia (Lange, 1970). However, our most detailed knowledge of fossil amentiferous flowers comes from studies of specimens from the Middle Eocene Claiborne Formation.

The Fagaceae are represented by compressed catkins closely allied to those produced by modern chestnuts (*Castanea*) (Crepet and Daghlian, 1980). These castaneoid catkins named *Castaneoidea* are up to 9 cm in length and consist of dichasia helically arranged around a robust axis. The dichasia are composed of

Figure 15.7: Reconstruction of a Mid-Cretaceous Catkin-like Inflorescence with Four-parted Florets Spirally Arranged around a Central Axis. (After Dilcher, 1979)

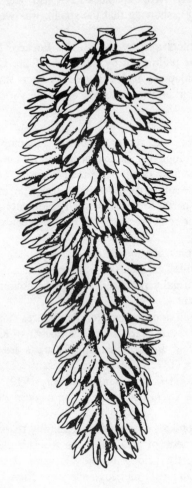

three florets subtended by bracts up to 1 mm in length. Florets have never been found open with exserted anthers, which suggested to Crepet and Daghlian (1980) that the studied catkins abscised prior to anthesis. The individual florets measure 0.8 mm × 1.0 mm and have a floral envelope terminating in at least five (probably six) lobes. Up to ten stamens occur within each floret and yield small (16 μm) tricolporate pollen.

Modern *Castanea* is usually insect pollinated in spite of producing small smooth pollen grains, normally associated with anemophily. It is therefore possible that *Castaneoidea* was also pollinated by insects. The stout axis of *Castaneoidea* also raises the possibility that the blossom was borne erect on the plant as in the extant genus.

In other members of the Fagaceae which are exclusively wind pollinated (for example *Fagus* and *Quercus*), we find ornamented pollen grains more typical of entomophilous blossoms, although their size is within the range best suited to wind dispersal. The fossil leaf and pollen record is a poor guide to taxonomic and phylogenetic status within the Fagaceae, but *Quercus* acorns are known in the Middle Eocene Clarno Formation (Manchester, personal communication cited in Crepet and Daghlian, 1980) and *Fagus* fruits are present in the Eocene London Clay (Chandler, 1964), which suggests that the family was well diversified by the Middle Eocene.

Wolfe (1973) suggested that the Fagaceae originated during the Upper Cretaceous in the Aquilapollenites Province (western North America and Asia). It is generally assumed that insect pollination within the family is derived from earlier adaptations to wind pollination.

Ulmaceous flowers from the Claiborne Formation have been described by Zavada and Crepet (1981). Borne in pairs, individual unisexual flowers were up to 9 mm in diameter and had a perianth composed of four tepals in a whorl. The tepals and the pedical were both covered in slender simple hairs. Each flower possessed 15 or more stamens bearing anthers filled with triporate pollen. The grains are somewhat large (40 μm) for wind pollination. Zavada and Crepet (1981) consider that on the basis of floral organisation and pollen characteristics the flowers are allied to the Celtidoideae of the Ulmaceae and named them *Eoceltis*. The main differences between the fossil *Eoceltis* and the flowers of modern members of the Celtidoideae are those that might be exhibited by an intermediate in a reduction sequence going from species with large bisexual entomophilous flowers (and large ornamented pollen) to species with small, unisexual flowers (and smaller pollen). *Eoceltis* has a larger perianth, numerous stamens, and pollen less well adapted for wind dispersal than the modern anemophilous Celtidoideae.

There can be no doubt that the Amentifereae were well represented by members of the Juglandaceae during Eocene times. Pollen and leaves of juglandaceous-like plants were widespread, but the most secure evidence for the family comes from inflorescences. *Eokachyra* (Crepet, Dilcher and Potter, 1975) is a staminate inflorescence with small helically arranged florets (3.3 × 2.7 mm) subtended by three-lobed bracts (Figure 15.8). Triporate pollen has been extracted from the anthers and is of a size (19.6 μm) and shape ideally suited to wind dispersal (Whitehead, 1969).

Another Eocene staminate catkin is *Engelhardia puryearensis* (Crepet, Daghlian and Zavada, 1980). With smaller florets, more acutely lobed bracts and smaller pollen (14.6 μm), these catkins are similar to those of the modern genus *Engelhardia*.

Juglandaceous winged fruits are known from both the Palaeocene (Hickey, 1977) and Eocene (Reid and Chandler, 1926; Dilcher, Potter and Crepet, 1976; Collinson, 1983) and, taken together with the pollen and leaf record, indicate

Figure 15.8: Reconstruction of *Eokachyra aeolia* Showing Part of a Staminate Inflorescence that Produced Triporate Pollen Ideally Suited to Wind Dispersal

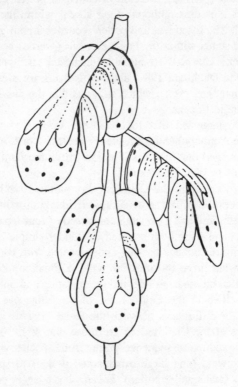

that the Juglandaceae diversified during the late Cretaceous. A number of authors postulated that, on the grounds of comparative pollen morphology, the plexus of pollen types referred to as the Normapolles complex could have been produced by plants that gave rise to some hamamelids (Doyle, 1969; Wolfe, 1973, 1975). The Normapolles complex first appeared in the middle Cenomanian, attained its maximum diversity in the late Santonian and became extinct by the end of the Eocene (Pacltova, 1966, 1978, 1981; Goczan *et al.*, 1967). Geographically the complex is restricted to Europe and eastern North America and for stratigraphic purposes over 70 genera have been recognised (Batten and Christopher, 1981). Although not all of these can be considered biologically distinct, the range of morphologies exhibited by the Normapolles group suggests they were produced by plants that were both wind and insect pollinated (Wolfe, 1975). Typically, Normapolles grains are triangular, triporate and somewhat smooth, and have a complex pore and wall structure (Figure 15.9).

The flowers bearing Normapolles pollen were completely unknown until recently. Friis (1983) described three genera of small, three-dimensionally

Figure 15.9: Three Examples of Normapolles-Type Pollen

preserved flowers, which she named *Manningia, Antiquocarya* and *Caryanthus*, from the Upper Cretaceous (Senonian) of Scania, southern Sweden, which appeared to have produced Normapolles grains.

Caryanthus (Figure 15.10A) flowers are small (0.86–1.36 mm in length), bisexual and bilaterally symmetrical. At the base of the ovary, subtending the flower, is a small bract with two lateral bractioles. The perianth consists of four tepals in two decussate pairs. The median tepals are narrow and external to somewhat broader lateral tepals. There are six to eight stamens, opposite the tepals, surrounding a gynoecium composed of two median carpels. The ovary and perianth have a hairy surface. Loose pollen grains within the specimens are all of the Normapolles type, about 13 μm in diameter, and are similar to the dispersed grains *Plicapolles*.

Manningia are also small (0.9–1.9 mm in length), perfect, actinomorphous flowers with five free tepals surrounding five stamens which in turn surround a uniloculate gynoecium. There is one style bearing three stigmatic branches (Figure 15.10B). Pollen recovered from anthers within the flowers (but not attached to them) is similar to the Normapolles genus *Trudopollis*. It is about 20 μm in diameter and triangular with convex sides.

Antiquocarya (Figure 15.10C) are again small (1.05–1.72 mm in length) bisexual flowers with six free tepals, six stamens, and a uniloculate gynoecium bearing three short styles. Individual pollen grains referable to the Normapolles complex have been found adhering to the flowers, but because no anthers were found the relationships between the Normapolles types and the *Antiquocarya* remain uncertain.

The close and distinct association of different Normapolles grains with the different flowers strongly suggests that the pollen was produced by the flowers. Friis (1983) compared the floral organisation of her specimens with flowers and fruits of modern angiosperms and concluded that there was a close relationship with members of the Hamamelideae, and in particular the Myricales (Myricaceae) and Juglandales (Juglandaceae and Rhoipteleaceae). Friis suggested that the modern hamamelids which mostly bear imperfect wind-pollinated flowers are derived from the Normapolles-producing complex. The ancestral flower type of the hamamelids was probably bisexual, with six tepals, six stamens and a three-loculed ovary, each ovary containing two anatropous seeds. It would be

Figure 15.10: Reconstruction of Three Upper Cretaceous Flowers
Containing Normapolles Pollen. (A) *Caryanthus*; (B) *Manningia*; (C)
Antiquocarya. (After Friis, 1983)

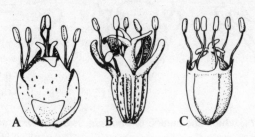

wrong to think of these flowers as being typical of all Normapolles producers.
Normapolles pollen was probably produced by a number of families occupying
a broad range of ecological niches (Batten, 1981; Pacltova, 1981; Zaklinskaya,
1981; Frederiksen, 1985), but the Cenomanian appearance of the Normapolles
complex would appear to suggest that the ancestors of some hamamelids
differentiated from the rest of the angiosperms early in flowering-plant evolution. However, if bisexual flowers are the basic 'form' for the group, it would
seem to support the view of others (for example Faegri and van der Pijl, 1979)
that anemophily is a derived condition within the hamamelids.

Early Angiosperm Reproductive Biology

Insect pollination or entomophily was apparently important very early in
angiosperm evolution and played a significant role in the subsequent success of
the group. Early angiosperm pollen possesses highly sculptured exines which
are to a certain degree correlated with insect vectors. One such example is the
Albian-Cenomanian *Afropollis*. The coarsely reticulate spheroidal palynomorph
has a wide environmental distribution and was apparently produced in limited
quantities by the as yet unknown parent plants (Doyle *et al.*, 1982). The
somewhat large size (40 μm diameter) of *Afropollis* and the coarsely ornamented
exine are typical of pollen grains produced by insect-pollinated flowers. Insect-
pollinated (entomophilous) flowers need only produce small quantities of pollen
because the behaviour of the insect vector in large part ensures that the pollen
is 'targeted' on to the stigma of a similar flower. The efficiency of the targeting
is a function of floral specialisation favouring only one group or even species
of insect and the behavioural patterns of that insect.

Even at the relatively low level of specialisation expected in early
angiosperms, targeting of pollen undoubtedly led to significant changes in plant
community structure that resulted in the eventual demise of gymnosperm
dominance in many areas. Insect pollination allows successful fertilisation and
outcrossing to take place, even when individuals of a species are widely
separated, or in isolated clusters, where wind pollination would be extremely
unreliable. Wind pollination, or anemophily, can only be successful when large

numbers of a species grow in close proximity, otherwise the probability of a suitable pollen grain landing on a stigma is very low. Regal (1977) points out that the dispersed distribution of angiosperms, particularly in forest environments, would be dependent upon widely foraging seed vectors, such as birds, which could carry seeds to new areas for colonisation, and widely scattered safe sites away from populations of specialised seed predators.

Insect pollination and a scattered distribution also increase the ability of angiosperms to survive and maintain genetic variability, even at low-density 'bottlenecks' brought about by changing environmental conditions. Here again, selection for rapid growth strategies, such as a rapid reproductive cycle, large leaves, and efficient vascular systems, would have been advantageous to the angiosperms.

When Normapolles pollen first appears in the Cenomanian, however, it is extremely well adapted for dispersal by wind in terms of size, shape, and ornamentation (Crepet, 1981). Other unrelated pollen types (the *Triorites*) occurring in equatorial Africa at about the same time also appeared to be ideally suited to wind dispersal. Both these groups of pollen types probably evolved from the Albian tricolpates, some of which also appeared to be adapted for wind dispersal. On this basis Crepet (1981) has suggested that there were strong selective pressures favouring anemophily early in angiosperm evolution, and, as we have seen (page 257), floral morphologies adapted for wind pollination were a feature of some Cenomanian angiosperms.

Walker and Walker (1984) envisage a pre-Barremian basement complex of entomophilous flowering plants, the living descendants of which are represented by the Magnoliales, Laurales and Winterales. From this basement complex a major line of anemophilous, apetalous angiosperms arose in the Barremian-Aptian interval which ultimately gave rise to the modern Chloranthaceae and primitive hamamelids (for example Trochodendrales, Cercidiphyllales and Hamamelidales). According to Walker and Walker (1984) the early selective pressure in favour of anemophily was the result of aridity in the South America/Africa rift system (Figure 13.4). Most modern dicotyledonous plants including the Dilleniidae, Rosideae and Asteridae may have been derived from the anemophilous group by a secondary return to entomophily.

Flower Cycles

The shortening of the life cycle opened up a range of competitive advantages for the angiosperms in that not only were they able to produce seeds quickly, but they were also able to stagger and vary flowering times. This in turn allowed them to compete successfully in otherwise unfavourable environments and/or develop complex communities.

In many gymnosperms a considerable time elapses between pollination and seed dispersal (in *Pinus*, for example, it is two years). This, of course, limits their ability to act as primary colonisers in unstable environments, and prior to the advent of angiosperms this niche was probably occupied by pteridophytes.

Reproduction by spores ensures that large populations of sporophytes can be built up quickly. However, as we have seen, conditions must exist to allow fertilisation to take place, and although plants like *Equisetum* have a very effective means of vegetative reproduction by rhizome fragmentation (most useful in riverside situations where bank collapse and water flow can transport pieces of rhizome to mud bars, for example), the success of the pteridophytes as primary colonisers is more limited.

In a sense the reproductive biology of early angiosperms mimicked that of the pteridophytes in that the shortening of the life cycle and the production of numerous small seeds ensured that large numbers of propagules were widely dispersed as quickly as possible. However, the sophisticated incompatibility mechanisms of the angiosperms also conferred on them the potential for rapid evolutionary development, and they soon diversified in unstable environments.

At first the pressures that constrained the time of flowering were probably those linked with fluctuating water availability because the seasonal fluctuations in temperature were likely to have been minimal in low latitudes during the Cretaceous. In today's herbaceous angiosperms that live in seasonally arid environments (typically Mediterranean-type climates) flowering takes place as soon as vegetative growth will allow following the advent of water availability, and pollination, fertilisation and seed development quickly follow. The situation with arborescent angiosperms in the seasonally arid tropics is somewhat different. Vegetative growth, food and water storage capacity enable the trees to delay flowering until the dry season when leaf fall has occurred and the flowers are more visible to insect pollinators. Under these circumstances the rapid angiosperm life cycle allows seed production and dispersal to take place before the next rainy season when the chances for germination are optimal (Janzen, 1967). Frankie *et al.* (1974a,b) and Frankie (1975) also observed that in these environments the flowering times of species were staggered to optimise the chances of being visited by insect vectors. This competition for vectors is a significant selective agent in modern environments (Levin and Anderson, 1970; Heinrich and Raven, 1972) and must have operated to a greater or lesser extent in the past.

Crepet (1981) notes that modern semi-arid tropical environments would seem to be an ideal environment for pollen dispersal by wind and yet insect pollination predominates. This would seem to argue against seasonal aridity as being a prime cause of the early development of anemophily in angiosperms. However, Crepet (1981) considers that the observations by Janzen (1967), Frankie (1975) and Frankie *et al.* (1976) that bees are the overwhelmingly important insect pollinators in modern tropical deciduous vegetation still allow room to postulate that aridity was the main cause of anemophily in early angiosperms. Bees first appear early in the Oligocene (Burnham, 1978) and it is unlikely that they were present in the Middle Cretaceous. The absence of such discriminating vectors might have put a premium on anemophily in seasonally arid environments at that time.

In warm humid environments where there is little seasonality either in water availability or temperature, the selective pressures favouring a rapid life cycle are not so strong. Here vegetative factors are probably more important (see Doyle *et al.*, 1982) and flowering may become, like leaf production and loss, more or less continuous. However, synchroneity of flowering time within a species, or population of that species, can confer certain advantages. If flowering of the various species within a complex community is staggered, this again optimises the probability that pollination will take place because immediate competition for vector availability is reduced. Here flowering time can be an important factor affecting gene flow, population structure and community complexity, both of the plants and of the vectors. Anemophily is not advantageous where the continual presence of leaves reduces its effectiveness, so in modern tropical rainforests it is practically non-existent.

The Cretaceous poleward expansion of the angiosperms brought them into contact with more seasonal climates, this time related not to water availability but to light and temperature. If we assume no major obliquity changes in the Earth's axis, then the Cretaceous polar light regime must have been similar to that of today with a long winter dark period and probably lower temperatures than in the summer (Spicer and Parrish, 1986). Deciduousness and early seasonal flowering evolved in response to aridity at lower latitudes would have been advantageous under these conditions and in a sense pre-adapted angiosperms to the polar regime. Here the premium was on completion of the life cycle and the accumulation of metabolites as quickly as possible before winter conditions returned.

Although insects were present at high late Cretaceous latitudes they may not have played an important role in pollination. In modern temperate and high-latitude environments, many angiosperm forest-forming tree species (for example those in the Amentiferae) are wind pollinated. Flowering takes place very early in the spring so that food reserves may be built up in the fruits and seeds before winter dormancy. Anthesis before insect activity has resumed forces the adoption of anemophily. Flowering before leaf expansion ensures that pollen dispersal is maximised.

Strongly seasonal environments also affect flowering in herbaceous angiosperms. Annual plants tend to be those that occupy open habitats and/or are primary colonisers. The openness of the situations they occupy enables them to grow, flower and complete their reproductive cycle relatively free from shading by trees. Pollination may be by wind (e.g. grasses) or by insects in temperate situations because flowering can afford to be delayed slightly until insects are active and only small seeds need be produced (hence minimising the time required for seed development). In harsher more polar or alpine environments there is a tendency towards anemophily as the growing season becomes shorter.

In temperate woodland herbs we see a slightly different phenomenon. Here the very early establishment of vegetative growth and flowering is advantageous before canopy leaves develop and reduce light levels on the woodland floor. It is probably in response to this that most woodland herbs have developed perenating organs such as bulbs, corms and rhizomes (as in daffodils, crocuses, etc.). These allow food reserves accumulated during the previous season to be used to establish growth quickly in the spring. For similar reasons many herbs are biennial and flower in the year after that in which germination took place. This allows food reserves to be accumulated in the first year which will later fuel the rapid and uninhibited production of seeds in the second year.

In addition to the poleward expansion, the development of early, and in some cases multiple, annual flowering cycles was probably favoured by global climatic deteriorations at the end of the Cretaceous and during the Tertiary. As we shall see in the final chapter, it was not until the later part of the Tertiary when climatic deterioration became severe and conditions approached their present state that open herb-rich environments and broad-leaved deciduous forests began to represent a large proportion of the Earth's vegetation. These deteriorations undoubtedly enhanced angiosperm evolution in that they favoured a rapid life cycle and flexibility of both vegetative and floral development.

16 THE EVOLUTION OF MODERN VEGETATION

Late Cretaceous Floral Provincialism

By Turonian times angiosperms had successfully outcompeted most elements of the existing non-angiospermous flora in lowland environments over much of the Northern Hemisphere. In the Southern Hemisphere the rise to dominance seems to have taken somewhat longer (J.A. Wolfe, personal communication). Even in polar regions a wide variety of angiosperm leaf and pollen types are found in fluvial margin, lacustrine margin, and interfluvial floodplain environments (Spicer and Parrish, 1986). Riparian vegetation continued to be dominated by platanoid leaf producers but the angiosperms had successfully invaded conifer-dominated lowland forests as a variety of understorey components. Swamp environments appear to have been the last to succumb to the angiosperm influx.

Although the late Cretaceous climate exhibited a lower equator-to-pole temperature gradient and was globally warmer than at present (Barron *et al.*, 1981; Barron and Washington, 1982; Barron, 1983), floras were not uniform. Vakhrameev (1978) has recognised several broadly latitudinal vegetational zones in the Cretaceous. Within a northern humid zone the Siberian-Canadian region was dominated by Czekanowskiales, ginkgos and conifers, with subordinate cycadophyte and fern elements, prior to the arrival of the angiosperms. All major elements of the flora were deciduous and were probably frost resistant. In the late Cretaceous angiosperm components are represented by leaves exhibiting a large size range and toothed margins. Overall the climate was humid and temperate to warm-temperate. Among the angiosperms hamamelid-like leaves are the most abundant, and although no floral structures have been recovered, it seems likely that most angiosperms were wind pollinated. Throughout most of the late Cretaceous the angiosperms were moderately diverse at high latitudes, and despite a decline during the latter part of the Cretaceous, formed the basis of a broad-leaved deciduous forest that Wolfe (1985) recognises as a distinct physiognomic unit in the Palaeocene.

During the late Cretaceous at high latitudes there are recognisable differences between coastal and inland floras. Coastal areas tend to have a greater diversity and contain elements (for example cycadophytes) that are better represented further south. This suggests that the coastal environment was ameliorated by oceanic influences.

Another phytogeographic province within Vakhrameev's (1978) northern humid zone is that of the more southerly Indo-European region. Here the climate is interpreted to have been warmer and frost-free. *Tempskya, Weichselia*, cycadoids and a predominance of cheirolepidiaceous conifers are characteristic elements in the early Cretaceous, whereas in the late Cretaceous most angiosperms have entire-margined leaves and tend to have a smaller mean size and size range than those in the North. Not only was the climate warmer than in the North, but it appears to have had a greater tendency towards seasonal aridity. Angiosperm floral structures indicate both biotic and abiotic pollination syndromes. In sub-equatorial regions of the Northern Hemisphere the climate appears to have been generally arid and the preservation of plant material is sparse. Cheirolepidiaceous conifers predominate, and angiosperms tend to be small-leaved.

During the mid-late Cretaceous the pollen floras of middle and high northern latitudes diverged into two more or less well-defined provinces. The pollen flora of eastern North America, Europe and western Asia was characterised by the Normapolles group, whereas pollen of the *Aquilapollenites* group was produced by plants growing in western North America and across Siberia. According to Srivastava (1975), an Austral province was characterised by angiospermous pollen apparently produced by proteaceous elements and *Nothofagus*-like plants.

Srivastava (1975) and Samoylovich (1977) identified the *Aquilapollenites* province as being circumpolar and separated from the Normapolles province by a sublatitudinal boundary. Tschudy (1981) reported generic differences within the Normapolles province and suggested that the Normapolles-producing plants arose in Europe and migrated to eastern North America prior to the opening up of the Atlantic in early late Cretaceous times.

During the Maastrichtian the *Aquilapollenites* and Normapolles provinces merged (Srivastava, 1981), apparently as the result of the disappearance of barrier continental seaways. Either competition or climatic deterioration seems to have decimated the *Aquilapollenites* producers at the end of the Cretaceous. The Austral province, however, remained intact and, compared with changes in the Northern Hemisphere, suffered little disturbance throughout the Tertiary.

The End of the Cretaceous

The end of the Cretaceous is marked by a number of apparent mass extinctions, the most well known being that of the dinosaurs. These significant changes occur in both plant and animal lineages at or about the Cretaceous/Tertiary boundary, and as far as terrestrial plants were concerned profoundly affected subsequent evolution by opening up niches and altering relationships with pollinators and seed vectors. The cause or causes of these extinctions have been the subject of much speculation. Among the more plausible explanations are a late Cretaceous climatic deterioration and/or a catastrophic 'asteroid' impact. It is of critical importance to determine whether the extinctions were the termination of a pre-existing gradual decline or were sudden and simultaneous across the broad spectrum of organisms involved. If the extinctions were sudden and simultaneous, then a single catastrophic causal factor was involved, but to demonstrate this demands extremely detailed stratigraphic analysis over wide areas.

Alvarez *et al.* (1980) reported finding an iridium-rich horizon at the Cretaceous/Tertiary (usually abbreviated to K/T) boundary position. Because iridium is comparatively rare on Earth but occurs in greater proportions in meteorites, they suggested that 65 million years ago a large (12 km diameter) asteroid collided with the Earth sending a dust cloud into the atmosphere that prevented any sunlight reaching the Earth's surface, drastically affected surface temperatures, and led to a near total collapse of the biosphere. The iridium-rich horizon was interpreted as the 'fallout' from the dust cloud.

Inevitably this exotic scenario has generated considerable debate and stimulated research efforts. Hickey (1981) interpreted the plant fossil record across the K/T boundary as evidence for a gradual rather than a catastrophic extinction pattern. Hickey suggested that K/T boundary extinctions were not exceptional and that a similar percentage of taxa became extinct across the Palaeocene/Eocene boundary where no catastrophe had been invoked. A distinct feature of the K/T floral changes was a post-boundary increase in the percentage of leaves with non-entire (toothed) margins which Hickey took to be indicative of a climatic cooling. Although Hickey's interpretations were based on assemblages from a number of North American localities that between them contained iridium-rich boundary clays and dinosaur remains, the stratigraphic resolution necessary to examine the catastrophe scenario was not achieved.

Elsewhere in North America, evidence for major vegetational disturbance at the K/T boundary has been demonstrated. In the Raton Basin of northern New Mexico and southern Colorado, intact sequences are found with iridium-rich boundary clays in coal-rich non-marine sediments (Pillmore *et al.*, 1984). Pollen and spore sequences across the boundary exhibit the same pattern at a number of sites. The latest Cretaceous angiosperm-rich pollen flora is decimated at the boundary and immediately replaced by assemblages consisting almost entirely of fern spores. Within 10–15 cm above the boundary this fern 'spike' subsides as the angiosperm component increases but with reduced diversity. A similar fern spike is seen in K/T boundary sections in Montana (Smit and van der Kaars, 1984). This pattern of plant succession parallels closely that seen in analogous modern vegetation that is destroyed catastrophically by, for example, volcanic eruptions (Spicer *et al.*, 1985). Of particular significance is the primary coloniser role of the ferns. Homospory, as we saw in the Silurian and early Devonian, is suited ideally to rapid colonisation of bare land surfaces because large numbers of wind-borne spores are produced, any of which, by self-fertilisation at the gametophyte stage, can give rise to a sporophyte. It seems likely therefore that the fern spike is evidence of short-term but intense vegetational disturbance. A volcanic cause for the disturbance has to be discounted for the Raton Basin sites because pollen/spore assemblages across known volcanic ash partings within the Basin do not exhibit the fern spike. It should be borne in mind that ferns only act as primary colonisers in volcanic terrains where destruction of the standing vegetation is severe by virtue of its close proximity to an active vent. There is no record of local contemporaneous volcanic activity in the Raton Basin.

In general the post K/T vegetational changes reflect normal ecological succession and recovery after a 'mass kill'. The relationship of such a mass kill to extinctions is, however, obscure. If there was an asteroid impact, the darkness produced by the resulting dust cloud may have lasted for several months to several years during which time photosynthetic activity would have ceased. Clearly this would have affected all green plants but it would not necessarily have caused widespread extinctions. For deciduous trees, weathering a post-impact dark period of a year or so's duration would have been relatively easy; they would have shed their leaves and entered dormancy. Dormancy and dying back to rhizome, corm, bulb and root systems would also allow perennial herbaceous forms to survive the event. If the dark period was of longer duration, many plants could have survived in seed form. Most terrestrial animals on the other hand, particularly the larger ones, would quickly have succumbed in the darkness, unable to find food.

In addition to darkness the lack of solar input would have led to a dramatic cooling particularly in the continental interiors. Such a post-impact winter would probably have had a greater effect on plant life than the darkness. In the equable Cretaceous climate few lower-latitude lineages are likely to have developed frost resistance and would have been greatly reduced in numbers. Today frost-sensitive plants tend to be evergreen and have leaves with entire margins. A combination of darkness and cold, therefore, could have produced the pattern of increase in toothed leaves observed by Hickey (1981).

Detailed analysis of leaf assemblages in the Raton Basin also seems to support the asteroid impact theory. J.A. Wolfe and G.R. Upchurch (personal communication) examined 66 leaf megafossil and 50 cuticle assemblages in palynologically dated latest Cretaceous to Palaeocene sediments including those containing the iridium-rich boundary layer. Below the boundary, angiosperm leaves tend to be small with thick hairy cuticles. Some evergreen conifers are also present. Overall the leaf flora is diverse and seems to represent evergreen plants. Wolfe and Upchurch take this as indicative of subhumid (semi-dry) vegetation extending up to the boundary layer. Immediately above the boundary the megafossil and cuticle flora consists almost exclusively of one type of fern. Angiosperms are represented only by a few pieces of cuticle similar to that produced by herbaceous forms. Within 2 m or so above the boundary the leaf flora begins to diversify as leaves physiognomically typical of early successional vegetation (thin, cordate, compound or lobed) appear. In contrast to those of the latest Cretaceous the earliest Palaeocene leaves are large and have attenuated 'drip tips' and smooth cuticles, all of which suggests high precipitation. Thereafter the early Palaeocene diversity gradually increases but assemblages remain characteristically those of humid conditions. The post-boundary recovery succession occupies a sedimentary thickness of over 350 m, which probably represents a time span of approximately 1.5 million years. Wolfe and Upchurch (1987) have termed this long-term pattern of assemblage change 'macrosuccession'.

The change fiom a latest Cretaceous subhumid vegetation to a Palaeocene 'wet' vegetation does not appear to be restricted to the Raton Basin, and precipitation appears to have been generally higher in the Palaeocene than in the Cretaceous (J.A. Wolfe, personal communication, 1985). Like Hickey (1981), Wolfe and Upchurch have observed a post K/T boundary increase in the abundance of toothed-margined deciduous leaves. The warm-loving (thermophilous or megathermal) broad-leaved evergreen taxa of the late Cretaceous suffer the greatest number of extinctions at the K/T boundary and are replaced by a low-diversity deciduous forest analogous to successional vegetation in mesothermal environments.

The scenario of Wolfe and Upchurch then is one of intense selection at the K/T boundary in favour of broad-leaved deciduous elements and a change from a subhumid to a wet environment. These effects appear to have been more strongly felt in the Northern Hemisphere at middle and low latitudes. To some extent modern vegetation appears to echo this selection in that the natural vegetation of the Northern Hemisphere is relatively depleted in entire-margined evergreen elements compared with the Southern Hemisphere. The post-boundary deciduous aspect of the vegetation continues throughout the remainder of the Palaeocene over much of the Northern Hemisphere. The polar broad-leaved deciduous flora, which became established in the late Cretaceous, was disturbed at the K/T boundary and suffered some extinctions possibly due to extreme cold, but its overall aspect was little altered and it undoubtedly played a part in supplying taxa to the more severely disrupted lower-latitude vegetation. Wolfe and Upchurch argue that the persistence of the deciduous floras suggests an absence of post-boundary mesothermal evergreen refugia, otherwise broad-leaved evergreens would rapidly have reinvaded the impoverished areas. Later, slow replenishment of broad-leaved evergreens in mesothermal vegetation must have come about by evolutionary adaptation from a severely disrupted megathermal flora. Gymnosperms as well as angiosperms would have suffered during the terminal Cretaceous event, and in particular remnants of low-latitude conifer-dominated vegetation were likely to have been the most severely affected. The loss of those forest ecosystems may well have paved the way for the first angiosperm-dominated multistratal forests typical of modern low-latitude lowlands.

The K/T boundary impact scenario is not universally accepted to have taken place but evidence that supports some kind of global biotic catastrophe leading to severe ecological disturbance, climate change and extinctions is accumulating.

The Evolution of Angiosperm Fruits and Seeds

So far we have considered angiosperm vegetative organs, floral morphology and pollen and have said little about fruits and seeds or to use a more general term 'diaspores'. After pollen and leaves, diaspores are the next most abundant angiosperm megafossils and their role in propagation and relation to floral morphology mean that they contain a great deal of ecological and evolutionary information. Following fertilisation the embryo and endosperm rapidly develop as the floral organs responsible for enhancing pollen arrival at the stigma wither away. Usually the integuments become sclerified and form the seed-coat or testa. The ovary may ripen into a fruit with the remnants of the style and stigma or the ovary wall becoming modified to enhance seed protection and dispersal. Both the lignification of the testa or ovary wall and the propensity for dispersal increase the fossilisation potential of diaspores.

The most common early angiosperm fruit type was a dry dehiscent follicle (derived from a single carpel) which split open down one side, although some capsules with more than one line of opening are known. Neither fruits nor seeds exhibit significant modifications for dispersal. Fleshy fruits are unknown in Cenomanian or older rocks. Overall these features suggest that the diaspores of early angiosperms were abiotically dispersed (Retallack and Dilcher, 1981c; Tiffney, 1984).

Late Cretaceous fruits and seeds are slightly more diverse as nuts and drupes (fleshy fruits with a seed or seeds surrounded by a stony layer) occur in addition to follicles and capsules. The continued abundance of small unspecialised diaspores suggests the continuing importance of abiotic dispersal mechanisms but nuts and drupes also indicate the potential for dispersal by animals (Tiffney, 1984).

In general the fruits and seeds of herbaceous and early successional plants tend to be small and are produced in large numbers, whereas diaspores of forest trees in stable environments tend to be large and produced in fewer numbers (Salisbury, 1942; Harper *et al.*, 1970). This relationship appears to have been remarkably robust throughout the history of terrestrial plant evolution (Chaloner and Sheerin, 1981) and throws a particularly illuminating light on angiosperm vegetational history in the Cretaceous and Tertiary. Tiffney (1984) observed that in general the diaspore size of Cretaceous angiosperms tends to be small with limited variation. Such a finding correlates well with the weedy opportunistic nature of the early angiosperms deduced from vegetative morphology and sedimentological evidence. However, throughout the late Cretaceous, angiosperm diaspore size remains small in spite of increasing floral and vegetative sophistication. This suggests that the ecological role of the angiosperms remained confined to that associated with an early successional status at least until the Tertiary.

In contrast to the late Cretaceous, early Tertiary and subsequent angiosperm diaspores are morphologically diverse and include both small and large forms. Average diaspore size increases in the Tertiary not as a result of an overall increase in fruit and seed size but because of an increase in abundance of large diaspores (Tiffney, 1984). Tiffney does not consider that these size changes are due to taphonomic factors or ecological sampling differences. Instead he considers that the large diaspores represent the advent of forest-forming physiognomically dominant angiosperms occupying stable environments. Until the Tertiary, gymnosperms were the physiognomic dominants in stable environments.

The evolution of large diaspores may also indicate an increase in the prevalence of biotic dispersal. Though there are exceptions (e.g. *Cocos nucifera*, the coconut), large diaspores can be only poorly dispersed by abiotic means (wind and water), and thus the development of closed forests in stable environments depends in large part on the presence of biotic seed vectors. There is considerable variation in seed size among Mesozoic gymnosperms, many of which were probably dispersed by animals (e.g. cycadeoids). It is possible that reptiles accounted for most biotic dispersal of Mesozoic gymnosperms, and the drastic reptilian decline at the end of the Cretaceous would have been yet another factor in post-Cretaceous vegetational change. Today birds and mammals are the most important biotic angiosperm diaspore vectors and their Tertiary radiations parallel that of the angiosperms. Fossil seeds and fruits show an increase in morphological adaptations to biotic dispersal, and as with flower and pollen morphology were part of a co-evolutionary relationship with vectors which further enhanced angiosperm diversification. Birds in particular are excellent long-distance vectors and play an important role in bringing about and maintaining the heterogeneity of modern angiosperm forest communities. This heterogeneity, coupled with the co-evolutionary development of sophisticated biotic pollination syndromes, sealed the fate of many Cretaceous gymnosperm wind-pollinated communities.

First Appearances of Extant Taxa

Although in many respects the angiosperm geographic expansion and late Cretaceous floral provincialism laid the foundations of modern vegetation, many angiosperms at this time bore little resemblance to flowering plants of modern taxa. Many of the families of extant taxa did not arise until the Tertiary, and even then our knowledge of them is often confined to the morphology of only their pollen. Other aspects of their overall morphology, reproductive biology and ecology may not have been typical of those living today.

It may be argued that the study of isolated organs is likely to give a distorted, perhaps even erroneous, impression of the evolution of groups of entire organisms, and even more tenuous is the relationship to ancient communities likely to be. As we have seen, mozaic evolution where different plant organs evolve at different rates appears to be the norm in plant evolution. It has to be borne in mind, however, that in the case of pollen its abundance and role in the reproduction process mean that it has the potential of acting as a yardstick against which evolutionary data from less abundant organs may be compared. The pollen record is more or less continuous through time, and subtle changes in pollen morphology can be related to changes in the reproductive process and floral morphology upon which angiosperm classification is based.

Using the pollen record Müller (1981, 1984) has traced the late Cretaceous and Tertiary first occurrences of extant taxa. His findings generally confirm the view that angiosperms with magnoliid characters were among the first to appear. Pollen with features characteristic of the Laurales first occurs in the Aptian followed by magnolialean pollen by Albian times. The pattern of first occurrences of extant orders is as follows.

Tertiary

Miocene — Lamiales, Plumbaginales, Dilleniales, Ranunculales, Capparales, Primulales, Alismales, Najales, Arales
Oligocene — Nelumboniales, Violales, Cucurbitales, Salicales, Rosales, Rhamnales, Oleales, Elaeagnales, Campanulales, Asterales
Eocene — Theales, Thymeliales, Nepenthales, Polemoniales, Dipsacales, Scrophulariales, Cyperales, Eupteliales, Saxifragales, Liliales
Palaeocene — Didymeliales, Hamamelidales, Eucommiales, Casuarinales, Polygonales, Rutales, Polygalales, Cornales, Gentianales, Restionales, Poales, Typhales

Cretaceous

Maastrichtian — Nymphaeales, Illicales, Caryophyllales, Ericales, Ebenales, Malvales, Fabales, Geraniales, Santalales, Proteales, Arecales, Pandanales
Campanian — Myrtales, Cercidiphylales, Juglandales, Euphorbiales
Santonian — Fagales, Betulales, Myricales
Turonian — Urticales, Hippuridales, Celastrales

The number of living orders is relatively small until the end of the Cretaceous, and most living orders appear to have arisen during the Tertiary. It would be wrong, however, to assume that this pattern represents an evolutionary 'explosion'. These are first occurrences and mark a level of organisation attained rather than rates of evolutionary change. Until the Maestrichtian many now extinct taxa remained at least locally dominant and their diversity is not represented in the same way as extant taxa. For example, Müller (1984) considered that the Normapolles should be accorded ordinal rank within the Hamamelideae whereas the other widespread pollen type, *Aquillapollenites*, probably represents an extinct monotypic family in the Santalales.

There are also some interesting differences in the pattern of evolution between pollen types typically dispersed by wind and those dispersed by insects. Wind-dispersed pollen tends to exhibit a pattern of evolutionary development that is more or less time-stable due to the constancy of selection pressure inherent in the abiotic vector. Insect-dispersed pollen on the other hand shows more dynamic evolutionary development that Müller (1984) attributes to the effect of co-evolutionary factors between pollen and vector. The rapid changes in morphology seen in insect-dispersed pollen also mean that there is considerable uncertainty in plotting the time of origin and early evolution of the predominantly insect-pollinated groups. The rates of acquisition of features characteristic of a particular order are highly variable depending on the dispersal syndrome, and pollen morphology may reflect evolution of the vector (or a change of vectors), rather than that of the plant *per se*. This probably explains the disparity between the time of origin of some groups (for example the Rosales) as determined by leaf studies and that determined by the pollen record.

If morphological similarities to modern taxa are reliable indicators of true relationships (and as we have seen with leaves they may not be), other patterns of evolutionary significance are present in the pollen record. Among the dicotyledons typical herbaceous taxa appear relatively late, whereas in the monocotyledons woody palms are a secondary development. The herbaceous habit is difficult to infer, however, from taxonomic affinity and it is possible that some Cretaceous angiosperms could have been herbaceous. According to Wolfe *et al.* (1975) occurrences of well-documented Cretaceous angiosperm wood are rare, suggesting that wood evolution lagged behind that of other aspects of angiosperms or that angiosperms were minimally woody in the Cretaceous.

Much of the older palaeobotanical literature exaggerates the modernity of Palaeogene (Palaeocene, Eocene and Oligocene) taxa by applying the names of extant genera to fragmentary and incomplete fossil material, often with little justification. Nevertheless, as the Tertiary progressed, the morphological similarity between the fossils and extant plants increased and the problem of when to apply the generic (and later species) names of extant plants (thereby imbuing the ancient plant with a biology identical to that of the modern descendant) becomes more acute. This problem is discussed at length in Collinson (1986).

Even when the use of modern names appears to be justified on morphological grounds, it should not be assumed that the ecological tolerances and community associations of today can be applied to the Tertiary. For example, the tolerances of extant *Platycarya* and *Nypa* are mutually exclusive and yet in Tertiary assemblages they are consistently found in association. In his review of early Tertiary sporomorph palaeoecology, Frederiksen (1985) cites instances (for example *Ephedra, Nypa, Platycarya, Pinus*) in which ecological tolerances and ranges of extant genera have apparently changed with time, but the overall impression is that unless there is evidence to the contrary, many palynologists assume ecological stasis. Global vegetational and environmental interpretations are made more reliably on physiognomic grounds, which to a large extent are independent of taxonomy and nomenclature (Wolfe, 1978, 1979).

Vegetational Changes in the Tertiary and Quaternary

Throughout the Palaeocene and Eocene, vegetation on a global scale became more modern in appearance in that taxa with modern affinities became forest dominants for the first time. However, the global climate was warmer than at present, and floras with a tropical and subtropical aspect extended to latitudes of 50° or so, confining broad-leaved deciduous and coniferous forests to higher latitudes (Wolfe, 1980). Broad-leaved evergreens and palms extended to as far north as 60° palaeolatitude.

Using the extensive Tertiary leaf floras of the Northern Hemisphere, Wolfe (1978, 1980) has documented a sudden vegetational change at the end of the Eocene. The broad-leaved deciduous forest expanded southwards, and at middle and high latitudes diversity declined. Similar terminal Eocene floristic changes were reported by Collinson, Fowler and Boulter (1981) based on fruit, seed, pollen and spore assemblages in southern England but over a longer time scale. Later in the Oligocene diversity began to rise again, perhaps due to climatic warming or evolutionary adaptation. There were minor temperature fluctuations in the Miocene, but overall there appears to have been a decrease in mean annual temperatures (Wolfe, 1978). There is evidence to suggest that this cooling affected even equatorial regions, as J. Müller (1966) reports finding alder and spruce pollen in Borneo. Extensive coniferous forests became established at high northern latitudes, but it was not until the Pliocene that tundra appeared.

The first doubtful records of grass pollen occur in the Campanian but there are no firm records until the Palaeocene. Thereafter grass pollen becomes increasingly abundant, with megafossils dating from the early Eocene (Daghlian, 1981; Müller, 1984). Open grassland probably first appeared in the Miocene at middle to high latitudes. The average size of angiosperm diaspores decreased in the latter half of the Tertiary as the number of large diaspores per assemblage decreased (Tiffney, 1984). This also suggests a decline in closed forest ecosystems and an increase in open, perhaps savannah-type, vegetation.

The Quaternary saw the Northern Hemisphere experiencing a series of major glaciations. Many tropical species suffered extinction or a severe restriction of range. During interglacial periods repopulation of the tropics by vegetation was probably from a number of refugia. At higher latitudes taxa migrated in response to the changing climate and the waxing and waning of the ice sheets. Numerous extinctions occurred at the species level, particularly in Europe, as taxa were limited in their southerly migrations by the east-west trending mountain ranges and the Mediterranean. In North America no such barriers existed, and consequently extinction levels were much lower. Changing geographic ranges, isolation of populations, subsequent re-expansion from refugia, mixing of previously separated species, and the opening up of niches by the demise of species resulted in a series of new speciation events. There is no evidence that any major new groups arose during that time. If events leading to the natural evolution of such groups were set in train, we shall probably never know because man's destruction of much of the natural environment has introduced a whole range of evolutionary pressures that previously did not exist.

REFERENCES

Addicott, F.T. (1982) *Abscission*, University of California Press, Berkeley, 369 pp

Alvarez, L.W., Alvarez, W., Asaro, F. and Michel, H.V. (1980) 'Extraterrestrial Cause for the Cretaceous-Tertiary Extinction', *Science*, *208*, 1095

Alvin, K.L. (1960) 'Further Conifers of the Pinaceae from the Wealden Formation of Belgium', *Mem. Inst. r. Sci. nat. Belg.*, *146*, 1–39

—— (1971) 'The Spore-bearing Organs of the Cretaceous Fern *Weichselia* Stiehler', *Bot. J. Linn. Soc.*, *61*, 87–92

—— (1974) 'Leaf Anatomy of *Weichselia* Based on Fusainized Material', *Palaeontology*, *17*, 587–98

—— (1982) 'Cheirolepidaceae: Biology, Structure and Paleoecology', *Rev. Palaeobot. Palynol.*, *37*, 71–98

—— (1983) 'Reconstruction of a Lower Cretaceous Conifer', *Bot. J. Linn. Soc. 86*, 169–76

Andrews, H.N. (1963) 'Early Seed Plants', *Science*, *142*, 925–31

—— (1974) 'Paleobotany 1947–1972', *Ann. Missouri Bot. Gdn*, *61*, 179–202

—— and Boureau, E. (1970) 'VI Classe des Leptosporangiopsida', in E. Boureau (Ed.), *Traité de Paleobotanique*, *IV(1) Filicophyta*, Masson et Cie, Paris, pp. 257–406

——, Gensel, P.G. and Forbes, W.H. (1974) 'An Apparently Heterosporous Plant from the Middle Devonian of New Brunswick', *Palaeontology*, *17*, 387–408

——, Kasper, A.E., Forbes, W.H., Gensel, P.G. and Chaloner, W.G. (1977) 'Early Devonian Flora of the Trout Valley Formation of Northern Maine', *Rev. Palaeobot. Palynol.*, *23*, 255–85

—— and Kern, E.M. (1947) 'The Idaho Tempskyas and Associated Fossil Plants', *Ann. Missouri Bot. Gnd*, *34*, 119–86

—— and Phillips, T.L. (1968) '*Rhacophyton* from the Upper Devonian of West Virginia', *Bot. J. Linn. Soc.*, *61*, 37–64

Arber, E.A.N. and Parkin, J. (1907) 'On the Origin of Angiosperms', *Bot. J. Linn. Soc.*, *38*, 29–80.

Archangelsky, S. (1965) 'Fossil Ginkgoales from the Tico Flora, Santa Cruz Province, Argentina', *Bull. Br. Mus. (Nat. Hist.) Geol.*, *10*, 121–37

Arnold, C.A. and Daugherty, L.H. (1963) 'The Fern Genus *Acrostichum* in the Eocene Clarno Formation of Oregon', *Univ. Michigan Contrib. Mus. Paleontol.*, *18*, 205–27

Asama, K. (1970) 'Evolution and Classification of Sphenophyllales in Cathaysia Land', *Bull. Nat. Sci. Tokyo*, *3*, 291–317

Ash, S.R. (1969) 'Ferns from the Chinle Formation (Upper Triassic) in the Fort Wingate Area, New Mexico', *US Geol. Surv. Prof. Paper*, 613-D, D1–D52

—— (1976) 'The Systematic Position of *Eoginkgoites*', *Amer. J. Bot.*, *63*, 1327–31

——, Litwin, R.J. and Traverse, A. (1982) 'The Upper Triassic Fern *Phlebopteris smithii* (Daugherty) Arnold', *Palynology*, *6*, 203–19

Atkinson, L.R. (1973) 'The Gametophyte and Family Relationships', in A.C. Jermy, J.A. Crabbe and B.A. Thomas (Eds), *The Phylogeny and Classification of the Ferns*, Bot. J. Linn. Soc., *67*, Suppl. No. 1., pp. 73–90

Axelrod, D.I. (1952) 'A Theory of Angiosperm Evolution', *Evolution*, *6*, 29–60

—— (1959) 'Poleward Migration of Early Angiosperm Flora', *Science*, *130*, 203–7

—— (1960) 'The Evolution of Flowering Plants', *Evolution after Darwin*, *1*, 277–305

Bailey, I.W. and Swamy, B.G.L. (1951) 'The Conduplicate Carpel of Dicotyledons and its Initial Trend of Specialization', *Amer. J. Bot.*, *38*, 373–9

Banks, H.P. (1968) 'The Early History of Land Plants', in E.T. Drake (Ed.), *Evolution and Environment*: A Symposium Presented on the Occasion of the 100th Anniversary of the Peabody Museum of Natural History at Yale University, Yale University Press, New Haven, Conn., pp. 73–107

—— (1970) *Evolution and Plants of the Past*, Wadsworth, Belmont, CA, 170 pp

——— (1975a) 'Reclassification of Psilophyta', *Taxon, 24* (4), 401–13

——— (1975b) 'Early Vascular Land Plants: Proof and Conjecture', *Biol. Science, 25,* 730–7

——— (1975c) 'The Oldest Vascular Land Plants: a Note of Caution', *Rev. Palaeobot. Palynol., 20,* 13–25

——— and Davis, M.R. (1969) '*Crenaticaulis,* a New Genus of Devonian Plants Allied to *Zosterophyllum,* and its Bearing on the Classification of Early Land Plants', *Amer. J. Bot., 56,* 436–49

———, Leclercq, S. and Hueber, F.M. (1975) 'Anatomy and Morphology of *Psilophyton dawsonii,* sp.n., from the Late Lower Devonian of Quebec (Gaspe), and Ontario, Canada', *Palaeontographica Amer., 8,* 77–127

Barron, E.J. (1983) 'A Warm, Equable Cretaceous: the Nature of the Problem', *Earth Science Reviews, 19,* 305–38

———, Thompson, S.L. and Schneider, S.M. (1981) 'An Ice-free Cretaceous? Results from Climate Model Simulations', *Science, 212,* 501–8

——— and Washington, W.M. (1982) 'Atmospheric Circulation during Warm Geological Periods. Is the Equator-to-pole Surface Temperature Gradient the Controlling Factor?', *Geology, 10,* 633–6

Barthel, M. (1980) 'Calamiten aus dem Oberkarbon und Rotliegenden des Thuringen Walds', *100 Jahre Arboretum (1879–1979), Berlin,* 237–58

Basinger, J.F. (1976) '*Paleorosa similkameensis* gen. et sp. nov. Permineralized Flowers (Rosaceae) from the Eocene of British Columbia', *Can. J. Bot., 54,* 2293–305

——— and Dilcher, D.L. (1980) 'Bisexual Flowers from the Mid-Cretaceous of Nebraska', *Bot. Soc. Amer., Misc. Series, 158,* 10

Batten, D.J. (1981) 'Stratigraphic, Palaeogeographic and Evolutionary Significance of Late Cretaceous and Early Tertiary Normapolles Pollen', *Rev. Palaeobot. Palynol., 35,* 125–37

——— and Christopher, R.A. (1981) 'Key to the Recognition of Normapolles and Some Morphologically Similar Pollen Genera', *Rev. Palaeobot. Palynol., 35,* 359–83

Battenburg, L.H. (1977) 'The *Sphenophyllum* Species in the Carboniferous Flora of Holz (Westphalian D, Saar Basin, Germany)', *Rev. Palaeobot. Palynol., 24,* 69–99

Baxter, R.W. (1955) '*Palaeostachya andrewsii,* a New Species of Calamite Cone from the American Carboniferous', *Amer. J. Bot., 42,* 342–51

Beck, C.B. (1960) 'Connection between *Archaeopteris* and *Callixylon*', *Science, 131,* 1524–5

——— (1962) 'Reconstruction of *Archaeopteris* and Further Consideration of its Phylogenetic Position', *Amer. J. Bot., 49,* 373–82

——— (1967) '*Eddya sullivanensis* gen. et. sp. nov., a Plant of Gymnosperm Morphology from the Upper Devonian of New York', *Palaeontographica B, 121,* 1–22

——— (1971) 'On the Anatomy and Morphology of Lateral Branch Systems of *Archaeopteris*', *Amer. J. Bot., 58,* 758–84

——— (1976) 'Current Status of the Progymnospermopsida', *Rev. Palaeobot. Palynol., 21,* 5–23

——— (1981) '*Archaeopteris* and its Role in Vascular Plant Evolution', in K.J. Niklas (Ed.), *Paleobotany, Paleoecology and Evolution,* Vol. 1, Praeger, New York, 193–330

———, Coy, K. and Schmid, R. (1982) 'Observations on the Fine Structure of Callixylon Wood', *Amer. J. Bot., 69,* 54–76

Beerbower, R. (1985) 'Early Development of Continental Ecosystems', in B.M. Tiffney (Ed.), *Geological Factors and the Evolution of Plants,* Yale University Press, Newhaven, CT, pp. 47–91

Bell, P.R. (1979) 'The Contributions of the Ferns to an Understanding of the Life Cycles of Vascular Plants', in A.F. Dyer (Ed.), *The Experimental Biology of Ferns,* Academic Press, London, pp. 57–85

Bessey, C.E. (1908) 'The Taxonomic Aspect of the Species Question', *Amer. Natur., 42,* 218–24

Best, L.S. and Bizychudek, P. (1981) 'Pollinator Foraging on Foxglove (*Digitalis purpurea*). A Test of a New Model', *Evolution, 36,* 70–9

Bierhorst, D.W. (1971) *Morphology of Vascular Plants,* Macmillan, New York, 560 pp

——— (1973) 'Non-appendicular Fronds in the Filicales', in A.C. Jermy, J.A. Crabbe and B.A. Thomas (Eds), *The Phylogeny and Classification of the Ferns, Bot. J. Linn. Soc., 67,* Suppl. No. 1, pp. 45–57

Bock, W. (1969) 'The American Triassic Flora and Global Distribution', *Geol. Center Research Ser.,* 3 & 4

Boltenhagen, E. (1967) 'Spores et Pollen du Cretace Superieur du Gabon', *Pollen Spores, 9*, 335–55
Bornman, C.H. (1972) '*Welwitschia mirabilis*: Paradox of the Namib Desert', *Endeavour, 31*, 95–9
Boulter, M.C. (1968) 'On a Species of Compressed Lycopod Sporophyll from the Upper Coal Measures of Somerset', *Palaeontology, 11*, 445–57
Bower, F.O. (1926) '*The Ferns (Filicales)* II', Cambridge University Press, Cambridge, 344 pp
Brack, S.D. (1970) 'On a New Structurally Preserved Arborescent Lycopsid Fructification from the Lower Pennsylvanian of North America', *Amer. J. Bot., 57*, 317–30
Brack-Hanes, S.D. and Thomas, B.A. (1983) 'A Re-examination of *Lepidostrobus Brongniart*', *Bot. J. Linn. Soc., 86*, 125–33
Brack-Hanes, S.D. and Vaughan, J.C. (1978) 'Evidence of Paleozoic Chromosomes from Lycopod Microgametophytes', *Science, 200*, 1383–5
Brenner, G.J. (1963) 'The Spores and Pollen of the Potomac Group of Maryland', *Md. Dep. Geol. Mines Water Resour. Bull., 27*, 1–215
———— (1968) 'Middle Cretaceous Spores and Pollen from Northeastern Peru', *Pollen Spores, 10*, 341–83
———— (1976) 'Middle Cretaceous Floral Provinces and Early Migrations of Angiosperms', in C.B. Beck (Ed.), *Origin and Early Evolution of Angiosperms*, Columbia Press, New York, 23–47
Briggs, D. and Walters, S.M. (1969) *Plant Variation and Evolution*, McGraw-Hill, New York and Toronto, 256 pp
———— (1984) *Plant Variation and Evolution*, 2nd edn, Cambridge University Press, Cambridge, 412 pp
Brokaw, C.J. (1958) 'Chemotaxis of Bracken Spermatozoids. The Role of Bimalate Ions', *J. Exp. Biol., 35*, 192–6
Brough, P. and Taylor, M.H. (1940) 'An Investigation of the Life Cycle of *Macrozamia spiralis* Miq.', *Proc. Linn. Soc. NSW, 65*, 494–524
Brown, J.T. (1975) '*Equisetum clarnoi*, a New Species Based on Petrifactions from the Eocene of Oregon', *Amer. J. Bot., 62*, 410–15
Burnham, L. (1978) 'Survey of Social Insects in the Fossil Record', *Psyche, 85*, 85–133
Burnham, R.J. (1986) 'Morphological Systematics and the Ulmoideae', in R.A. Spicer and B.A. Thomas (Eds), *Systematic and Taxonomic Approaches in Palaeobotany*, Systematic Association Special Volume 31, pp. 105–21

Chaloner, W.G. (1958) 'Isolated Megaspore Tetrads of *Stauropteris burntislandica*', *Ann. Bot., 22*, 197–204
———— (1967) 'Spores and Land-plant Evolution', *Rev. Palaebot. Palynol., 1*, 83–93
———— (1968) 'The Cone of *Cyclostigma kiltorkense* Haughton, from the Upper Devonian of Ireland', *Bot. J. Linn. Soc., 61*, 25–36
———— (1970) 'The Rise of the First Land Plants', *Biol. Rev., 45*, 353–77
———— and Collinson, M.E. (1975) 'Applications of SEM to Sigillarian Impression Fossil', *Rev. Palaeobot. Palynol., 20*, 85–101
————, Hill, A.J. and Lacey, W.S. (1977) 'First Devonian Platyspermic Seed and its Implications in Gymnosperm Evolution', *Nature, 265*, 233–5
———— and Macdonald, P. (1980) *Plants Invade the Land*, HMSO, London, 16 pp
————, Mensah, M.K. and Crane, M.D. (1974) 'Nonvascular Land Plants from the Devonian of Ghana', *Palaeontology, 17*, 925–47
———— and Meyer-Berthaud, B. (1983) 'Leaf and Stem Growth in the Lepidodendrales', *Bot. J. Linn. Soc., 76*, 135–48
———— and Sheerin, A. (1979) 'Devonian Macrofloras', in *The Devonian System, Spec. Papers Palaeontol., 23*, 145–61
———— and Sheerin, A. (1981) 'The Evolution of Reproductive Strategies in Early Land Plants', in G.G.E. Scudder and J.L. Reveal (Eds), *Evolution Today*, Proceedings of the Second International Congress of Systematic and Evolutionary Biology, Hunt Institute, Pittsburgh, pp. 93–100
Chamberlain, C.J. (1935) *Gymnosperms, Structure and Evolution*, University of Chicago Press, Chicago, 484 pp
Chandler, M.E.J. (1964) *The Lower Tertiary Floras of Southern England IV. A Summary and Survey Findings in the Light of Recent Botanical Observations*, British Museum (Natural History) London, 151 pp

Chiarugi, a. (1960) 'Tavole Chromosomihe delle Pteridoyphyta', *Caryologia, 13,* 27–150

Church, A.M. (1968) *Thalassiophyta and the Subaerial Transmigration* (Facsimile of the 1919 edition), Hafner, New York

Clement-Westerhof, J.A. (1984) 'Aspects of Permian Palaeobotany and Palynology. IV. The Conifer *Ortiseia* Florin from the Val Gardena Formation of the Dolomites and the Vincentinian Alps (Italy) with Special Reference to a Revised Concept of the Walchiaceae (Göppert) Schimper', *Rev. Palaeobot. Palynol., 41,* 51–166

Collinson, M.E. (1978) 'Dispersed Fern Sporangia from the British Tertiary', *Ann. Bot., 42,* 233–50

—— (1980) 'A New Multiple-floated *Azolla* from the Eocene of Britain with a Brief Review of the Genus', *Palaeontology, 23,* 213–29

—— (1983) *Fossil Plants of the London Clay,* Palaeontological Association, London, 121 pp

—— (1986) 'Use of Modern Generic Names for Plant Fossils', in R.A. Spicer and B.A. Thomas (Eds), *Systematic and Taxonomic Approaches in Palaeobotany,* Systematics Association Special Volume 31, pp 91–104

——, Fowler, K. and Boulter, M.C. (1981) 'Floristic Changes Indicate a Cooling Climate in the Eocene of Southern England', *Nature, 291,* 315–17

Corna, O. (1970) 'Plant Remains in the Ordovician of the Bohemian Massif', *Geol. Zbornik-Geol. Carpathica, 21,* 183–6

Couper, R.A. (1958) 'British Mesozoic Microspores and Pollen Grains', *Palaeontographica B, 103,* 75–179

Crepet, W.L. (1974) 'Investigations of North American Cycadeoids: the Reproductive Biology of *Cycadeoidea*', *Palaeontographica B, 148,* 144–69

—— (1978) 'Investigations of Angiosperms from the Eocene of North America: an Aroid Inflorescence', *Rev. Palaeobot. Palynol., 25,* 241–52

—— (1979) 'Some Aspects of the Pollination Biology of Middle Eocene Angiosperms', *Rev. Palaeobot. Palynol., 27,* 213–38

—— (1981) 'The Status of Certain Families of the Amentiferae during the Middle Eocene and Some Hypotheses Regarding the Evolution of Wind Pollination in Dicotyledonous Angiosperms', *Paleoecol. Evol., 1,* 103–28

—— and Daghlian, C.P. (1980) 'Castaneoid Inflorescences from the Middle Eocene of Tennessee and the Diagnostic Value of Pollen (at the Subfamily Level) in the Fagaceae', *Amer. J. Bot., 67,* 739–57

——, Daghlian, C.P. and Zavada, M. (1979) 'Investigations of Angiosperms from the Eocene of North America: a New Juglandaceous Catkin', *Rev. Palaeobot. Palynol., 30,* 361–70

—— and Dilcher, D.L. (1977) 'Investigations of Angiosperms from the Eocene of North America: a Mimosoid Inflorescence', *Amer. J. Bot., 64,* 714–25

——, Dilcher, D.L. and Potter, F.W. (1975) 'Investigations of Angiosperms from the Eocene of North America: a Catkin with Juglandaceous Affinities', *Amer. J. Bot., 62,* 813–23

Cridland, A.A. (1964) '*Amyelon* in American Coal-balls', *Palaeontology, 7,* 186–209

Cronquist, A. (1968) *Evolution and Classification of Flowering Plants,* Houghton Mifflin, Boston, 296 pp

—— (1978) 'Once Again, What is a Species?', in *Biosystematics in Agriculture,* Allenheld & Osmun, Montclair, NJ, pp. 3–20

Cutler, D.F., Alvin, K.L. and Price, C.E. (1982) *The Plant Cuticle,* Linnean Society Symposium Series No 10, London, 461 pp

Daghlian, C.P. (1981) 'A Review of the Fossil Record of Monocotyledons', *Bot. Rev., 47,* 517–55

Dawson, J.W. (1859) 'On Fossil Plants from the Devonian Rocks of Canada', *Quart. J. Geol. Soc. London, 15,* 477–88

Delevoryas, T. (1953) 'A New Male Cordaitean Fructification from the Kansas Carboniferous', *Amer. J. Bot., 40,* 144–50

—— (1962) *Morphology and Evolution of Fossil Plants,* Holt, Rinehart & Winston, New York

—— (1968) 'Investigations of North American Cycadeoids: Structure, Ontogeny and Phylogenetic Considerations of Cones of Cycadeoidea', *Palaeontographica B, 121,* 122–33

—— (1971) 'Biotic Provinces and the Jurassic-Cretaceous Floral Transition', *Proc. N. Amer. Paleontol. Conv.* (1), 1660–74

—— (1982) 'Perspectives on the Origin of Cycads and Cycadeoids', *Rev. Palaebot. Palynol.*, *37*, 115–32

—— and Hope, R.C. (1971) 'A New Triassic Cycad and its Phyletic Implications', *Postilla*, *150*, 1–21

—— and —— (1973) 'Fertile Coniferophyte Remains from the Late Triassic Deep River Basin, North Carolina', *Amer. J. Bot.*, *60*, 810–18

—— and —— (1981) 'More Evidence from Conifer Diversification in the Upper Triassic of North Carolina', *Amer. J. Bot.*, *68*, 1003–7

Dennis, R.L. and Eggert, D.A. (1978) '*Parasporotheca*, gen. nov., and its Bearing on the Interpretation of the Morphology of Permineralized Medullosan Pollen Organs', *Bot. Gaz.*, *139*, 117–39

Dettmann, M.E. (1973) 'Angiospermous Pollen from Albian to Turonian Sediments of Eastern Australia', *Geol. Soc. Aust. Spec. Publ.*, *4*, 3–34

Dilcher, D.L. (1969) '*Podocarpus* from the Eocene of North America', *Science*, *164*, 299–301

—— (1971) 'A Revision of the Eocene Flora of Southeastern North America', *Palaeobotanist*, *20*, 7–18

—— (1974) 'Approaches to the Identification of Angiosperm Leaf Remains', *Bot. Rev.*, *40*, 1–157

—— (1979) 'Early Angiosperm Reproduction: an Introductory Report', *Rev. Palaeobot. Palynol.*, *27*, 291–328

—— and Crane, P.R. (1984) '*Archaeanthus*: an Early Angiosperm from the Cenomanian of the Western Interior of North America', *Ann. Missouri Bot. Gdn*, *71*, 351–83

——, Potter, F.W. and Crepet, W.L. (1976) 'Investigations of Angiosperms from the Eocene of North America: Juglandaceous Winged Fruits', *Amer. J. Bot.*, *63*, 532–44

Dimichele, W.A. and Phillips, T.L. (1985) 'Arborescent Lycopod Reproduction and Paleocology in a Coal-swamp Environment of Late Middle Pennsylvanian Age (Herrin Coal, Illinois, U.S.A)', *Rev. Palaeobot. Palynol.*, *44*, 1–26

Dobzhansky, Th. (1970) *Genetics of the Evolutionary Process*, Columbia University Press, New York, 555 pp

Dover, G. (1982) 'Molecular Drive: a Cohesive Mode of Species Evolution', *Nature*, *299*, 111–17

Doyle, J.A. (1969) 'Cretaceous Angiosperm Pollen of the Atlantic Coastal Plain and its Evolutionary Significance', *J. Arnold Arbor.*, *50*, 1–35

—— (1978) 'Origin of Angiosperms', *Ann. Rev. Ecol. Syst.*, *9*, 365–92

——, Biens, P., Doerenkamp, A. and Jardiné, S. (1977) 'Angiosperm Pollen from the Pre-Albian Lower Cretaceous of Equatorial Africa', *Bull. Centres Rech. Explor. Prod. Elf Aquitaine*, *1*, 451–73

—— and Donaghue, M.J. (1986) 'Relationship of Angiosperms and Gnetales: a Numerical Cladistic Analysis', in R.A. Spicer and B.A. Thomas (Eds), *Systematic and Taxonomic Approaches in Palaeobotany*, Systematics Association Special Volume 31, pp. 177–99

—— and Hickey, L.J. (1976) 'Pollen and Leaves from the Mid-Cretaceous Potomac Group and their Bearing on Early Angiosperm Evolution', in C.B. Beck (Ed.), *Origin and Early Evolution of Angiosperms*, Columbia University Press, New York, pp. 139–206

——, Jardine, S. and Doerenkamp, A. (1982) '*Afropollis*, a New Genus of Early Angiosperm Pollen, with Notes on the Cretaceous Palynostratigraphy and Paleoenvironments of Northern Gondwana', *Bull. Centres Rech. Explor. Prod. Elf Aquitaine*, *6*, 39–117

—— and Robbins, E.I. (1977) 'Angiosperm Pollen Zonation of the Continental Cretaceous of the Atlantic Coastal Plain and its Application to Deep Wells in the Salisbury Embayment', *Palynology*, *1*, 43–78

Duckett, J.G. (1973) 'Comparative Morphology of the Gametophytes of the Genus *Equisetum*, Subgenus *Equisetum*', *Bot. J. Linn Soc.*, *66*, 1–22

—— and Pang, W.C. (1984) 'The Origins of Heterospory: a Comparative Study of Sexual Behaviour in the Fern *Platyzoma microphyllum* R. Br. and the Horsetail *Equisetum giganteum* L.', *Bot. J. Linn. Soc.*, *88*, 11–34

Dufek, D. and Stidd, B.M. (1981) 'The Vascular System of *Dolerotheca* and its Phylogenetic Significance', *Amer. J. Bot.*, *68*, 897–907

Dyer, A.F. (1979) (Ed.), *The Experimental Biology of Ferns*, Academic Press, London, 657 pp

Eames, A.J. (1952) 'Relationships of the Ephedrales', *Phytomorphology*, *2*, 79–100

Eckenwalder, J.E. (1980) 'Cycads: The Prime of their Lives', *Fairchild Trop. Gdn, Bull.*, Jan. 1980, 11–18

Edhorn, A. St. (1977) 'Early Cambrian Algae Croppers', *Can. J. Earth Sci.*, *14*, 1014–20

Edwards, D. (1970) 'Further Observations on the Lower Devonian Plant, *Gosslingia breconensis* Heard', *Phil. Trans. R. Soc. Lond. B*, *258*, 225–43

—— (1979) 'A Late Silurian Flora from the Lower Old Red Sandstone of South-west Dyfed', *Palaeontology*, *22*, 23–52

—— and Davies, E.C.W. (1976) 'Oldest Recorded *in situ* Tracheids', *Nature*, *263*, 494–5

—— and Edwards, D.S. (1986) 'A Reconsideration of the Rhyniophytina Banks', in R.A. Spicer and B.A. Thomas (Eds), *Systematic and Taxonomic Approaches in Palaeobotany*, Systematics Association Special Volume 31, Oxford University Press, Oxford, pp. 201–22

—— and Fanning, U. (1985) 'Evolution and Environment in the Late Silurian-Early Devonian: the Rise of the Pteridophytes', *Phil. Trans. R. Soc. Lond. B*, *309*, 147–65

Edwards, D.S. (1980) 'Evidence for the Sporophytic Status of the Lower Devonian Plant *Rhynia gwynne-vaughanii* Kidston and Lang', *Rev. Palaeobot. Palynol.*, *29*, 177–88

—— (1986) '*Aglaophyton major* a non-vasculer land-plant from the Devonian Rhynie Chert', *Bot. J. Linn. Soc.*, *93*, 173–204

Eggert, D.A. (1961) 'The Ontogeny of the Carboniferous Arborescent Lycopsida', *Palaeontographica B*, *108*, 43–92

—— (1972) 'Petrified *Stigmaria* of Sigillarian Origin from North America', *Rev. Palaeobot. Palynol.*, *14*, 85–99

—— (1974) 'The Sporangium of *Horneophyton lignieri* (Rhyniophytina)', *Amer. J. Bot.*, *61*, 405–13

—— and Gaunt, D.D. (1973) 'Phloem of *Spenophyllum*', *Amer. J. Bot.*, *60*, 755–70

Ehrlich, P.R. and Holm, R.W. (1962) 'Patterns and Populations', *Science*, *137*, 652

El-Saadaway, W.E. and Lacey, W.S. (1979a) 'Observations on *Nothia aphylla* Lyon ex Hoeg', *Rev. Palaebot. Palynol.*, *27*, 119–47

—— and —— (1979b) 'The Sporangia of *Horneophyton lignieri* (Kidston and Lang) Barghoorn and Darrah', *Rev. Palaeobot. Palynol.*, *28*, 137–44

Erdtman, G. (1963) 'Palynology', in R.D. Preston (Ed.), *Recent Advances in Botanical Research*, *1*, Academic Press, New York, pp. 149–208

Faegri, K. and Iversen, J. (1950) *Textbook of Pollen Analysis*, Munksgaard, Copenhagen, 237 pp

—— and van der Pijl, L. (1979) *The Principles of Pollination Ecology*, 3rd edn, Pergamon Press, Oxford, 244 pp

Ferguson, D.K. (1985) 'The Origin of Leaf-assemblages — New Light on an Old Problem', *Rev. Palaeobot. Palynol.*, *46*, 117–88

Fisher, R.A. (1930) *The Genetical Theory of Natural Selection*, Clarendon Press, Oxford (rev. edn 1958), Dover, New York, 291 pp

Florin, R. (1933) 'Studien über die Cycadales des Mesozoikums', *Kungl. Svenska Vetenskapsakademiens Handlingar III*, *12*, 1–134

—— (1944) 'Die Koniferen des Oberkarbons und des Untern Perms', *Palaeontographica B*, *85*, 457–654

—— (1950a) 'On Female Reproductive Organs in the Cordaitinae', *Acta Horti Bergiana*, *15*, 111–34

—— (1950b) 'Upper Carboniferous and Lower Permian Conifers', *Bot. Rev.*, *16*, 258–82

—— (1951) 'Evolution in Cordaites and Conifers', *Acta Horti Bergiana*, *15*, 285–388

—— (1958) 'On Jurassic Taxads and Conifers from Northeastern Europe and Eastern Greenland', *Acta Horti Bergiana*, *17*, 259–388

Fowler, K. (1975) 'Megaspores and Massulae of *Azolla prisca* from the Oligocene of the Isle of Wight', *Palaeontology*, *18*, 483–507

Francis, J.E. (1984) 'The Seasonal Environment of the Purbeck (Upper Jurassic) Fossil Forests', *Palaeogeog., Palaeoclimatol., Palaeoecol.*, *48*, 285–307

Frankenberg, J.M. and Eggert, D.A. (1969) 'Petrified *Stigmaria* from North America: Part 1. *Stigmaria ficoides*, the Underground Portions of Lepidodendraceae', *Palaeontographica B*, *128*, 1–47

Frankie, G.W. (1975) 'Tropical Forest Phenology and Pollinator Plant Co-evolution', in L.E. Gilbert and P.H. Raven (Eds), *Co-evolution of Animals and Plants*, University of Texas Press,

Austin
——, Baker, H.G. and Opler, P.A. (1974a) 'Comparative Phenological Studies of Trees in Tropical Wet and Dry Forests of Costa Rica', *J. Ecol.*, *62*, 881–919
——, —— and —— (1974b) 'Tropical Plant Phenology: Applications for Studies in Community Ecology', in M. Lieth (Ed.), *Phenology and Seasonality Modelling*, Springer-Verlag, New York, 444 pp
——, Opler, P.A. and Bawa, K.S. (1976) 'Foraging Behaviour of Solitary Bees: Implications for Outcrossing of a Neotropical Forest Tree Species', *J. Ecol.*, *64*, 1049–57
Frederiksen, N.O. (1985) 'Review of Early Tertiary Sporomorph Paleoecology', *AASP Contributions Series*, No. 15, 92 pp
Freudenbert, K. (1968) 'The Constitution and Biosynthesis of Lignin', in K. Reendenberg and A.C. Neish (Eds), *Constitution and Biosynthesis of Lignin*, Springer-Verlag, New York, pp. 47–129
Friis, E.M. (1983) 'Upper Cretaceous (Senonian) Floral Structures of Juglandalean Affinity Containing Normapolles Pollen', *Rev. Palaeobot. Palynol.*, *39*, 161–88

Galtier, J. (1966) 'Observations nouvelles sur le genre *Clepsydropsis*', *Naturalia Monspeliensia ser. Botanique*, *17*, 111–32
—— (1981) 'Structure foliaires de fougères et pteridospermales du Carbonifère inferieur et leur signification evolutive', *Palaeontographica B*, *180*, 1–38
Garratt, M.J., Tims, J.D., Rickards, R.B., Chambers, T.C. and Douglas, J.G. (1984) 'The Appearance of *Baragwanathia* (Lycophytina) in the Silurian', *Bot. J. Linn. Soc.*, *89*, 355–8
Gastaldo, R.A. (1981) 'Taxonomic Considerations for Carboniferous Coalfield Compression Equisetalean Strobili', *Amer. J. Bot.*, *68*, 1319–24
Gensel, P.G. (1976) '*Renalia hueberi*, a New Plant from the Lower Devonian of Gaspé', *Rev. Palaeobot. Palynol.*, *22*, 19–37
—— (1979) 'Two *Psilophyton* Species from the Lower Devonian of Eastern Canada with Discussion of Morphological Variation within the Genus', *Palaeontographica B*, *168*, 81–9
—— (1980) 'Devonian *in situ* Spores: a Survey and Discussion', *Rev. Palaebot. Palynol.*, *30*, 101–32
——, Andrews, H.N. and Forbes, W.H. (1975) 'A New Species of *Sawdonia* with Notes on the Origin of Microphylls and Lateral Sporangia', *Bot. Gaz.*, *136*, 50–62
—— and Skog, J.E. (1977) 'Two Early Mississipian Seeds from the Price Formation of Southwestern Virginia', *Brittonia*, *29*, 332–51
Gillespie, W.H., Rothwell, G.W. and Scheckler, S.E. (1981) 'The Earliest Seeds', *Nature*, *293*, 462–4
Gilmour, J.S.L. and Gregor, J.W. (1939) 'Demes: a Suggested New Terminology', *Nature*, *144*, 333–4
—— and Heslop-Harrison, J. (1954) 'The Deme Terminology and the Units of Micro-evolutionary Change', *Genetica*, *27*, 147–61
Góczán, F., Groot, J.J., Krutzsch, N. and Pacltová, B. (1967) 'Die Gattungen des "Stemma Normapolles Pflug. 1953b" (Angiospermae), Neubeschreibungen und Revision Europäisher (Oberkreide bis Eozän)', *Palaeontol. Abh. B.*, *3*, 429–539
Good, C.W. (1975) 'Pennsylvanian-age Calamite Cones, Elater-bearing Spores, and Associated Vegetative Organs', *Palaeontographica B*, *153*, 28–99
Goswami, H.K. and Ayra, B.S. (1968) 'Heterosporous Sporangia in *Isoetes*', *Brit. Fern. Gaz.*, *10*, 39–40
Gould, R.E. and Delevoryas, T. (1977) 'The Biology of *Glossopteris*: Evidence from Petrified Seed-bearing and Pollen-bearing Organs', *Alcheringa*, *1*, 387–99
Gould, S.J. and Eldridge, N. (1977) 'Punctuated Equilibria: the Tempo and Mode of Evolution Reconsidered', *Paleobiology*, *3*, 115–51
Grand'Eury, F.C. (1877) 'Flore Carbonifère du Department de la Loire et du Centre de la France', *Memoire de l'Academie des Sciences, Paris*, *24*, 624 pp
Granoff, J.A., Gensel, P.G. and Andrews, H.N. (1976) 'A New Species of *Pertica* from the Devonian of Eastern Canada', *Palaeontographica B*, *155*, 119–28
Gray, J. and Boucot, A.J. (1977) 'Early Vascular Land Plants: Proof and Conjecture', *Lethaia*, *10*, 145–74
Grierson, J.D. and Bonamo, P.M. (1979) '*Leclercqia complexa*: Earliest Ligulate Lycopod (Middle Devonian)', *Amer. J. Bot.*, *66*, 474–6

Gunnison, D. and Alexander, M. (1975) 'Basis for the Resistance of Several Algae to Microbial Decomposition', *Appl. Microbiol.*, *29*, 729–38

Harper, J.L., Lovell, P.M. and Moore, K.C. (1970) 'The Shapes and Sizes of Seeds', *Ann. Rev. Ecol. Syst.*, *1*, 327–57

Harris, T.M. (1931) 'The Fossil Flora of Scoresby Sound, East Greenland. I. Cryptograms (Exclusive of Lycopodiales)', *Meddr. Gronland*, *85*, 1–104

———— (1933) 'A New Member of the Caytoniales', *New Phytol.*, *32*, 97–114

———— (1937) 'The Fossil Flora of Scoresby Sound, East Greenland. Part 5: Stratigraphic Relations of the Plant Beds', *Meddr. Gronland*, *112*, 1–114

———— (1940) 'On *Caytonia* Thomas', *Ann. Bot. (NS)*, *4*, 713–34

———— (1942) 'Notes on the Jurassic Flora of Yorkshire. 1. *Ptilophyllum caytonense* sp. n. 2. *Deltolepis credipata* gen. et. sp. n. 3. *Nilssonia compta* and its Reference to *Beania gracilis*', *Ann. Mag. Nat. Hist.*, Ser. II, 9, 568–87

———— (1958) 'The Seed of *Caytonia*', *Palaeobotanist*, *7*, 93–106

———— (1961a) 'The Fossil Cycads', *Palaeontology*, *4*, 313–23

———— (1961b) *The Yorkshire Jurassic Flora. I: Thallophyta-Pteridophyta*, British Museum (Natural History), London, 212 pp

———— (1964) *The Yorkshire Jurassic Flora. II. Caytoniales, Cycadales and Pteridosperms*, British Museum (Natural History), London, 191 pp

———— (1969) *The Yorkshire Jurassic Flora. III. Bennettitales*, British Museum (Natural History), London, 186 pp

———— (1973a) 'The Strange Bennettitales', the 19th Sir Albert Charles Seward Memorial Lecture, 1970, Birbal Sahni Institute of Palaeobotany, Lucknow, 11 pp

———— (1973b) 'What Use Are Fossil Ferns?', in A.C. Jermy, J.A. Crabbe and B.A. Thomas (Eds), *The Phylogeny and Classification of the Ferns, Bot. J. Linn. Soc.*, *67*, Suppl. No. 1, pp. 41–4

———— (1974) '*Williamsoniella lignieri*: its Pollen and the Compression of Spherical Pollen Grains', *Palaeontology*, *17*, 125–48

———— (1976a) 'A Slender Upright Plant from Wealden Sandstones', *Proc. Geol. Assn*, *87*, 413–22

———— (1976b) 'Two Neglected Aspects of Fossil Conifers', *Amer. J. Bot.*, *63*, 902–10

———— (1976c) 'The Mesozoic Gymnosperms', *Rev. Palaeobot. Palynol.*, *21*, 119–34

———— (1979) *The Yorkshire Jurassic Flora. V. Coniferales*, British Museum (Natural History), London, 165 pp

Hebant, C. (1977) *The Conducting Tissues of Bryophytes*, J. Cramer, Germany, 157 pp

Heer, O. (1868) 'Die fossile Flora der Polarlander', in *Flora Fossils Arctica*, Band 1: Zurich, 152 pp

Heinrich, B. (1975) 'Energetics of Pollination', *Ann Rev. Ecol. Syst.*, *6*, 139–79

———— (1976) 'Resource Partitioning among Some Social Insects: Bumblebees', *Ecology*, *57*, 874–89

———— (1979) 'Resource Heterogeneity and Patterns of Movement in Foraging Bumblebees', *Oecologia*, *140*, 235–45

———— and Raven, P.M. (1972) 'Energetics and Pollination Ecology', *Science*, *176*, 597–602

Hennig, W. (1966) *Phylogenetic Systematics*, University of Illinois Press, Chicago, Ill., 263 pp

Herngreen, G.F.W. (1973) 'Palynology of Albian-Cenomanian strata of borehole 1-QS-1-MA State of Maranhao, Brazil', *Pollen Spores*, *15*, 515–55

———— (1974) 'Middle Cretaceous Palynomorphs from Northeastern Brazil', *Bull. Sci. Geol. Strasbourg*, *27*, 101–16

Heslop-Harrison, J. (1971) 'Sporopollenin in the Biological Context', in J. Brooks, P.R. Grant, M. Muir, P. van Gijzel and G. Shaw (Eds), *Sporopollenin*, Academic Press, New York, pp. 1–30

———— (1976) 'The Adaptive Significance of the Exine', in I.K. Ferguson and J. Muller (Eds), *The Evolutionary Significance of the Exine*, Linn. Soc. Symp. Series No 1., Academic Press, New York, pp. 27–38

Heywood, V.H. (1980) 'The Impact of Linnaeus on Botanical Taxonomy — Past, Present and Future', *Veroffentlichungen der Joachim Jungius — Gesellschaft der Wissenschaften, Hamburg*, *43*, 97–115

Hickey, L.J. (1973) 'Classification of the Architecture of Dicotyledonous Leaves', *Amer. J. Bot.*, *80*, 17–33

———— (1977) 'Stratigraphy and Palaeobotany of the Golden Valley Formation (Early Tertiary) of

Western North Dakota', *Geol. Soc. Amer. Memoir, 150*, 183 pp

―――― (1979) 'A Revised Classification of the Architecture of Dicotyledonous Leaves', in C.R. Metcalfe and L. Chalk (Eds), *Anatomy of the Dicotyledons*, Clarendon Press, Oxford

―――― (1981) 'Land Plant Evidence Compatible with Gradual, not Catastrophic Change at the End of the Cretaceous', *Nature, 292*, 529–31

―――― and Doyle, J.A (1977) 'Early Cretaceous Fossil Evidence for Angiosperm Evolution', *Bot. Rev., 43*, 3–104

―――― and Wolfe, J.A. (1975) 'The Basis of Angiosperm Phylogeny, Vegetation, Morphology', *Ann. Missouri Bot. Gdn, 62*, 538–89

Hill, C.R. and Crane, P.R. (1982) 'Evolutionary Cladistics and the Origin of Angiosperms', in K.A. Joysey and A.E. Friday (Eds), *Problems of Phylogenetic Reconstruction*, Academic Press, London and New York, pp. 269–361

Hirmer, M. (1927) *Handbuch der Paläobotanik*, R. Oldenbourg, Munich, 708 pp

Holmes, J.C. (1981) 'The Carboniferous Fern *Psalixochlaena cylindrica* as Found in Westphalian 'A' Coal Balls from England. Part II, the Frond and Fertile Parts', *Palaeontographica B, 176*, 147–73

Hopkins, W.S. (1974) 'Some Spores and Pollen from the Christopher Formation (Albian) of Ellef and Amund Ringes Island, and Northwestern Melville Island, Canadian Arctic Archipelago', *Canad. Geol. Surv. Paper, 74–30*, pp. 1–39

Hueber, F.M. (1971) '*Sawdonia ornata*: a New Name for *Psilophyton princeps* var. *ornatum*', *Taxon, 16*, 641–2

―――― and Banks, H.P. (1979) '*Serrulacaulis furcatus* gen. et sp. nov., a New Zosterophyll from the Lower Upper Devonian of New York State', *Rev. Palaeobot. Palynol., 28*, 169–89

Hughes, N.F. (1976) *Palaeobiology of Angiosperm Origins*, Cambridge University Press, Cambridge, 242 pp

―――― (1977) 'Palaeo-succession of Angiosperm Evolution', *Bot. Rev., 43*, 105–27

――――, Drewry, G.E. and Laing, J.E. (1979) 'Barremian Earliest Angiosperm Pollen', *Palaeontology, 22*, 513–35

Ishchenko, T.A. and Shlyakov, R.N. (1979) 'Middle Devonian Liverworts (Marchantiidae) from Podolia', *Paleontol. J., 13*, 369–80

Jain, R.K. (1971) 'Pre-Tertiary Records of Salviniaceae', *Amer. J. Bot., 58*, 487–97

―――― and Hall, J.W. (1969) 'A Contribution to the Early Tertiary Fossil Record of the Salviniaceae', *Amer. J. Bot., 56*, 527–39

Janzen, D.H. (1967) 'Synchronization of Sexual Reproduction of Trees within the Dry Season in Central America', *Evolution, 21*, 620–37

Janzen, D.M. and Norris, G. (1975) 'Evolutionary Significance and Botanical Relationships of Cretaceous Angiosperm Pollen of the Western Canadian Interior', *Geosci. Man, 11*, 47–60

Jardiné, S., Kieser, G. and Reyre, Y. (1974) 'L'individualisation progressive du continent africain vue a travers les données palynologiques de l'ere secondaire', *Bull. Sci. Geol. Strasbourg, 27*, 69–85

―――― and Magloire, L. (1965) 'Palynologie et stratigraphie du Cretace des bassins du Senegal et de Cote-d'Ivoire', *Mem. Bur. Rech. geol. min., 32*, 187–245

Jennings, J.R. (1975) '*Protostigmaria*, a New Plant Organ from the Lower Mississippian of Virginia', *Palaeontology, 18*, 19–24

―――― and Eggert, D.A. (1977) 'Preliminary Report on Permineralized *Senftenbergia* from the Chester Series of Illinois', *Rev. Palaeobot. Palynol., 24*, 221–5

Jermy, A.C., Crabbe, J.A. and Thomas, B.A. (Eds) (1973) *The Phylogeny and Classification of the Ferns, Bot. J. Linn. Soc., 67*, Suppl. No. 1, 284 pp

Jongmans, W.J. (1954) 'The Carboniferous Flora of Peru', *Bull. Br. Mus. (Nat. Hist.), Geol., 2*, 191–223

Jordan, K. (1905) 'Der Gegensatz zwischen geographischer und nicht geographischer Variation', *Z. wiss. Zool., 83*, 151–210

Jung, W.W. (1968) '*Hirmeriella muensteri* (Schenk) Jung nov. comb., eine bedeutsame Konifere des Mesozoikums', *Palaeontographica B, 122*, 56–93

Kaspar, A. and Andrews, H.N. (1972) '*Pertica*, a New Genus of Devonian Plants from Northern

Maine', *Amer. J. Bot.*, *59*, 897–911

Keely, J.E., Osmond, C.B. and Raven, J.A. (1984) *'Stylites*, a Vascular Land Plant without Stomata Absorbs CO_2 via its Roots', *Nature*, *310*, 694–5

Kemp, E.M. (1968) 'Probable Angiosperm Pollen from British Barremian to Albian Strata', *Palaeontology, 11*, 421–34

Kerp, J.H.F. (1983) 'Aspects of Permian Palaeobotany and Palynology. I *Sobernheimia jonkeri* nov. gen., nov sp. a New Fossil Plant of Cycadalean Affinity from the Waderner Gruppe of Sobernheim', *Rev. Palaeobot. Palynol.*, *38*, 173–83

Kevan, P.G., Chaloner, W.G. and Saville, D.B.O. (1975) 'Interrelationships of Early Terrestrial Arthropods and Plants', *Palaeontology*, *18*, 391–417

Kidston, R. and Lang, W.H. (1917–1921) 'On Old Red Sandstone Plants Showing Structure, from the Rhynie Chert Bed, Aberdeenshire. Parts I–V', *Trans. R. Soc. Edinb.*, *51*, 761–84; *52*, 603–27, 643–80, 831–54, 855–902

Kimura, T. and Sekido, S. (1975) '*Nilssoniocladus* N. Gen. (Nilssoniaceae N. Fam.) Newly Found from the Early Lower Cretaceous of Japan', *Palaeontographica B, 153*, 111–18

Konijnenburg-van Cittert, J.H.A. (1971) '*In situ* Gymnosperm Pollen from the Middle Jurassic of Yorkshire', *Bot. Neerland.*, *20*, 1–96

Kovach, W.L. and Dilcher, D.L. (1984) 'Dispersed Cuticles from the Eocene of North America', *Bot. J. Linn. Soc.*, *88*, 63–104

Krassilov, V.A. (1970) 'Approach to the Classification of Mesozoic "Ginkgoalean" Plants from Siberia', *Palaeobotanist*, *18*, 12–19

—— (1977) 'Contributions to the Knowledge of the Caytoniales', *Rev. Palaeobot. Palynol.*, *24*, 115–78

—— (1978) 'Araucariaceae as Indicators of Climate and Palaeolatitudes', *Rev. Palaeobot. Palynol.*, *26*, 113–24

—— (1981) '*Orestovia* and the Origin of Vascular Plants', *Lethaia*, *14*, 235–50

—— (1984) 'New Paleobotanical Data on Origin and Early Evolution of Angiospermy', *Ann. Missouri Bot. Gdn*, *71*, 577–92

Krausel, R. and Weyland, H. (1935) 'Neue Pflanzenfunds in Rheinischen Unterdevon', *Palaeontographica B, 80*, 171–90

Lacey, W.S., Van Dijk, D.E. and Gordon-Gray, K.D. (1974) 'New Permian Glossopteris Flora from Natal', *S. Afr. J. Sci.*, *70*, 154–6

Laing, J.F. (1975) 'Mid-Cretaceous Angiosperm Pollen from Southern England and Northern France', *Palaeontology*, *18*, 775–808

Lang, W.H. (1937) 'On the Plant Remains from the Downtonian of England and Wales', *Phil. Trans. R. Soc. Lond. B*, *227*, 245–91

Lange, R.T. (1970) 'The Maslin Bay Flora, South Australia. 2. The Assemblage of Fossils', *Neues Jahrb. Geol. Palaeontol. Monash, 8*, 486–90

Lebedev, Y.L. (1976) 'Evolution of Albian-Cenomian Floras of Northeast U.S.S.R. and the Association between their Composition and Facies Conditions', *Int. Geol. Rev.*, *19*, 1183–90

Lee, B. and Priestly, J.M. (1924) 'The Plant Cuticle. I. Its Structure, Distribution and Function', *Ann. Bot.*, *38*, 525–45

Leisman, G.A. (1964) '*Mesidophyton paulus* gen. et. sp. nov., a New Herbaceous Sphenophyll', *Palaeontographica B, 114*, 135–46

Lele, K.M. and Walton, J. (1962) 'Contributions to the Knowledge of *Zosterophyllum myretonianum* Penhallow from the Lower Old Red Sandstone of Angus', *Trans. R. Soc. Edinb.*, *64*, 469–75

Lemoigne, Y. (1967) 'Paleoflore a Cupressales dans le Trias-Rhetien du Contentin', *C.R. Acad. Sci. Paris, 264, D*, 715–18

—— (1971) 'Nouvelles Diagnoses du Genre *Rhynia* et de l'Espece *Rhynia gwynne-vaughanii*', *Bull. Soc. Bot. Fr.*, *117*, 307–20

Leppik, E.E. (1957) 'A New System for Classification of Flower Types', *Taxon, 6*, 64–7

Lesquereux, L. (1892) 'The Flora of the Dakota Group', *US Geol. Surv. Monogr.*, *17*, 400 pp

Levin, D.A. (1978) 'Pollinator Behaviour and the Breeding Structure of Plant Populations', in A.J. Richards (Ed.), *The Pollination of Flowers by Insects*, Linnean Society Symposium Series 6, Academic Press, London, 133–50

—— (1979) 'The Nature of Plant Species', *Science, 204*, 381–4

—— and Anderson, W.W. (1970) 'Competition for Pollination between Simultaneously Flowering Species', *Amer. Natur., 104*, 455–67

Logan, K.T. and Thomas, B.A. (1987) 'Distribution of Lignin Derivatives in Fossil Plants', *New Phytol.*, in press

Long, A.G. (1975) 'Further Observations on Some Lower Carboniferous Seeds and Cupules', *Trans. R. Soc. Edinb., 69*, 267–93

—— (1977) 'Lower Carboniferous Pteridosperm Cupules and the Origin of Angiosperms', *Trans. R. Soc. Edinb., 70*, 13–35

Longman, K.A. and Jenik, J. (1974) *Tropical Forest and its Environment*, Longmans, London, 196 pp

Lovis, J.D. (1975) '*Aspidistes thomasii* — a Jurassic Member of the Thelypteridaceae', *Fern Gaz., 11*, 137–40

Lyon, A.G. (1957) 'Germinating Spores in the Rhynie Chert', *Nature, 180*, 1219

Mägdefrau, K. (1956) *Palaobiologie der Pflanzen*, 3rd edn, G. Fischer, Jena, 443 pp

Maheshwari, P. and Vasil, V. (1961) 'The Stomata of *Gnetum*', *Ann. Bot., 25*, 313–19

Mamay, S.H. (1950) 'Some American Carboniferous Fern Fructifications', *Ann. Missouri Bot. Gdn, 37*, 409–59

—— (1969) 'Cycads: Fossil Evidence of a Late Paleozoic Origin', *Science, 164*, 295–6

—— (1976) 'Paleozoic Origin of Cycads', *US Geol. Surv. Prof. Paper, 934*, 1–48

Mapes, G. and Rothwell, G.W. (1984) 'Permineralized Ovulate Cones of *Lebachia* from the Late Palaeozoic Limestones of Kansas', *Palaeontology, 21*, 69–94

Martin, J.T. and Juniper, B.E. (1970) *The Cuticles of Plants*, Edward Arnold, London, 347 pp

Matten, L.C. and Lacey, W.S. (1981) 'Cupule Organization in Early Seed Plants', in R.C. Romans (Ed.), *Geobotany II*, Plenum, New York, pp. 221–34

Mayr, E. (1940) 'Speciation Phenomena in Birds', *Amer. Natur., 74*, 249–78

—— (1957) 'The Species Problem', in E. Mayr (Ed.), *Amer. Assn Adv. Sci. Publ., 50*, 1–22

—— (1963) *Animal Species and Evolution*, Harvard University Press, Cambridge, Mass, 797 pp

Melville, R. (1960) 'A New Theory of the Angiosperm Flower', *Nature, 188*, 14–18

—— (1962) 'A New Theory of the Angiosperm Flower. 1. The Gynoecium', *Kew Bulletin, 16*, 1–50

—— (1983) 'Glossopteridae, Angiospermidae and the Evidence for Angiosperm Origin', *Bull. J. Linn. Soc., 86*, 279–323

Merker, H. (1958) 'Zum Fehlenden Gleide der Rhynienflora', *Bot. Not., 111*, 608–18

—— (1959) 'Analyse der Rhynien-Basis und Nechweis des Gametophyton', *Bot. Not., 112*, 441–52

Meyen, S.V. (1969) 'The Angara Members of Gondwana Genus *Barakaria* and its Systematic Position', *Argumenta Palaeobotanica, 3*, 1–14

—— (1971) 'Phyllotheca-like Plants from the Upper Palaeozoic Flora of Angaraland', *Palaeontographica B, 133*, 1–33

—— (1972) 'Are there Ligula and Parichnos in Angara Carboniferous Lepidophytes?', *Rev. Palaeobot. Palynol., 14*, 149–57

—— (1973) 'Plant Morphology in its Nomothetical Aspects', *Bot. Rev., 39*, 205–60

—— (1976) 'Carboniferous and Permian Lepidophytes of Angaraland', *Palaeontographica B, 157*, 112–57

—— (1984) 'Basic Features of Gymnosperm Systematics and Phylogeny as Evidenced by the Fossil Record', *Bot. Rev., 50*, 1–113

—— and Menshikova, L.V. (1983) 'Systematics of the Upper Palaeozoic Articulates of the Family Tchernoviaceae', *Bot. J., 68*, 721–9

Mierzejewski, P. (1982) 'The Nature of Pre-Devonian Tracheid-like Tubes', *Lethaia, 15*, 148

Millay, M.A. (1976) 'Synangia of North American Pennsylvanian Petrified Marattialeans', PhD thesis, University of Illinois

—— (1977) '*Acaulangium* gen. n. A fertile Marattialean from the Upper Pennsylvanian of Illinois', *Amer. J. Bot., 64*, 223–9

—— (1979) 'Studies of Paleozoic Marattialeans. A Monograph of the American Species of *Sclocecopteris*', *Palaeontographica B, 169*, 1–69

—— (1982) 'Studies of Paleozoic Marattialeans: the Morphology and Probable Affinities of *Telangium pygmaeum* Graham', *Amer. J. Bot., 69*, 1566–72

—— and Eggert, D.A. (1979) '*Idanothekion* gen. nov. A Synangiate Pollen Organ with Saccate Pollen from the Middle Pennsylvanian of Illinois', *Amer. J. Bot.*, *57*, 50–61

—— and Taylor, T.N. (1979) 'Paleozoic Seed Fern Pollen Organs', *Bot. Rev.*, *45*, 301–75

Miller, C.N. (1971) 'Evolution of the Fern Family Osmundaceae Based on Anatomical Studies', *Cont. Mus. Paleontol. Univ. Mich.*, *23*, 105–69

—— (1977) 'Mesozoic Conifers', *Bot. Rev.*, *43*, 217–80

—— (1982) 'Current Status of Paleozoic and Mesozoic Conifers', *Rev. Palaeobot. Palynol.*, *37*, 99–114

—— and Brown, J.T. (1973) 'Paleozoic Seeds with Embryos', *Science*, *179*, 184–5

Morgan, J. (1959) 'The Morphology and Anatomy of American Species of the Genus *Psaronius*', *Ill. Biol. Monogr.*, *27*, 1–108

Muhammad, A.F. and Sattler, R. (1982) 'Vessel Structure of *Gnetum* and the Origin of Angiosperms', *Amer. J. Bot.*, *69*, 1004–21

Muir, M.D. and Sutton, J. (1970) 'Some Fossiliferous Precambrian Chert Pebbles within the Torridonian of Britain', *Nature*, *226*, 443–5

Müller, H. (1966) 'Palynological Investigations of Cretaceous Sediments in northeastern Brazil', *Proc. 2nd West Afn Micropaleontol. Coll.* (Ibadan, 1965) Brill, Leiden, 123–36

Müller, J. (1966) 'Montane Pollen from the Tertiary of North West Borneo', *Blumea*, *14*, 231–5

—— (1970) 'Palynological Evidence on Early Differentiation of Angiosperms', *Biol. Rev. Cambridge Phil. Soc.*, *45*, 417–50

—— (1981) 'Fossil Pollen Records of Extant Angiosperms', *Bot. Rev.*, *47*, 1–142

—— (1984) 'Significance of Fossil Pollen for Angiosperm History', *Ann. Missouri Bot. Gdn*, *71*, 419–43

Namboodiri, K.K. and Beck, C.B. (1968) 'A Comparative Study of the Primary Vascular System of Conifers. 3 parts', *Amer. J. Bot.*, *55*, 447–72

Nathorst, A.G. (1911) 'Palaobotanische Mitteilungen', *Kungl. Svenska Vetenskapsakademiens Handlingar*, *46*, 1–33

Neish, A.C. (1968) 'Cinnamic Acid Derivatives as Intermediates in the Biosynthesis of Lignin and Related Compounds', in M.M. Zimmermann (Ed.), *The Formation of Wood in Forest Trees*, Academic Press, New York, pp. 219–39

Niklas, K.J. (1976) 'Morphological and Ontogenetic Reconstruction of *Parka decipiens* Fleming and *Pachytheca* Hooker from the Lower Old Red Sandstone, Scotland', *Trans. R. Soc. Edinb.*, *69*, 483–99

—— (1981) 'Airflow Patterns around Some Early Seed Plant Ovules and Cupules: Implications Concerning Efficiency in Wind Pollination', *Amer. J. Bot.*, *68*, 635–50

—— (1983) 'The Influence of Paleozoic Ovule and Cupule Morphologies on Wind Pollination', *Evolution*, *37*, 968–86

—— and Phillips, T.L. (1976) 'Morphology of *Protosalvinia* from the Upper Devonian of Ohio and Kentucky', *Amer. J. Bot.*, *63*, 9–29

—— and Pratt, L.M. (1980) 'Evidence for Lignin-like Constituents in Early Silurian (Llandoverian) Plant Fossils', *Science*, *209*, 396–7

Norris, G., Janzen, D.M. and Awai-Thorne, B.V. (1975) 'Evolution of Cretaceous Terrestrial Palynoflora in Western Canada', *Geol. Assn Canada, Special Paper 13*, 333–64

Norstog, K. (1982) 'Pollination in Cycads', *Bot. Soc. Amer. (Abstracts) Misc. Publ.*, *162*, 20 pp

Oliver, F.W. and Scott, D.H. (1904) 'On the Structure of the Palaeozoic Seed *Lagenostoma lomaxi*, with a Statement of the Evidence upon which it is Referred to *Lyginodendron*', *Phil. Trans. R. Soc. Lond. B*, *197*, 193–247

Pacltova, B. (1976) 'Notes on the Evolution and Distribution of Angiosperms during the Upper Cretaceous', *Evol. Biol.*, *9*, 133–8

—— (1978) 'Significance of Palynology for the Biostratigraphic Division of the Cretaceous Bohemia', *Paleontologicka Konference*, *77*, 243–51

—— (1981) 'The Evolution and Distribution of Normapolles Pollen during the Cenophytic', *Rev. Palaebot. Palynol.*, *35*, 175–208

Page, C.N. (1972a) 'An Interpretation of the Morphology and Evolution of the Cone and Shoot of *Equisetum*', *Bot. J. Linn. Soc.*, *65*, 359–97

────── (1972b) 'An Assessment of Inter-specific Relationships in *Equisetum* Subgenus *Equisetum*', *New Phytol.*, *71*, 355–69

────── (1973) 'Ferns, Polyploids, and their Bearing on the Evolution of the Canarian Flora', *Monogr. Biol. Canar.*, *4*, 83–8

────── (1979) 'The Diversity of Ferns. An Ecological Perspective', in A.F. Dyer (Ed.), *The Experimental Biology of Ferns*, Academic Press, London, pp. 9–56

Pant, D.D. (1962) 'The Gametophyte of the Psilophytales', in P. Maheshwari, B.M. Johri and I.K. Vasil (Eds), *Proceedings of the Summer School of Botany, Darjeeling 1960*. Ministry of Scientific Research and Cultural Affairs, New Delhi, pp. 276–301

────── (1977) 'Early Conifers and Conifer Allies', *J. Indian Bot. Soc.*, *56*, 23–37

────── (1982) 'The Lower Gondwana Gymnosperms and their Relationships', *Rev. Palaeobot. Palynol.*, *37*, 55–70

────── and Basu, N. (1979) 'Some Further Remains of Fructifications from the Triassic of Nidpur, India', *Palaeontographica B*, *168*, 129–46

────── and Nautiyal, D.D. (1967) 'On the Structure of *Buriadia heterophylla* (Feistmantel) Seward and Sahni and its Fructification', *Phil. Trans. R. Soc. Lond. B*, *252*, 27–48

────── and Singh, R.S. (1974) 'On the Stem and Attachment of *Gangamopteris* and *Glossopteris* Leaves, Part II. Structural Features', *Palaeontographica B*, *147*, 42–73

Parrish, J.T. (1982) 'Upwelling and Petroleum Source Beds, with Reference to Paleozoic', *Amer. Assn Petroleum Geol. Bull.*, *66*, 750–74

────── (1985) 'Global Paleogeography, Atmospheric Circulation and Rainfall in the Barremian Age (Late Early Cretaceous)', *US Geol. Surv. Open-file Report*, *85–728*, 14 pp

──────, Ziegler, A.M. and Scotese, C.R. (1982) 'Rainfall Patterns and the Distribution of Coals and Evaporites in the Mesozoic and Cenozoic', *Palaeogeog. Palaeoclimatol. Palaeoecol.*, *40*, 67–101

Pettitt, J.M. and Beck, C.B. (1968) '*Archaeosperma arnoldii* — a Cupulate Seed from the Upper Devonian of North America', *Cont. Mus. Paleontol. Univ. Mich.*, *22*, 139–54

────── and Chaloner, W.G. (1964) 'The Ultrastructure of the Mesozoic Pollen *Classopollis*', *Pollen Spores*, *6*, 611–20

Phillips, T.L. (1979) 'Reproduction of Heterosporous Arborescent Lycopods in the Mississippian-Pennsylvanian of Euramerica', *Rev. Palaeobot. Palynol.*, *27*, 239–89

Picket-Heaps, J. (1976) 'Cell Division in Eukaryotic Algae', *Bioscience*, *26*, 445–50

Pigg, K.B. and Rothwell, G.W. (1983) '*Chaloneria* gen. nov., Heterosporous Lycophyte from the Pennsylvanian of North America', *Bot. Gaz.*, *144*, 132–47

Pillmore, C.L., Tschudy, R.H., Orth, C.J., Gimore, J.S. and Knight, J.D. (1984) 'Geologic Framework of Nonmarine Cretaceous-Tertiary Boundary Sites, Raton Basin, New Mexico and Colorado', *Science*, *223*, 1180–3

Pirozynski, K.A. and Malloch, D.W. (1975) 'The Origin of Land Plants: a Matter of Mycotrophism', *Biosystems*, *6*, 153–64

Plumstead, E.P. (1952) 'Descriptions of Two New Genera and Six New Species of Fructification Borne on *Glossopteris* Leaves', *Trans. Geol. Soc. South Africa*, *55*, 281–328

Potonié, H. (1899) *Lehrbuch der Pflanzenpaleontologie mit besondere Rucksicht auf die Bedurfnisse des Geologen*, Pt 4, Dummber, Berlin, pp. 289–402

Pratt, L.M., Phillips, T.L. and Dennison, J.M. (1978) 'Evidence of Non-vascular Land Plants from the Early Silurian (Llandoverian) of Virginia, U.S.A.', *Rev. Palaeobot. Palynol.*, *25*, 121–49

Ramanujan, C.G.K., Rothwell, G.W. and Stewart, W.N. (1974) 'Probable Attachment of the *Dolerotheca campanulum* to a *Myeloxylon-Alethopteris* Type Frond', *Amer. J. Bot.*, *61*, 1057–66

Raven, J.A. (1977) 'The Evolution of Vascular Land Plants in Relation to Supracellular Transport Processes', *Adv. Bot. Res.*, *5*, 153–219

Regal, P.J. (1977) 'Ecology and Evolution of Flowering Plant Dominance', *Science*, *196*, 622–9

Reid, E.M. and Chandler, M.E.J. (1926) *Catalogue of Cainozoic Plants in the Department of Geology. 1. The Bembridge Flora*, British Museum (Natural History), London, 206 pp

Remy, W. and Remy, R. (1977) *Die Floren des Erdalterums*, Gluckauf GmbH, Essen, 468 pp

────── and ────── (1980a) 'Devonian Gametophytes with Anatomically Preserved Gametangia', *Science*, *208*, 295–6

────── and ────── (1980b) '*Lyonophyton rhyniensis* nov. gen. et nov. spec., ein Gametophyt aus dem Chert von Rhynie (Unterdevon, Schottland)', *Argumenta Palaeobot.*, *6*, 37–72

——— and ——— (1980c) '*Sciadophyton steinmann* — ein Gametophyt aus dem Siegen', *Argumenta Palaeobot.*, *6*, 73–94

Retallack, G.J. (1975) 'The Life and Times of a Triassic Lycopod', *Alcheringa*, *1*, 3–29

——— and Dilcher, D.L. (1981a) 'Arguments for a Glossopterid Ancestry of Angiosperms', *Paleobiology*, *7*, 54–67

——— and ——— (1981b) 'Early Angiosperm Reproduction: *Prisea reynoldsii* gen et sp. nov. from Mid-Cretaceous Coastal Deposits in Kansas, U.S.A.', *Palaeontographica B*, *179*, 103–51

——— and ——— (1981c) 'A Coastal Hypothesis for the Dispersal and Rise to Dominance of Flowering Plants', in K.J. Niklas (Ed.), *Paleoecology and Evolution*, vol. 2, Praeger, New York, pp. 27–77

Reymanova, M. (1974) 'On Anatomy and Morphology of *Caytonia*', in *Symposium on Origins and Phytogeography of Angiosperms*, B. Sahni Inst. Palaeobot. Spec. Publ. 2, Lucknow, India, pp. 50–7

Reznik, M. (1960) 'Vergleichende Biochemie der Phenylpropane', *Ergeb. Biol.*, *23*, 14–46

Ribbins, M.M. and Collinson, M.E. (1978) 'Further Notes on Pyritised Fern Rachides from the London Clay', *Tert. Res.*, *2*, 47–50

Richardson, J.B. and Ionnides, N. (1973) 'Silurian Palynomorphs from the Tanezzuft and Acacus Formations, Tripolitania, North Africa', *Micropalaeontology*, *19*, 257–307

Rigby, J.F. (1978) 'Permian Glossopterid and other Cycadopsid Fructifications from Queensland', *Geol. Surv. Queensland, Palaeontol. Paper*, *41*, 1–21

Rothwell, G.W. (1972) 'Evidence of Pollen Tubes in Paleozoic Pteridosperms', *Science*, *175*, 772–4

——— (1975) 'The Callistophytaceae (Pteridospermopsida): 1. Vegetative Structures', *Palaeontographica B*, *151*, 171–96

——— (1977) 'Evidence for a Pollination-drop Mechanism in Paleozoic Pteridosperms', *Science*, *198*, 1251–2

——— (1981) 'The Callistophytales (Pteridospermopsida): Reproductively Sophisticated Paleozoic Gymnosperms', *Rev. Palaeobot. Palynol.*, *32*, 103–21

——— (1982a) '*Cordaianthus duquesnensis* sp. nov., Anatomically Preserved Ovulate Cones from the Upper Pennsylvanian of Ohio', *Amer. J. Bot.*, *69*, 239–47

——— (1982b) 'New Interpretation of the Earliest Conifers', *Rev. Palaeobot. Palynol.*, *37*, 7–28

——— (1984) 'The Apex of *Stigmaria* (Lycopsida) Rooting Organ of Lepidodendrales', *Amer. J. Bot.*, *71*, 1031–4

——— and Warner, S. (1984) '*Cordaixylon dumusum* n. sp. (Cordaitales). 1. Vegetative Structures', *Bot. Gaz.*, *145*, 275–91

Sahni, B.D. (1932) 'A Petrified *Williamsonia* (*W. sewardiana*, sp. nov.) from the Rajmahal Hills, India', *Palaeontol. Ind.*, *20*, 1–19

Salisbury, E.J. (1942) *The Reproductive Capacity of Plants, Studies in Quantitative Biology*, G. Bell & Sons, London

Samoylovich, S.R. (1977) 'A New Outline of the Floristic Zoning of the Northern Hemisphere in the Late Senonian', *Paleontol. J.*, *3*, 118–27

Saporta, G. de (1877) *L'ancienne vegetation polaire*, Congres International de Science et Geographie Compte Vendu

——— (1894) *Flora fossile du Portugal*, Direction de Travaux Geologiques du Portugal, Lisbon

Sarkanen, K.V. and Ludwig, C.H. (1971) 'Definition and Nomenclature', in K.V. Sarkanen and G.M. Ludwig (Eds), *Lignins: Occurrence, Formation, Structure and Reactions*, Wiley-Interscience, New York, pp. 1–18

Scheckler, S.E. (1975) *Rhymokalon*, a New Plant with Cladoxylalean Anatomy from the Upper Devonian of New York State', *Can. J. Bot.*, *53*, 25–38

——— (1978) 'Ontogeny of Progymnosperms, Part II: Shoots of Upper Devonian Archaeopteridales', *Can. J. Bot.*, *56*, 3136–70

——— and Banks, H.P. (1971a) 'Anatomy and Relationships of some Devonian Progymnosperms from New York', *Amer. J. Bot.*, *58*, 737–51

——— and ——— (1971b) '*Proteokalon* a New Genus of Progymnosperms from the Devonian of New York State and its Bearing on Phylogenetic Trends in the Group', *Amer. J. Bot.*, *58*, 874–84

Schiehing, M.M. and Pfefferkorn, H.W. (1984) 'The Taphonomy of Land Plants in the Orinoco Delta: a Model for the Incorporation of Plant Parts in Clastic Sediments of Late Carboniferous Age of Euramerica', *Rev. Palaeobot. Palynol., 41*, 205–40

Schlanker, C.M. and Leisman, G.A. (1969) 'The Herbaceous Carboniferous Lycopod *Selaginella fraiponti* comb. nov.', *Bot. Gaz., 130*, 35–41

Schlesinger, W.H. (1977) 'The Use of Water and Minerals by Evergreen and Deciduous Shrubs in Okefenokee Swamp', *Bot. Gaz., 138*, 490–7

—— (1978) 'Community Structure, Dynamics and Nutrient Cycling in the Okefenokee Cypress Swamp-forest', *Ecol. Monogr., 48*, 43–65

Schopf, J.M. (1971) 'Notes on Plant Tissue Preservation and Mineralization in a Permian Deposit of Peat from Antarctica', *Amer. J. Sci., 271*, 522–43

—— (1975) 'Modes of Fossil Preservation', *Rev. Palaeobot. Palynol., 20*, 27–35

—— (1976) 'Morphological Interpretations of Fertile Structures in Glossopterid Gymnosperms', *Rev. Palaeobot. Palynol., 21*, 25–64

—— (1978) '*Foerstia* and Recent Interpretations of Early Vascular Plants', *Lethaia, 11*, 139–43

—— and Schwietering, J.F. (1970) 'The *Foerstia* Zone of the Ohio and Chattanooga Shales', *US Geol. Surv. Bull., 1294-H*, 1–15

Schopf, J.W. (1978) 'The Evolution of the Earliest Cells', *Scient. Amer., 239*, 110–38

—— and Oehler, D.Z. (1976) 'How Old Are the Eukaryotes?', *Science, 193*, 47–9

Schuster, R.M. (1966) *The Hepaticae and Anthoceratae of North America East of the Hundredth Meridian*, Vol. 1. Columbia, New York, 802 pp

Schweitzer, H.-J. (1963) 'Der weibliche Zipfen von *Pseudovoltzia liebeana* und seine Bedeutung für die Phylogenie der Koniferen', *Palaeontographica B, 113*, 1–29

—— (1978) 'Die Rato-Jurassischen Floren des Iran und Afghanistans. 5. *Todites princeps, Thaumatopteris brauniana* und *Phlebopteris polypodioides*', *Palaeontographica B, 168*, 17–60

—— (1981) 'Der Generationswechsel Rheinischer Psilophyten', *Bonner Paläobot. Mitteil., 8*, 1–19

Scott, A.C. and Chaloner, W.G. (1983) 'The Earliest Conifer from the Westphalian B of Yorkshire', *Proc. R. Soc. Lond. B, 220*, 163–82

Scott, A.C. and Collinson, M.C. (1983) 'Investigating Fossil Plant Beds Parts 1 and 2', *Geology Teaching, 7*, 114–22; *8*, 12–26

Scott, D.H. (1909) *Studies in Fossil Botany*, 2nd edn, Black, London, 683 pp

Scott, R.A. (1960) 'Pollen of *Ephedra* from the Chinle Formation (Upper Triassic) and the Genus *Equisetosporites*', *Micropalaeontology, 6*, 271–6

—— and Smiley, C.J. (1979) 'Some Cretaceous Plant Megafossils and Microfossils from the Nanushuk Group, Northern Alaska, a Preliminary Report', *US Geological Survey Circular 749*, 89–111

Sellards, E.H. (1903) '*Codonotheca*, a New Type of Spore-bearing Organ from the Coal Measures', *Amer. J. Sci., 16*, 87–95

Setlik, J. (1956) 'Contribution to the Study of *Noeggerathia foliosa* Sternberg', *Rozpravy Ustred. Ustavu Geol., 21*, 1–106 (Czech with French summary)

Shivas, M.G. (1958) 'Contributions to the Cytology and Taxonomy of Species of *Polypodium* in Europe and America. I Cytology; II Taxonomy', *Bot. J. Linn. Soc., 370*, 13–36

Singh, C. (1975) 'Stratigraphic Significance of Early Angiosperm Pollen in the Mid-Cretaceous Strata of Alberta', *Geol. Assn Can. Spec. Paper 13*, 365–89

Singh, H. (1961) 'The Life History and Systematic Position of *Cephalotaxus drupaceae* Sieb. et Zucc.', *Phytomorphology, 11*, 153–97

Skog, J.E. and Banks, H.P. (1973) '*Ibyka amphikoma* gen. et sp. n., a New Protoarticulate Precursor from the Late Middle Devonian of New York State', *Amer. J. Bot., 60*, 366–80

Sleep, A. (1970) 'An Introduction to the Ferns of Japan', *Brit. Fern Gaz., 10*, 127–41

Smart, J. (1971) 'Palaeoecological Factors Affecting the Original Winged Insects', *Proc. XIII Int. Congr. Ent., Moscow 1968, 1*, 304–6

Smiley, C.J. (1967) 'Paleoclimatic Interpretations of Some Mesozoic Floral Sequences', *Amer. Assn Petroleum Geol. Bull., 51*, 849–63

Smit, J. and Van der Kaars, S. (1984) 'Terminal Cretaceous Extinctions in the Hell Creek Area, Montana: Compatible with Catastrophic Extinction', *Science, 223*, 1177–9

Smith, D.L. (1962) 'Three Fructifications from the Scottish Lower Carboniferous', *Palaeontology, 5*, 225–37

Spicer, R.A. (1977) 'Predepositional Formation of Leaf Impressions', *Palaeontology, 20*, 907–12
———— (1980) 'The Importance of Depositional Sorting to the Biostratigraphy of Plant Megafossils',
in D.L. Dilcher and T.N. Taylor (Eds.), *Biostratigraphy of Fossil Plants*, Dowden, Hutchinson
and Ross, Stroudsburg, Pennsylvania, pp. 171–83
———— (1981) 'The Sorting and Deposition of Allochthonous Plant Material in a Modern Environment at Silwood Lake, Silwood Park, Berkshire, England', *US Geol. Surv. Prof. Paper* 1143,
77 pp
———— (1986a) 'Pectinal Veins: a New Concept in Terminology for the Description of
Dicotyledonous Leaf Venation Patterns', *Bot. J. Linn. Soc.*, in press
———— (1986b) 'Comparative Leaf Architectural Analysis of Cretaceous Radiating Angiosperms',
in R.A. Spicer and B.A. Thomas (Eds), *Systematic and Taxonomic Approaches in Palaeobotany*,
Systematics Association Special Volume 31, pp. 223–34
———— (1986c) 'Albian — Paleocene Plant Megafossils from Alaska', *US Geol. Surv. Prof. Paper*,
in press
———— (1987) 'The Significance of the Cretaceous Flora of Northern Alaska for the Reconstruction
of the Climate of the Cretaceous', in E. Kemper (Ed.), *Das Klima der Kreide*, Geologisches
Jahrbuch Series A, in press
————, Burnham, R.J., Grant, P. and Glicken, M. (1985) '*Pityrogramma calomelanos*, the
Primary Post-eruption Colonizer of Volcan Chichonal, Chiapas, Mexico', *Amer. Fern J., 75*,
1–5
———— and Hill, C.R. (1979) 'Principal Components and Correspondence Analyses of Quantitative
Data from a Jurassic Plant Bed', *Rev. Palaeobot. Palynol., 28*, 273–99
———— and Parrish, J.T. (1986) 'Paleobotanical Evidence for Cool North Polar Climates in the Mid-
Cretaceous (Albian-Cenomanian)', *Geology, 14*, 703–6
Sporne, K.R. (1974) *The Morphology of Angiosperms*, Hutchinson, London, 207 pp
Srivastava, S.K. (1975) 'Maestrichtian Microspore Assemblages from the Interbasaltic Lignites of
Mull, Scotland', *Palaeontographica B, 150*, 125–56
———— (1981) 'Evolution of Upper Cretaceous Phytogeoprovinces and their Pollen Flora', *Rev.
Palaeobot. Palynol., 35*, 155–73
Stebbins, G.L. (1957) 'Self-fertilization and Population Variability in Higher Plants', *Amer. Natur.,
91*, 337–54
———— (1965) 'The Probable Growth Habit of the Earliest Flowering Plants', *Ann. Missouri Bot.
Gdn, 52*, 457–68
———— (1971) 'Relationships between Adaptive Radiation, Speciation and Major Evolutionary
Trends', *Taxon, 20* (1), 3–16
———— (1974) *Flowering Plants. Evolution above the Species Level*, Belknap Press of Harvard
University Press, Mass., 399 pp
Stein, W.E. (1982) 'The Devonian Plant *Reimannia*, with a Discussion of the Class Progymno-
spermopsida', *Palaeontology, 25*, 605–22
Stewart, C.M. (1966) 'Excretion and Heartwood Formation in Living Trees', *Science, 153*,
1068–74
Stewart, K.D. and Mattox, K.R. (1975) 'Comparative Cytology, Evolution and Classification of the
Green Algae with Some Consideration of the Origin of other Organisms with Chlorophylls a and
b', *Bot. Rev., 41*, 105–35
———— and ———— (1978) 'Structural Evolution in the Flagellated Cells of Green Algae and Land
Plants', *Biosystems, 10*, 145–52
Stewart, W.N. (1960) 'More about the Origins of Vascular Plants', *Plant Sci. Bull., 6*, 1–4
———— (1983) *Paleobotany and the Evolution of Plants*, Cambridge University Press, Cambridge,
405 pp
———— and Delevoryas, T. (1956) 'The Medullosan Pteridosperms', *Bot. Rev., 22*, 45–80
Stidd, B.M. (1978) 'An Anatomically Preserved *Potoniea* with *in situ* Spores from the Pennsylvanian of Illinois', *Amer. J. Bot., 65*, 677–83
———— (1980) 'The Neotenous Origin of the Pollen Organ of the Gymnosperm *Cycadeoidea* and
Implications for the Origin of Higher Taxa', *Paleobiology, 6*, 161–7
———— (1981) 'The Current Status of Medullosan Seed Fern', *Rev. Palaeobot. Palynol., 32*,
63–101
———— and Cosentino, K. (1976) '*Nucellangium*: Gametophytic Structures and Relationship to
Cordaites', *Bot. Gaz., 137*, 342–9

Stockey, R.A. (1981) 'Some Comments on the Origin and Evolution of Conifers', *Can. J. Bot.*, *59*, 1932–40
—— (1982) 'The Araucariaceae: an Evolutionary Perspective', *Rev. Palaeobot. Palynol.*, *37*, 133–54
Storch, D. (1980) *Sphenophyllum tenerrimum* besass trilete Sporen. *Wiss. Zeit. d. Humboldt-Univ. z Berl., Math. Nat. R.*, *29*, 387–8
Surange, K.R. and Chandra, S. (1973) '*Denkania indica* gen. et sp. nov. — a Glossopteridian Fructification from the Lower Gondwana of India', *Palaeobotanist*, *20*, 264–8
—— and —— (1974) '*Lidgettonia mucronata* sp. nov. a Female Fructification from the Lower Gondwana of India', *Palaeobotanist*, *21*, 121–6
—— and Maheshwari, H.K. (1970) 'Some Male and Female Fructifications of the *Glossopteris* Flora and their Relationships', *Palaeontographica B*, *129*, 178–92

Takhtajan, A.L. (1969) *Flowering Plants: Origin and Dispersal*, Oliver & Boyd, Edinburgh, 310 pp
—— (1976) 'Neoteny and the Origin of Flowering Plants', in C.B. Beck (Ed.), *Origin and Early Evolution of Angiosperms*, Columbia University Press, New York, pp. 207–19
Taylor, T.N. (1965) 'Paleozoic Seed Studies: a Monograph of the American Species of *Pachytesta*', *Palaeontographica B*, *117*, 1–46
—— (1971) '*Halletheca reticulatus* gen. et sp. n.: a Synangiate Pennsylvanian Pteridosperm Pollen Organ', *Amer. J. Bot.*, *58*, 300–8
—— (1978) 'The Ultrastructure and Reproductive Significance of *Monoletes* (Pteridospermales) Pollen', *Can. J. Bot.*, 3105–18
—— (1981) *Palaeobotany: an Introduction to Fossil Plant Biology*, McGraw-Hill, New York, 589 pp
—— (1982a) 'Reproductive Biology in Early Seed Plants', *Bioscience*, *32*, 23–8
—— (1982b) 'The Origin of Land Plants: a Paleobotanical Perspective', *Taxon*, *31*, 155–77
—— and Alvin, K. (1984) 'Ultrastructure and Development of Mesozoic Pollen: *Classopollis*', *Amer. J. Bot.*, *71*, 575–87
—— and Brauer, D.F. (1983) 'Ultrastructural Studies of *in situ* Devonian Spores *Barinophyton citrulliforme*', *Amer. J. Bot.*, *70*, 106–12
—— and Daghlian, C.P. (1980) 'The Morphology and Ultrastructure of *Gothania* (Cordaites) Pollen', *Rev. Palaeobot. Palynol.*, *29*, 1–14
—— and Millay, M.A. (1979) 'Pollination Biology in Early Seed Plants', *Rev. Palaeobot. Palynol. 27*, 329–55
—— and —— (1980) 'Morphologic Variability of Pennsylvanian Lyginopterid Seed Ferns', *Rev. Palaeobot. Palynol.*, *32*, 27–62
—— and —— (1981) 'Additional Information on the Pollen Organ *Halletheca* (Medullosales)', *Amer. J. Bot.*, *68*, 1403–7
Thomas, B.A. (1966) 'The Cuticle of the Lepidodendroid Stem', *New Phytol.*, *65*, 296–303
—— (1968) 'A Revision of the Carboniferous Lycopod Genus *Eskdalia* Kidston', *Palaeontology*, *11*, 439–44
—— (1970) 'A New Specimen of *Lepidostrobus binneyanus* from the Westphalian B of Yorkshire', *Pollen et Spores*, *12*, 217–34
—— (1974) 'The Lepidodendroid Stoma', *Palaeontology*, *17*, 525–39
—— (1977) 'Epidermal Studies in the Interpretation of *Lepidophloios* Species', *Palaeontology*, *20*, 273–93
—— (1978) 'Carboniferous Lepidodendraceae and Lepidocarpaceae', *Bot. Rev.*, *44*, 321–64
—— (1981) 'Structural Adaptations Shown by the Lepidocarpaceae', *Rev. Palaeobot. Palynol.*, *32*, 377–88
—— (1985a) 'Pteridophyte Success and Past Biota — a Palaeobotanist's Approach', *Proc. R. Soc. Edinb.*, *86B*, 423–30
—— (1985b) 'Stomatal Mechanism as the Basis of Evolution of Crassulacean Acid Metabolism', *Nature*, *314*, 200
—— and Brack-Hanes, S.D. (1984) 'A New Approach to Family Groupings in the Lycophytes', *Taxon*, *33*, 247–55
Thomas, H.H. (1925) 'The Caytoniales, a New Group of Angiospermous Plants from the Jurassic Rocks of Yorkshire', *Phil. Trans. R. Soc. Lond. B*, *213*, 299–363
—— (1955) 'Mesozoic Pteridosperms', *Phytomorphology*, *5*, 177–85

Tiffney, B.H. (1984) 'Seed Size, Dispersal Syndromes, and the Rise of the Angiosperms: Evidence and Hypothesis', *Ann. Missouri Bot. Gdn, 71*, 551–76

Townrow, J.A. (1960) 'The Peltaspermaceae, a Pteridosperm Family of Permian and Triassic Age', *Palaeontology, 3*, 333–61

——— (1962) 'On *Pteruchus*, a Microsporophyll of the Corystospermaceae', *Bull. Brit. Mus. (Nat. Hist.) Geol., 6*, 289–320

——— (1967) 'On *Rissika* and *Mataia* Podocarpaceous Conifers from the Lower Mesozoic of Southern Lands', *Proc. R. Soc. Tasmania, 101*, 103–36

Tryon, R. and Gastony, G. (1975) 'The Biogeography of Endemism in the Cyatheaceae', *Fern Gaz., 11*, 73–80

——— and Tryon, A.F. (1982) *Ferns and Allied Plants. With Special Reference to Tropical America*, Springer, New York, 857 pp

Tschudy, R.H. (1981) 'Geographic Distribution and Dispersal of Normapolles Genera in North America', *Rev. Palaeobot. Palynol., 35*, 283–314

Tyler, S.A. and Barghoorn, E.S. (1954) 'Occurrence of Structurally Preserved Plants in Pre-Cambrian Rocks of the Canadian Shield', *Science, 119*, 606–8

Upchurch, G.R. (1984) 'Cuticle Evolution in Early Cretaceous Angiosperms from the Potomac Group of Virginia and Maryland', *Ann. Missouri Bot. Gdn, 71*, 522–50

Vakhrameev, V.A. (1973) 'Pokrytosemennye i granitsa nizhnego i verkhnego mela', in A.F. Khlonova (Ed.), *Palinologiya Mesofita*, Nauka, Moscow, pp. 131–5

——— (1978) 'The Climates of the Northern Hemisphere in the Cretaceous in the Light of Paleobotanical Data', *Paleontol. J., 2*, 3–17

Van Campo, M. (1971) 'Precisions nouvelles sur les structures comparées des pollens de Gymnospermes et d'Angiospermes', *Compt. Rend. Acad. Sci. Paris, D, 272*, 2071–4

Vishnu-Mittre (1958) 'Studies on Fossil Flora of Nipania (Rajmahal) Series, Bihar-coniferales', *Palaeobotanist, 6*, 82–122

Walker, J.W. (1976) 'Evolutionary Significance of the Exine in the Pollen of Primitive Angiosperms', in I.K. Ferguson and J. Muller (Eds), *The Evolutionary Significance of the Exine*, Linn. Soc. Symp. Series No. 1, Academic Press, London, pp. 251–308

——— and Walker, A.G. (1984) 'Ultrastructure of Lower Cretaceous Angiosperm Pollen and the Origin and Early Evolution of Flowering Plants', *Ann. Missouri Bot. Gdn, 71*, 464–521

Walker, T.G. (1958) 'Hybridization in Some Species of *Pteris* L.', *Evolution, 12*, 82–92

——— (1973) 'Evidence from Cytology in the Classification of Ferns', in A.C. Jermy, J.A. Crabbe and B.A. Thomas (Eds), *The Phylogeny and Classification of the Ferns, Bot. J. Linn. Soc., 67, Suppl. No. 1*, pp. 91–110

Walton, J. (1964) 'On the Morphology of *Zosterophyllum* and Some Other Early Devonian Plants', *Phytomorphology, 14*, 155–60

Watson, E.V. (1964) *The Structure and Life of Bryophytes*, Hutchinson University Library, London, 192 pp

Watson, J. (1982) 'The Cheirolepidiaceae: a Short Review', in D.D. Nautiyal (Ed.), *Phyta, Studies on Living and Fossil Plants. Pant. Comm. Vol.*, Allahabad, India, 265–73

——— and Alvin, K.L. (1976) 'Silicone Rubber Casts of Silicified Plants from the Cretaceous of Sudan', *Palaeontology, 19*, 641–50

Weiland, C.R. (1906) *American Fossil Cycads*, Carnegie Institute, Washington, DC

——— (1916) *American Fossil Cycads*, Vol. II, Carnegie Institute, Washington, DC

Weiss, C.E. (1876) 'Steinkohlen — Calamarien. Atlas', *Abh. geol. Spec. Preussen, II* (1) 19 pls

White, D. and Stadnichenko, T. (1923) 'Some Mother Plants of Petroleum in the Devonian Black Shales', *Econ. Geol., 18*, 238–52

White, M.E. (1978) 'Reproductive Structures of the Glossopteridales in the Plant Fossil Collection of the Australian Museum', *Records Austral. Mus., 31*, 473–505

White, M.J.D. (1978) *Modes of Speciation*, Freeman, San Francisco, 455 pp

Whitehead, D.R. (1969) 'Wind Pollination in the Angiosperms: Evolutionary and Environmental Considerations', *Evolution, 23*, 28–35

Wigglesworth, V.B. (1973) 'Evolution of Insect Wings and Flight', *Nature, 246*, 127–9

Wolfe, J.A. (1972) 'Significance of Comparative Foliar Morphology to Paleobotany and

Neobotany' (Abstract), *Amer. J. Bot., 59*, 664

—— (1973) 'Fossil Forms of Amentiferae', *Brittonia, 25*, 334–55

—— (1975) 'Some Aspects of Plant Geography of the Northern Hemisphere during the Late Cretaceous and Tertiary', *Ann. Missouri Bot. Gdn, 62*, 264–79

—— (1978) 'A Paleobotanical Interpretation of Tertiary Climates in the Northern Hemisphere', *Amer. Sci., 66*, 694–703

—— (1979) 'Temperature Parameters of Humid to Mesic Forests of Eastern Asia and Relation to Forests of Other Regions of the Northern Hemisphere and Australasia', *US Geol. Surv. Prof. Paper 1106*, 37 pp

—— (1980) 'Tertiary Climates and Floristic Relationships at High Latitudes in the Northern Hemisphere', *Palaeogeog., Palaeoclimatol., Palaeoecol., 30*, 313–25

—— (1985) 'The Distribution of Major Vegetational Types during the Tertiary', in *The Carbon Cycle and Atmospheric CO_2 Natural Variations Archean to Present*, American Geophysical Union, Geophysical Monograph 32, pp. 357–75

——, Doyle, J.A. and Page, V.M. (1975) 'The Bases of Angiosperm Phylogeny: Paleobotany', *Ann. Missouri Bot. Gdn, 62*, 801–24

—— and Pakiser, H.M. (1971) 'Stratigraphic Interpretations of Some Cretaceous Microfossil Floras of the Middle Atlantic States', *US Geol. Surv. Prof. Paper 750-B*, 35–47

—— and Upchurch, G.R. (1987) 'Mid-Cretaceous to Early Tertiary Vegetation and Climate: Evidence from Fossil Leaves and Woods', in E.M. Friss, W.G. Chaloner and P.R. Crane (Eds) *Origin of the Angiosperms and Their Biological Consequences*, Cambridge University Press, Cambridge, in press

Zaklinskaya, E.D. (1981) 'Phylogeny and Classification of the Normapolles', *Rev. Palaeobot. Palynol., 35*, 139–47

Zavada, M.S. and Crepet, W.L. (1981) 'Investigations of Angiosperms from the Middle Eocene of North America; Flowers of the Celtidoideae', *Amer. J. Bot., 68*, 924–33

Zdebska, D. (1982) 'A New Zosterophyll from the Lower Devonian of Poland', *Palaeontology, 25*, 247–63

Zimmermann, W. (1930) *Die Phylogenie der Pflanzen. 1*, Gustav Fischer, Jena

—— (1952) 'Main Results of the "Telome Theory" ', *Palaeobotanist, 1*, 456–70

INDEX